Descriptive Physical Oceanography
An Introduction

Other titles of interest

THE OPEN UNIVERSITY
The Ocean Basins—Their Structure and Evolution
Seawater—It Composition, Properties and Behaviour
Ocean Circulation
Waves, Tides and Shallow-water Processes
Ocean Chemistry and Deep-sea Sediments
Case Studies in Oceanography and Marine Affairs

PARSONS et al
Biological Oceanography Processes, 3rd Edition

POND & PICKARD
Introductory Dynamical Oceanography, 2nd Edition

TCHERNIA
Descriptive Regional Oceanography

WILLIAMS & ELDER
Fluid Physics for Oceanography and Physicists

Descriptive Physical Oceanography

An Introduction

FIFTH ENLARGED EDITION (in SI units)

by

George L. Pickard
M.A., D. Phil., D.M.S. (Hon.), F.R.S.C.
Professor Emeritus of Oceanography and Physics
Former Director of the Department of Oceanography
University of British Columbia

and

William J. Emery
B.Sc., Ph.D; *Professor*
Colorado Center for Astrodynamical Research
Aerospace Engineering Science Department
University of Colorado

Butterworth-Heinemann
Linacre House, Jordan Hill, Oxford OX2 8DP
A division of Reed Educational and Professional Publishing Ltd

 A member of the Reed Elsevier plc group

OXFORD BOSTON JOHANNESBURG
MELBOURNE NEW DELHI SINGAPORE

First published 1964
Reprinted 1966, 1968, 1970
Second edition 1975
Third edition 1979
Fourth enlarged edition 1982
Reprinted 1984, 1985, 1986, 1988, 1989
Fifth edition 1990
Reprinted 1995, 1996

© G. L. Pickard and W. J. Emery 1990

All rights reserved. No part of this publication may be reproduced in
any material form (including photocopying or storing in any medium by
electronic means and whether or not transiently or incidentally to some
other use of this publication) without the written permission of the
copyright holder except in accordance with the provisions of the Copyright,
Designs and Patents Act 1988 or under the terms of a licence issued by the
Copyright Licensing Agency Ltd, 90 Tottenham Court Road, London,
England W1P 9HE. Applications for the copyright holder's written
permission to reproduce any part of this publication should be addressed
to the publishers

British Library Cataloguing in Publication Data
Pickard, George L. (George Lawson)
  Descriptive physical oceanography – 5th ed.
  1. Physical oceanography
  I. Title II. Emery William J.
  551.4601

Library of Congress Cataloguing in Publication Data
Pickard, George L.
  Descriptive physical oceanography: an introduction/
  George L. Pickard and William J. Emery – 5th ed.
  p. cm.
  Includes bibliographical references
  1. Oceanograph I. Emery William J. II. Title
  GC 150.5.P52   1990   551.46–dc20   90–6814

ISBN 0 7506 2759 X

Printed and bound in Great Britain by BPC Wheatons, Exeter

Preface to the Fifth Edition

THE development of interest in oceanography in recent years has led to an increased demand from students for information on the subject. The texts available hitherto have been either more elementary in treatment or more comprehensive and extensive than may be desirable for an introduction to the subject for the undergraduate. The present text is an attempt to supply information on the Synoptic or Descriptive aspects of Physical Oceanography at a level suitable as an introduction for graduate students, for undergraduates in the sciences and possibly for senior school students who wish to learn something of the aims and achievements in the field of scientific study.

This fifth edition has been prepared to bring the material up-to-date in the rapidly developing field of physical oceanography. Some of the aspects which have been enlarged include descriptions of newer observational methods (e.g. instruments, satellite applications) and modern methods of handling data using computers rather than just paper and pencil, the use of tracers for investigating the circulation, the properties and distributions of sea-ice, equatorial circulation details, and more on the El Niño—Southern Oscillation (ENSO) phenomenon. The treatment of the Ekman circulation has been moved to Chapter 7 to relate it more closely to the main ocean circulations, and additional features have been included in Chapter 8 on Coastal Oceanography. Many additional references to the published literature have been added.

The International System of Units (SI units) is used as a basis, since this has been recommended by the International Association for the Physical Sciences of the Ocean. For brevity, the term "sverdrup" (Sv), which is not an SI unit, will still be used for volume transports; dissolved oxygen will be expressed in millilitres per litre (mL/L) as most of the existing literature uses this unit, and dynamic metres will be used in the geostrophic method discussion for the same reason. As it may be some time before all oceanographers adopt the SI and to assist in relating to the wealth of literature using the previous mixed system of units, an Appendix describing the two systems is included.

In presenting the synoptic approach it must be emphasized that this represents only one aspect of physical oceanography. The other, and complementary, one is the dynamical approach through the laws of mechanics. This is described in other texts listed in the Suggestions for Further Reading such as those by Bowden (1983, on coastal oceanography) Gill (1982), Officer (1976, on estuaries), Pedlosky (1974), Pond and Pickard (1983), Stommel ("The Gulf Stream", 1965). A valuable reference volume is that edited by Warren and Wunsch (1981). The student who requires a full introduction to physical oceanography must study both aspects.

This text is intended to be introductory to the subject. For the student in the biological or geological sciences it may provide sufficient information on physical oceanography as background for their studies. For a student in physics or mathematics it should serve to present a basic description of the oceans before he or she proceeds to the more advanced texts and original literature making free use of mathematical methods. The text by Tchernia provides a more detailed coverage of the oceans by regions although it must be appreciated that considerably more information, particularly on the Equatorial regions, Gulf Stream, Southern Ocean, etc. has been accumulated since that volume was prepared.

The Bibliography at the end of the book is in two sections. The "Suggestions for Further Reading" lists a number of texts which would be helpful to a student wishing to read further in descriptive oceanography together with some sources of tables of use to physical oceanographers and a list of some journals which contain articles in the field of oceanography. The "References to Journal and Review Articles" section provides the specific references made by author and date in the body of the text for the convenience of the student who wishes to examine these items in more detail. Suggestions are also offered for more extensive lists of references.

The text is based on a course presented by the authors and colleagues for thirty years at the University of British Columbia to introduce undergraduate and graduate students to physical oceanography. It owes much to the more comprehensive text *The Oceans* by Sverdrup, Johnson and Fleming, and G. L. Pickard wishes to acknowledge this and also the stimulation received during a year at the Scripps Institution of Oceanography. He is particularly indebted to Dr. J. P. Tully of the former Pacific Oceanographic Group for initiating him into oceanography and for encouragement since, and to Dr. R. W. Burling and others for constructive comments on the text. He is also indebted to the Directors and staffs of the Section d'Océanographie, ORSTOM, New Caledonia and of the Australian Institute of Marine Science for providing facilities and encouragement to obtain practical experience of oceanography in the equatorial Pacific and in the Great Barrier Reef regions respectively. W. J. Emery would like to acknowledge the guidance of Dr. Klaus Wyrtki in introducing him to physical oceanography and providing opportunities for study. Helpful suggestions were offered by Dr. Aas, University of Oslo, on the optical aspects and by Mr. Sydney Levitus, Princeton University, on surface temperatures, and these have been incorporated.

In addition, it will be realized that although the authors have personal experience of some aspects of physical oceanography and of some regions, they have relied very much on the results and interpretations of others in order to present an adequate coverage of the subject. They therefore gratefully acknowledge their indebtedness to the many oceanographers whose works they have consulted in texts and journals in assembling the material for this book.

If the reader concludes the text with a feeling that our knowledge of the sea is incomplete at present, one of our objectives will have been achieved. This was to indicate to the student that there is still much to be learned of the ocean and that if he or she is interested in observing the marine world and interpreting it there are still many opportunities to do so.

Acknowledgements

PERMISSION from authors and publishers to reproduce or adapt the following material is acknowledged with sincere thanks.

Fig. 4.5 Adapted from Pollard, 1977, Fig. 1, *Deep-Sea Research*, Pergamon Press.
Fig. 4.6 Data from the Pacific Oceanographic Group, Fisheries Research Board of Canada.
Fig. 4.7 Adapted from Tchernia, 1980, Fig. 7.10, Pergamon Press.
Fig. 4.16 Adapted from Lewis et al., 1988, *Journal of Geophysical Research*, American Geophysical Union.
Fig. 5.7 b, c, d, e, f Adapted from Wyrtki, 1965, Figs. 1–5, *Journal of Geophysical Research*, American Geophysical Union.
Fig. 5.9 From Stommel, 1981, unpublished manuscript.
Fig. 6.6 From Wyrtki, 1974, University of Hawaii.
Figs. 6.13, 6.14 Adapted from Figs. 22 and 29 in L. V. Worthington, 1981, by permission of the MIT Press, Cambridge, Massachusetts. Copyright © 1981 by the Massachusetts Institute of Technology.
Fig. 6.15 From Pickard, 1977, Australian Institute of Marine Science.
Figs. 7.2, 7.3 Based on Levitus, 1988, Figs. 1, 2, 3, 6, 7, 8, *Journal of Physical Oceanography*, American Meteorological Society.
Fig. 7.6 From Deacon, 1937, Cambridge University Press.
Fig. 7.8 From Sverdrup et al., 1946, Fig. 164, Prentice Hall Inc.
Fig. 7.11 From Cornillon et al., 1987, *Journal of Geophysical Research*, American Geophysical Union.
Fig. 7.13 From Parker, 1971, Fig. 1, *Deep-Sea Research*, Pergamon Press.
Fig. 7.14 Adapted from Richardson et al., 1978, Figs. 23, 29, *Journal of Geophysical Research*, American Geophysical Union.
Fig. 7.17 Adapted from Fugilister, 1960, Figs. 23, 29, Woods Hole Oceanographic Institution.
Figs. 7.19, 7.20 Adapted from Sverdrup et al., 1946, Fig. 209B, Prentice Hall Inc.

Figs. 7.22, 7.42   From Emery and Dewar, 1981, Pergamon Press.
Fig. 7.24   Adapted from Wüst, 1961, *Journal of Geophysical Research*, American Geophysical Union.
Fig. 7.25   Adapted from J. W. Murray and E. Izdar, 1989, *Oceanography Magazine*, Vol. 2, No. 1, 15–21, Figs. 3, 4 & 5.
Figs. 7.28, 7.29, 7.30   Adapted from Coachman, L. K., 1962, On the water masses of the Arctic Ocean, Department of Oceanography, University of Washington, Report M. 62–11.
Figs. 7.33, 7.34, 7.35, 7.38   From Wyrtki and Kilonsky, 1984, Fig. 2, 5, 4 & 7, *Journal of Physical Oceanography*, American Meteorological Society.
Fig. 7.40   Adapted from Wooster and Reid, 1963, Fig. 8, Wiley-Interscience.
Fig. 7.43   Adapted from Tsuchiya, 1962, Johns Hopkins Press.
Plate 4   Courtesy Aanderaa Instruments Ltd.
Plate 7   Courtesy General Oceanics Inc.

Contents

Chapter 1 Introduction   1

Chapter 2 Ocean Dimensions, Shapes and Bottom Materials   5
2.1 *Dimensions*   5
2.2 *Sea-floor dimensions*   7
   2.21 *Scales*   7
   2.22 *Shore*   7
   2.23 *Continental shelf*   8
   2.24 *Continental slope and rise*   8
   2.25 *Deep-sea bottom and sounding*   9
   2.26 *Sills*   10
2.3 *Bottom material*   10

Chapter 3 Physical Properties of Sea-Water   12
3.1 *Vocabulary*   12
3.2 *Properties of pure water*   12
3.3 *Temperature*   13
3.4 *Salinity and conductivity*   13
3.5 *Density*   17
   3.51 *Units for density*   17
   3.52 *Effects of temperature and salinity on density*   17
   3.53 *Effect of pressure on density and temperature*   19
   3.54 *Specific volume and anomaly*   20
   3.55 *Tables and formulae for density and specific volume anomaly*   21
3.6 *Other characteristic properties*   23
3.7 *Sound in the sea*   23
3.8 *Light in the sea*   30
3.9 *Colour of sea-water*   32

Chapter 4 Typical Distributions of Water Characteristics in the Oceans   34
4.1 *Introduction*   34
   4.11 *General*   34
   4.12 *Collection and analysis of data*   34
   4.13 *General statistics and area descriptions*   36
4.2 *Temperature distribution*   37
   4.21 *Surface temperature*   37
   4.22 *Upper layers and the thermocline*   40
   4.23 *Langmuir circulations*   42

4.24	Temporal variations of temperature in the upper layer	44
4.25	Deep water; potential temperature	47
4.3	Salinity distribution	49
4.31	Surface salinity	49
4.32	Upper-layer salinity	51
4.33	Deep-water salinity	52
4.34	Temporal variations of salinity	53
4.4	Density distribution	53
4.41	Density at the surface	53
4.42	Subsurface density and the pycnocline	53
4.43	Static stability	54
4.44	Geographic distribution of density	56
4.5	Dissolved oxygen distribution	57
4.6	Other water-motion tracers	59
4.61	Introduction	59
4.62	Nutrients	59
4.7	Transparency of the upper ocean from Secchi disc measurements	62

Chapter 5 Water, Salt and Heat Budgets of the Oceans 64

5.1	Conservation of volume	64
5.2	Conservation of salt	65
5.21	Principle	65
5.22	Two examples of applications of the two conservation principles	67
5.221	The Mediterranean Sea	67
5.222	The Black Sea	67
5.3	Conservation of Heat Energy; the Heat Budget	68
5.31	Heat-budget terms	68
5.32	Short- and long-wave radiation; elements of radiation theory	70
5.33	Short-wave radiation (Q_s)	71
5.331	Incoming solar radiation	71
5.332	Effects of atmospheric absorption, solar elevation, clouds, etc.	72
5.333	Spatial and temporal variations	75
5.34	Long-wave radiation (Q_b)	75
5.341	Determining factors	75
5.342	Temporal and spatial variations	76
5.343	Effect of clouds	77
5.344	Effect of ice and snow cover	77
5.35	Evaporation (Q_e)	78
5.36	Heat conduction (Q_h)	81
5.361	Eddy conduction	81
5.362	Convection	81
5.363	Bowen's Ratio	81
5.37	Some comments on the calculation of the heat budget terms	83
5.38	Geographic distribution of the heat-budget terms	84
5.381	Pacific Ocean	84
5.382	Atlantic Ocean	87
5.383	World ocean; northern hemisphere	88
5.384	Large-scale programmes on the heat budget	89

Chapter 6 Instruments and Methods 92

6.1	Introduction	92
6.2	Instruments	94
6.21	Winches, wire, etc.	94
6.22	Depth measurement	95

Contents

6.23 *Water properties*	95
6.231 *Water-sampling bottles*	95
6.232 *Temperature measurement*	96
6.233 *Salinity measurement*	98
6.234 *Density measurement*	101
6.24 *Expendable instruments*	102
6.25 *Current measurement*	102
6.251 *Lagrangian methods*	103
6.252 *Eulerian methods; drag and propeller-type meters*	107
6.253 *Eulerian methods; non-propeller-type meters*	110
6.254 *Mooring current meters and other instruments*	112
6.255 *Geostrophic method and dynamic topography*	115
6.256 *Acoustic tomography*	123
6.26 *Radiation measurement*	124
6.3 *Platforms*	127
6.31 *Sea- and airborne*	127
6.32 *Satellites*	129
6.321 *Satellite remote sensing*	129
6.322 *Weather satellites*	129
6.323 *The Coastal Zone Colour Scanner*	132
6.324 *SEASAT*	132
6.325 *Other sensors and derived products*	134
6.326 *Other satellite applications in oceanography*	136
6.327 *The future*	137
6.4 *Age of ocean water*	137
6.5 *Graphical presentation of data*	139
6.51 *Variation in space; profiles and sections*	139
6.511 *Vertical direction*	139
6.512 *Horizontal direction*	140
6.52 *Variation in time*	140
6.521 *Time-series plots*	140
6.522 *Current data plots*	142
6.53 *Isentropic analysis*	143
6.54 *Characteristic diagrams*	144
6.541 *Two characteristics, e.g.* T–S, T–O_2	144
6.542 *Global* T–S *characteristics of ocean waters*	146
6.543 *Core method*	147
6.544 *Three characteristics, e.g.* T–S–V, T–S–t	148
6.55 *Conclusion*	153

Chapter 7 Circulation and Water Masses of the Oceans   154

7.1 *Introduction; mechanisms*	154
7.11 *Thermohaline circulation*	154
7.12 *Wind-driven circulation*	158
7.13 *Mixing processes*	165
7.14 *Circulation and water masses*	170
7.2 *Southern Ocean*	173
7.21 *Divisions of the Southern Ocean*	173
7.22 *Southern Ocean circulation*	175
7.23 *Southern Ocean water masses*	177
7.231 *Antarctic zone*	177
7.232 *Subantarctic zone*	179
7.3 *Atlantic Ocean*	180
7.31 *Atlantic Ocean as a whole*	180
7.32 *South Atlantic Ocean*	181
7.321 *South Atlantic circulation*	181
7.322 *South Atlantic volume transports*	181

7.33	Equatorial Atlantic circulation	184
7.34	North Atlantic circulation: general	184
	7.341 Gulf Stream system	185
	7.342 Gulf Stream volume transport	190
	7.343 Gulf Stream temperature and salinity distributions	190
	7.344 Gulf Stream rings and ocean eddies	192
7.35	Atlantic Ocean water masses	195
	7.351 Atlantic Ocean upper waters	196
	7.352 Atlantic Ocean deep-water masses and circulation	196
	7.353 T–S characteristics of subsurface waters of the Atlantic Ocean	201
7.4	North Atlantic Adjacent Seas	208
	7.41 Mediterranean Sea	208
	7.42 Black Sea	211
	7.43 Baltic Sea	213
	7.44 Norwegian and Greenland Seas	214
	7.45 Labrador Sea, Baffin Bay and Hudson Bay	216
	7.46 Adjacent seas; inflow and outflow characteristics	218
7.5	Arctic Sea	219
	7.51 Arctic Sea: upper-layer circulation	220
	7.52 Arctic Sea water masses	220
	7.521 Arctic Water	221
	7.522 Atlantic Water	223
	7.523 Deep or Bottom Water	224
	7.53 Arctic Sea budgets	225
	7.54 Ice in the sea	226
	7.541 Ice physics	227
	7.542 Distribution of sea-ice	231
	7.543 Build-up and break-up of sea-ice in several regions	232
	7.544 Icebergs	233
7.6	Pacific Ocean	235
	7.61 Pacific Ocean circulation	235
	7.611 Pacific Ocean: equatorial circulation	235
	7.612 A meridional section across the upper equatorial Pacific	240
	7.613 Zonal and temporal variations of the Pacific equatorial current system	242
	7.614 Convergences and divergences associated with current systems	248
	7.615 Deeper currents in the equatorial Pacific	251
	7.62 North Pacific circulation	252
	7.621 North Pacific gyre and the Kuroshio	252
	7.622 Influence of the North Pacific Sea-Surface Temperature on the North American climate	254
	7.63 South Pacific circulation	255
	7.64 Eastern boundary currents; Peru Current and El Niño	256
	7.641 Peru Current	256
	7.642 El Niño and the Southern Oscillation (ENSO) and La Niña	258
	7.65 Pacific Ocean water masses	263
	7.651 Pacific Ocean upper waters	263
	7.652 Pacific Ocean deep waters	268
7.7	Indian Ocean	270
	7.71 Indian Ocean circulation	270
	7.72 Indian Ocean water masses	273
7.8	Red Sea and Persian Gulf	274

Chapter 8 Coastal Oceanography   276

8.1 Introduction   276
8.2 Coastal Upwelling   279
8.3 Ekman circulation in shallow water   280

8.4 *Estuaries*	281
8.41 *Types of estuaries*	281
8.42 *Estuarine circulation*	284
8.5 *Coral reefs*	286
8.51 *Introduction*	286
8.52 *Topography*	286
8.53 *Water properties*	287
8.54 *Tides*	288
8.55 *Currents*	288
8.56 *Circulation in lagoons*	289

Chapter 9   Some Directions for Future Work   291

APPENDIX   Units used in descriptive physical oceanography   298

BIBLIOGRAPHY   301

■ Suggestions for Further Reading   301
■ References to Journal and Review Articles   303

INDEX   313

CHAPTER 1

Introduction

OCEANOGRAPHY is the general name given to the scientific study of the oceans, with an emphasis on their character as an environment. It is conveniently divided in terms of the basic sciences into physical oceanography, biological oceanography, chemical oceanography and geological oceanography. This book is concerned primarily with one aspect of the first of these.

The basic goal of oceanographic study is to obtain a clear and systematic description of the oceans, sufficiently quantitative to permit us to predict their behaviour in the future with some certainty. While we can do this in a general fashion for some characteristics and in some regions, we are a long way from being able to predict details with confidence. In other words, there is still a great deal of scientific study of the oceans to be done and understanding to be achieved.

Generally the individual scientist studying the ocean devotes himself to investigations in one of the sciences, but very often supporting information may be obtained from observations in other sciences. In fact, one of the intriguing aspects of oceanography is that it is not yet too highly compartmented or specialized, and there is much co-operation between those working in the different sciences.

There are many reasons for developing our knowledge of the oceans. As sources of food, of chemicals and of power, they are as yet only exploited to a very minor degree. They are still a vitally important avenue of transportation and are always likely to be. They form a sink into which industrial waste is dumped, but they do not form a bottomless pit into which material like radioactive waste can be thrown without due thought being given to where it may be carried by currents. The vast heat capacity of the oceans exerts a significant effect on the climate of the land, while the continuous movement of the currents and waves along the coast must be taken into account when piers, breakwaters and other structures are built.

In all of these applications, and in many others, a knowledge of the circulation of the oceans is needed. The goal of the physical oceanographer is to obtain a systematic quantitative description of the character of the ocean waters and of their movements. The latter include the major ocean currents which circulate continuously but with fluctuating velocity and position, medium and small scale circulation features, the variable coastal currents, the reversing tidal currents, the rise and fall of the tide, and the waves generated by wind or earthquake. The character of the ocean waters includes those aspects, such as temperature and salt content, which together determine density and hence vertical movement, and also includes other dissolved substances or biological species in so far as they yield information about the currents.

The physical study of the oceans is approached in two ways. In what is called the synoptic or descriptive approach, observations are made of specific features and these are reduced to as simple a statement as possible of the character of the features themselves and of their relations to other features. The dynamical or theoretical approach is to apply the already known laws of physics to the ocean, regarding it as a body acted upon by forces, and to endeavour to solve the resulting mathematical equations to obtain information on the motions to be expected from the forces acting. In practice there are limitations and difficulties associated with both methods, and our present knowledge of the oceans has been developed by a combination of the synoptic and the dynamical approaches. Ideally, the method is as follows. Preliminary observations give one some idea what features of the ocean require explanation. The basic physical law which is considered to apply to the situation is then used to set up an equation between the forces acting and the motions observed. A solution of this equation, even an approximate one, will give some indication of how the motions may vary in time or space. It may also suggest further observations which may be made to test whether or not the law selected or the features entered into the equation are adequate or not. If not, the theory is modified in the light of the test observations and the procedure of alternate observation of nature and development of the theory is pursued until a satisfactory theory is obtained. The method is typical of scientific research.

Our present knowledge in physical oceanography represents an accumulation of data, most of which have been gathered during the past hundred years. The purpose of this book is to summarize some of these data to give an idea of what we now know about the distribution of the physical characteristics of the ocean waters and of their circulation. The achievements of the alternate but parallel approach through the laws of mechanics are described in other texts such as those mentioned in the Preface.

During its history physical oceanography has gone through several phases. Presumably ever since man started to sail the oceans he has been concerned with ocean currents as they affect the course of his ship. This distinctly practical approach is more a branch of the related field of hydrography, which includes the preparation of navigation charts and of current and tide tables, than of oceanography, but out of it came the study of the currents for the purpose of determining *why* they behave in the way they do as well as *how*. Many of the earlier navigators, such as Cook and Vancouver, made valuable scientific observations during their voyages in the late 1700s, but it is generally considered that Mathew Fontaine Maury (1855) started the systematic large-scale collection of ocean current data, using ship's navigation logs as his source of information. Many physical data on surface currents and winds were collected, and still are, from this source. The first major expedition designed expressly to study all the scientific aspects of the oceans was that of H.M.S. *Challenger* which circumnavigated the globe from 1872 to 1876. The first large-scale expedition organized primarily to gather physical oceanographic data was the German *Meteor* expedition to study the Atlantic Ocean from 1925 to 1927. Expeditions in increasing numbers in the following years have added to our knowledge of the oceans, both in single ship and in multi-ship operations including the loosely coordinated worldwide International Geophysical Year projects in 1957–58, the International Indian Ocean Expedition in 1962–65, the oceanographic aspects of GATE in 1974 (GARP Atlantic Tropical Experiment where GARP = Global Atmospheric Research Programme) and, in the late 1970s, POLYGON, MODE and POLYMODE in the Atlantic (see Section 7.344), the Coastal Upwelling Ecosystems projects in the Pacific and Atlantic, NORPAX (North Pacific Experiment) and ISOS (International Southern Ocean Study) and many other projects. Nevertheless, there are still areas such as the Arctic, the Southern Ocean and the southern Pacific and Indian Oceans for which we have very limited information on which to base our large-scale, steady-state description. Only in a few selected regions do sufficient data exist to allow study of the significant variations in space and time; most of the world's ocean remains a very sparsely sampled environment.

Some of the earliest theoretical studies of the sea were of the surface tides by Newton (1687) and Laplace (1775), and of waves by Gerstner (1847) and Stokes (1874), Following this, about 1896, some of the Scandinavian meteorologists started to turn their attention to the ocean, since dynamical meteorology and dynamical oceanography have much in common. The present basis for dynamical oceanography owes much to the early work of Bjerknes *et al.* (1933), Ekman (1905, 1953), Helland-Hansen (1934) and others.

In recent years attention has been given to other phases, including the circulation and water properties at the ocean boundaries, along the coasts and in estuaries, and also in the deep and bottom waters of the oceans. The coastal waters are more accessible for observation than the open ocean but show large fluctuations in space and time, which present difficulties for theoretical study. For these reasons, the earlier studies tended to be of the open ocean but more detailed studies in recent years have revealed a hitherto unexpected wealth of detail in the form of eddies and shorter-scale time and space variations in the open ocean. The deep and bottom waters are very difficult to observe: this makes it hard to acquire information to start the theoretical studies, and also to test them.

The plan followed in this book will be to describe briefly the ocean basins and something of their topography as it affects ocean circulation, and then to introduce some of the terminology of physical oceanography. After a brief summary of the properties of fresh-water and sea-water, a general description of the distribution of water characteristics both in the vertical and in the horizontal will be presented to give the reader some feeling for typical conditions. A discussion of the sources of gain or loss of heat and water to the ocean follows and then a description of instruments and of methods for data analysis and presentation. After this there is a description of the water characteristics and of the currents in the individual oceans of the world and in coastal regions as such, and finally a few comments on the present state of our knowledge in descriptive physical oceanography.

One aspect of this study which will become apparent during our description is that there are strong interactions between the ocean and the atmosphere (e.g. the ENSO phenomenon, Section 7.642), and to understand one it is necessary to understand the other. In consequence, oceanographers and meteorologists work closely together in studying their respective regions and as larger-scale studies develop, both theoretical and observational, these are clearly joint studies of both the hydrosphere and the atmosphere.

A comment on the title of this book should be made before proceeding. Among oceanographers the term "synoptic oceanography" is understood to refer to the method of approach which starts with the observation of data and then continues with the preparation of a concise description, i.e. a "synopsis". (The adjective "synoptic" is also used in the term "synoptic data", meaning data for an area collected as quickly as possible to try to eliminate the effects of variations with time.) However, description and synopsis are only the start. The oceanographer then seeks regularities in the data and interprets the distributions of properties with the object of obtaining information on the circulation. Therefore a more exact title for this aspect would be "Interpretative Oceanography", but unfortunately this term is not in general use. "Synoptic Oceanography" was the first choice for a title and would be clear in its meaning for trained oceanographers. However, this book is not intended for them but for would-be oceanographers or for those who desire an introduction to this aspect of science. In the end, the title "Descriptive Physical Oceanography" was chosen in the hope that this would best indicate the character of its contents to those who were not yet familiar with the field. It is hoped, however, that the reader who completes its study will no longer be a novice but will by then appreciate that "synoptic" oceanography does not stop at description but continues with its main aim, interpretation.

CHAPTER 2

Ocean Dimensions, Shapes and Bottom Materials

2.1 Dimensions

The oceans are basins in the surface of the solid earth containing salt water. The purpose of this chapter is to introduce some of the nomenclature and to direct attention to features of the basins which have a close connection with the circulation and are of importance to the physical oceanographer. A more detailed description of the geology and geophysics of the ocean basins is given in *Submarine Geology* or in *Geological Oceanography*, both by Shepard, or in the articles by Menard and by Bullard in *Oceanography* listed in the Bibliography at the end of this book. A more recent and comprehensive review is presented by Kennett (1981).

In order to appreciate the shapes of the oceans and seas it is almost essential to examine them on a globe, since map projections on to flat paper always introduce distortions when large portions of the earth are to be represented. From the oceanographic point of view it is convenient to distinguish the various regions in terms of their oceanographic characteristics, particularly their circulations.

Anticipating the information to be given in later chapters, the major ocean areas will be defined now as the Southern Ocean, the Atlantic Ocean, the Pacific Ocean, the Indian Ocean and the Arctic Sea. The last four are clearly divided from each other by land masses but the divisions between the Southern Ocean and the others to its north are determined only by the characteristics of the ocean waters and by their circulations as will be described in Chapter 7. Then there are smaller bodies of water such as the (European) Mediterranean Sea, the Caribbean Sea, the Sea of Japan, the Bering Sea, etc., which are clearly bounded by land or by island chains. The term "sea" is also used for a portion of an ocean which is not divided off by land but has local distinguishing oceanographic characteristics. Examples are the Norwegian, the Labrador and the Tasman Seas.

Looking at a globe again, it is evident that more of the earth's surface is covered by sea than by land, about 71% compared with 29%. Furthermore, the proportion of water to land in the southern hemisphere is much greater (4:1) than in the northern hemisphere (1.5:1). In area, the Pacific Ocean is about as large as the Atlantic and Indian Oceans combined. If one includes the neighbouring sectors of the Southern Ocean with the three main oceans north of it, then the Pacific Ocean occupies about 46% of the total world ocean area, the Atlantic Ocean about 23%, the Indian Ocean about 20%, and the rest combined about 11%.

The average depth of the oceans is close to 4000 metres while the seas are generally about 1200 m deep or less. Relative to sea level the oceans are much deeper than the land is high. While only 11% of the land surface of the earth is more than 2000 m above sea level, 84% of the sea bottom is more than 2000 m deep. However, the maxima are similar: the height of Mt. Everest is about 8840 m while the maximum depth recorded in the oceans is 11,524 m by H.M.S. *Cook* in the Mindanao Trench in the western Pacific. Figure 2.1 shows the distributions of land elevations and of sea depths relative to sea level in 1000-m intervals as the percentage of the total area of the earth's surface. The land data are from Kossina (1921) and the ocean data from Menard and Smith (1966).

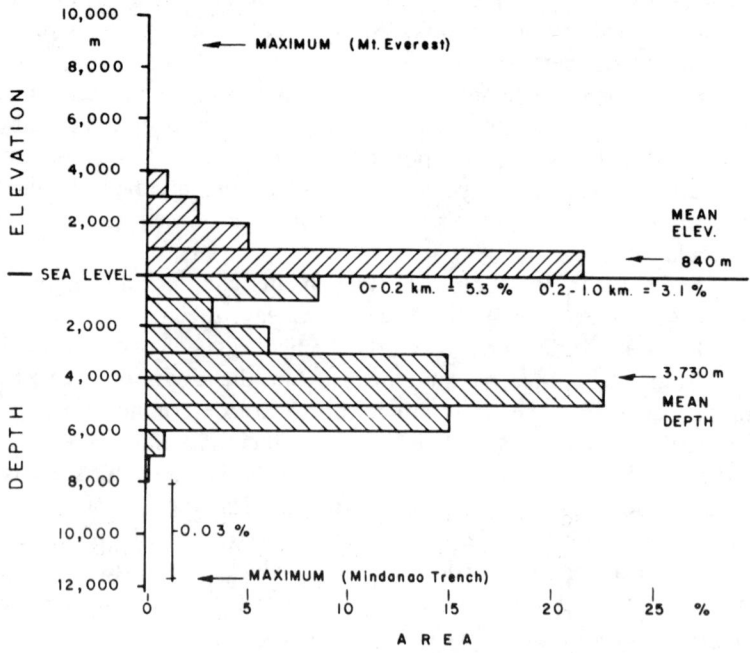

FIG. 2.1. Areas of earth's surface above and below sea-level as percentage of total area of earth (in 1000-m intervals).

Ocean Dimensions, Shapes and Bottom Materials

Although the average depth of the oceans, 4 km, is a considerable distance, it is small compared with the horizontal dimensions of the oceans, which are of the order of 5000 to 15,000 km. An idea of the relative dimensions of the Pacific Ocean may be obtained by stating that they are much the same as a sheet of very thin typing paper. This analogy makes the oceans appear as a very thin skin on the surface of the earth. Relative to the major dimensions of the earth they are thin, but there is a great deal of detail and structure in this thin layer between the sea surface and the bottom of the ocean.

2.2 Sea-floor Dimensions

2.21 Scales

Very often we wish to present some of these details by drawing a vertical cross-section of a part of the oceans. A drawing to true scale would have the relative dimensions of the edge of a sheet of paper and would be either too thin to show details or too long to be convenient. Therefore we usually have to distort our cross-section by making the vertical scale much larger than the horizontal one. For instance, we might use a scale of 1 cm on the paper to represent 100 km horizontally in the sea while depths might be on a scale of 1 cm to represent 100 m, i.e. 0.1 km. In this case the vertical dimensions on our drawing would be magnified by 1000 times compared with the horizontal ones. This gives us room to show the detail but exaggerates the slope of the sea bottom or of lines of constant property (*isopleths*) drawn on the cross-section, such as in Figs. 2.2, 7.12, etc. It is as well to remind oneself that such slopes are in reality far less than they appear on the cross-section drawings. For instance, a line of constant temperature (*isotherm*) with a real slope of 1 in 100 would be exceptionally steep in the ocean, one of 1 in 1000 very steep and of 1 in 10,000 more usual.

The continents form the major lateral boundaries to the oceans, and the detailed features of the shoreline and of the sea bottom are important in their effects on circulation. Starting from the land, the main divisions recognized are the shore, the continental shelf, the continental slope and rise, and the deep-sea bottom (shown schematically in Fig. 2.2).

2.22 Shore

The *shore* is defined here as part of the land mass close to the sea which has been modified by the action of the sea. It is as well to note in this connection that there is ample evidence to indicate that sea level in the past has varied over a range of about 100 m when glaciers were smaller or larger than they are now. The *beach* is the seaward limit of the shore and extends roughly from the highest to the lowest tide levels. Sandy beaches are often in a state of dynamic equilibrium. That is to say, they may be composed of sand all the time but it

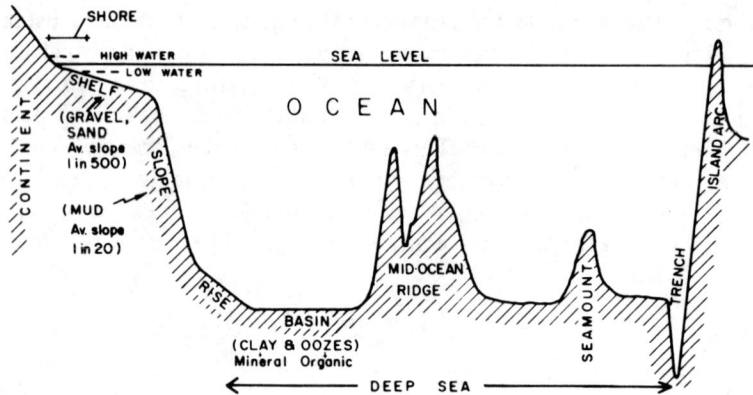

Fig. 2.2. Schematic section through ocean floor to show principal features.

may not always be the same sand. This may be continually moving along the shore under the influence of waves and nearshore currents. Evidence for this process can be seen in the way in which sand accumulates against new structures built on the shore, or by the way in which it is removed from a beach when a breakwater is built in such a way as to cut off the supply of sand beyond it. On some beaches, the sand may be removed by currents associated with high waves at one season of the year and replaced by different currents associated with lower waves at another season.

2.23 Continental shelf

The *continental shelf* extends seaward from the shore with an average gradient of 1 in 500. Its outer limit (the "break-in-slope" or "shelf-break") is set where the gradient increases to about 1 in 20 on the average to form the continental slope down to the deep sea bottom. The shelf has an average width of 65 km. In places it is much narrower than this, while in others, as in the north-eastern Bering Sea or the Arctic shelf off Siberia, it is as much as ten times this. The bottom material is dominantly sand, with rock or mud being less common. The division between the shelf and slope is made on the basis of the break-in-slope which is usually clearly evident when one examines a vertical cross-section of the sea bottom from the shore outward. The average depth at the shelf break is about 130 m. Most of the world's fisheries are located on the continental shelf.

2.24 Continental slope and rise

The *continental slope* averages about 4000 m vertically from the shelf to the deep sea bottom, but in places extends as much as 9000 m vertically, in a relatively short horizontal distance. In general, the continental slope is considerably steeper than the slopes from lowland to highland on land. The material of the slope is predominantly mud, with some rock outcrops. Very typical features of the shelf and slope are the submarine canyons which are of worldwide occurrence. They are valleys in the slope, either V-shaped or with vertical sides, and are usually found off coasts with rivers and never off desert areas. The lower part of the slope, where it grades into the deep-sea bottom is referred to as the *continental rise*.

2.25 Deep-sea bottom and sounding

From the bottom of the continental slope the gradient decreases down the continental rise to the *deep-sea bottom*, the last and most extensive area. Depths of 3000 to 6000 m are found over 74% of the ocean basins with 1% being deeper. Perhaps the most characteristic aspect of the deep-sea bottom is the variety of its topography. Before any significant deep ocean soundings were available the sea bottom was regarded as uniformly smooth. When detailed sounding started in connection with cable laying, it became clear that this was not the case and there was a swing to regarding the sea bottom as predominantly rugged. Neither view is exclusively correct, for we know now that there are mountains, valleys and plains on the ocean bottom just as on land. The characteristic features are, as on land, either basically long and narrow (welts and furrows) or of roughly equal lateral extent (swells and basins). The *Mid-Ocean Ridge* is the most extensive feature of the earth's topography. Starting south of Greenland it extends along the middle of the Atlantic from north to south and then through the Indian and Pacific Oceans. In the Atlantic it separates the bottom waters, as can be seen from their very different properties east and west of the ridges (Fig. 7.17). However, there are narrow gaps in this ridge at some of the "fracture zones". These are roughly vertical planes perpendicular to the ridge and on either side of which the crust has moved in opposite directions perpendicular to the ridge. This leaves a gap in the ridge through which water below the ridge top may leak from one side of the ridge to the other. One such is the Romanche Gap through the Mid-Atlantic Ridge close to the equator.

Individual mountains (*seamounts*) are widely distributed in the oceans. Some project above the surface to form islands, while the tops of others are below the surface.

In some of the large basins the sea floor is very smooth, possibly more so than the plains areas on land. Stretches of the abyssal plain in the western North Atlantic have been measured to be smooth within 2 m over distances of 100 km.

The deepest parts of the oceans are in the *trenches*. The majority of the deep ones are in the Pacific, including the Aleutian, Kurile, Philippine and Mariana Trenches, with a few in other oceans such as the Puerto Rico and the South Sandwich Trenches in the Atlantic and the Sunda Trench in the Indian Ocean.

In dimensions they are narrow relative to their length and have depths to 10,000 m. They are often shaped like an arc of a circle in plan form with an arc of islands on one side. Because of this they are asymmetrical in cross-section, the island side extending as much as 10,000 m from the trench bottom to the sea surface while the other side is only half as high as it terminates at the ocean depth of about 5000 m.

Our present knowledge of the shape of the ocean floor results from an accumulation of sounding measurements, most of which have been made within the last 60 years. The early measurements were made by lowering a weight on a measured line until the weight touched bottom. This method was slow; in deep water it was uncertain because it was difficult to tell when the weight touched the bottom, and to be certain that the line was vertical. Since 1920 most depth measurements have been made with *echo-sounders* which measure the time taken for a pulse of sound to travel from the ship to the bottom and be reflected back to the ship. One half of this time is multiplied by the average speed of sound in the sea-water under the ship to give the depth. With present-day equipment, the time can be measured very accurately and the main uncertainty over a flat bottom is in the value used for the speed of sound. This varies with water temperature and salinity (see Section 3.7) and if these are not measured at the time of sounding an average value must be used. This introduces a possibility of error. Over trenches or places where rapid changes with depth occur there may also be some uncertainty about whether the echo comes from directly under the ship (to give the true depth) or from one side (to give too small a value).

2.26 Sills

One other term used frequently in reference to bottom topography is *sill*. This refers to a ridge, above the average bottom level in a region, which separates one basin from another or, in the case of a fjord (Chapter 8), separates the landward basin from the sea outside. The *sill depth* is the maximum depth from the sea surface to the top of the ridge, i.e. the maximum depth at which direct flow across the sill is possible.

2.3 Bottom Material

On the continental shelf and slope, most of the bottom material comes directly from the land, either brought down by rivers or blown by the wind. The material of the deep-sea bottom is often more finely divided than that on the shelf or slope. Much of it is pelagic in character, i.e. it has been formed in the sea itself. The two major deep ocean sediments are inorganic "red" clay and the organic "oozes". The former has less than 30% of organic material and is mainly mineral in content. It consists of fine material from the land (which may have travelled great distances before finally settling out on to the bottom), volcanic material, and meteoric remains. The oozes are over 30% organic and originate from the remains of living organisms (plankton). The calcareous oozes have a high percentage of calcium carbonate from the shells of animal plankton, while the siliceous oozes have a high proportion of silica from the shells of planktonic plants and animals. The siliceous oozes are found mainly in the Southern Ocean and in the equatorial Pacific. In both of these regions, the distribution is clearly related to the water flow above it—around Antarctica in the first case and parallel to the equator in the second.

Except when turbidity currents (mud slides down the slope) deposit their loads on the ocean bed, the average rate of deposition of the sediments is from 0.1 to 10 mm per 1000 years, and much information on the past history of the oceans is stored up in them. Samples of bottom material are obtained with a "corer" which is a steel pipe 2 to 30 m long which is lowered vertically and forced to penetrate into the sediments by the heavy weight at its upper end. The "core" of sediment retained in the pipe may represent material deposited over a period from 100,000 to 10 million years per metre of length. Sometimes the material is layered, indicating stages of sedimentation of different materials. In some places, layers of volcanic ash can be related to historical records of eruptions; in others, organisms characteristic of cold or of warm waters are found in different layers and suggest changes in temperature of the overlying water during the period represented by the core. In some places gradations from coarse to fine sediments in the upward direction suggest the occurrence of turbidity currents bringing material to the region with the coarser material settling out first and the finer later.

The physical oceanographer looks at the sediments for the information which they give him on the movement of the water above and on its temperature history. This information may be of movements in the past. It may also give some idea of the mean flow, averaging out small fluctuations. The distribution of material around Antarctica is an example of this. The surface of the ocean bottom also reveals information. Photographs of the deep-sea bottom have been obtained in recent years and some of them show ripples such as one sees on a sand beach after the tide has gone out. Such ripples are only found on the beach where the water speed is quite considerable, such as in the backwash from waves. We conclude from the ripples on the deep-ocean bottom that currents of similar speed occur there. This discovery helped to dispel the earlier notion that all deep-sea currents are very slow.

CHAPTER 3

Physical Properties of Sea-Water

3.1 Vocabulary

In preparation for discussing the physical properties of sea-water in this chapter, and in anticipation of discussions of measurement techniques in Chapter 6, several words will be defined. They are:

Determination: the actual direct measurement of a variable, e.g. the length of a piece of wood with a ruler,

Estimation: a value for one variable derived from the determination of one or more others, e.g. the estimation of salinity from the determination of chlorinity or conductivity and temperature.

Accuracy: the difference between a result obtained and the true value.

Precision: the difference between one result and the mean of several obtained by the same method, i.e. reproducibility (includes random errors only).

Systematic error: one which results from a basic (but unrealized) fault in the method and which causes values to be consistently different from the true value (cannot be detected by statistical analysis of values obtained and affects Accuracy).

Random error: one which results from basic limitations in the method, e.g. the limit to the accuracy with which one can read the level of liquid in a burette. It is possible to determine a value for this type of error by statistical analysis of a sufficient number of measurements; it affects Precision. Truly random errors are identified by their Gaussian distribution.

3.2 Properties of Pure Water

Many of the unique characteristics of the ocean can be ascribed to the nature of water itself. Consisting of two positively charged hydrogen ions and a single negatively charged oxygen ion, water is arranged as a polar molecule having positive and negative sides. This molecular polarity leads to water's

Physical Properties of Sea-Water

high dielectric constant (ability to withstand an electric field) and its high solvent ability. Water is able to dissolve more substances than any other fluid. This property explains the abundance of ions in the ocean resulting in its salty character.

The polar nature of the water molecule causes it to form polymer-like chains of up to eight molecules. A certain amount of energy goes into linking these molecules which explains the ocean's ability to absorb heat energy which then may be transported by currents. This aspect of the ocean plays an important, but poorly understood, role in the interaction between sea and atmosphere and the determination of the global climate.

As water is heated, molecular activity increases and thermal expansion occurs. At the same time, added energy is available for the formation of molecular chains whose alignment causes the water to shrink. The combination of these effects results in pure water having a maximum density at 4°C rather than at its freezing point. In sea-water, these molecular effects are overshadowed by the presence of salt which affects the density as described in Section 3.52.

Another important effect of the chainlike molecular structure is the very high surface tension of water. In the ocean, one effect of this can be seen in the formation of surface capillary waves which depend on surface tension for a restoring force. Such capillary waves, despite their small size, are considered to play an important part in determining the friction between wind and water which is responsible for the generation of larger waves and for the major circulations of the upper layers of the oceans.

3.3 Temperature

An important physical characteristic of sea-water is its temperature. Easily measured with thermometers it was one of the first ocean parameters studied. Even today it continues to be easier to observe temperature and its vertical profile than other properties. In most of the mid- and lower-latitude upper ocean (between the surface and 500 m), temperature is the primary parameter determining density, and studies of temperature profiles alone have yielded valuable insights into circulation features (and also sound-speed distributions). Some of the factors determining temperature are described in Section 5.3 and techniques for its measurement are described in Section 6.232. Temperature (t) is expressed on the Celsius scale (°C) in oceanography (e.g. $t = 5°C$) (except for some expressions for thermal radiation); when referring to temperature *differences* (Δt) we shall express them in kelvins (K) (e.g. $\Delta t = 2.5 K$) to avoid confusion with actual temperature (°C). (See Appendix.)

3.4 Salinity and Conductivity

Sea-water is a complicated solution and contains the majority of the known elements. Some of the more abundant components are chlorine ion 55.0% of the total dissolved material, sulphate ion 7.7%, sodium ion 30.6%, magnesium ion 3.7% and potassium ion 1.1%. A significant feature of seawater is that while the total concentration of dissolved salts varies from place to place, the ratios of the more abundant components remain almost constant. This may be taken as evidence that over geologic time the oceans have become well mixed, i.e. while there are well-marked circulations within each ocean, water must also circulate, between the oceans. At the same time there are significant differences in total concentration of the dissolved salts from place to place and at different depths. This indicates that processes must be continually in action to concentrate or dilute sea-water in specific localities; these processes are features of the sea which oceanographers wish to understand.

The total amount of dissolved material in sea-water is termed the *salinity* and has been defined as "the total amount of solid materials in grams contained in one kilogram of sea-water when all the carbonate has been converted to oxide, the bromine and iodine replaced by chlorine and all organic matter completely oxidised". For example, the average salinity of ocean water is about 35 grams of salts per kilogram of sea-water (g/kg), previously written as "S = 35‰" or as "S = 35 ppt" and read as "thirty-five parts per thousand". The direct determination of salinity by chemical analysis or by evaporating sea-water to dryness is too difficult to carry out routinely. The method which was used from the beginning of the century until recently was to determine the amount of chlorine ion (plus the chlorine equivalent of the bromine and iodine), called *chlorinity*, by titration with silver nitrate and then to scale up to the salinity by a relation based on the measured ratio of chlorinity to total dissolved substances (e.g. see Wallace (1974) or Wilson (1975) for a full account). Later, the current definition of salinity was established as "the mass of silver required to precipitate completely the halogens in 0.328 523 4 kg of the sea-water sample". The relation between salinity and chlorinity was redetermined in the early 1960s and since then has been taken as:

$$\text{Salinity} = 1.806\ 55 \times \text{Chlorinity}.$$

This is now referred to as the *absolute salinity*, symbol S_A.

However, this definition of salinity has now been replaced by one (called *practical salinity*, symbol S) based on the *electrical conductivity* of sea-water, because almost all salinity estimations are now made by determining this quantity which depends on salinity and temperature (see Lewis and Perkin, 1978; Lewis and Fofonoff, 1979). Such values are now written as, for example, $S = 35.00$ or occasionally as $S = 35.00$ psu (practical salinity units). Because the dependence of electrical conductivity on temperature is significant this has either to be controlled very accurately by a thermostat or measured and corrected for during the measurement. The real advances which permitted the change from chemical titration to the electrical salinometer were refinements in the electrical circuits to permit accurate compensation for temperature.

Physical Properties of Sea-Water 15

Then the inductive (electrodeless) method (Section 6.233) was developed to avoid problems with electrochemical effects at the electrodes in a conductance cell but recently techniques to eliminate the effects of electrochemical changes have resulted in a return to the electrode-type conductance cell for both laboratory and *in situ* salinometers. The precision of the titration method in routine use is about ± 0.02 in salinity while by electrical conductivity it is about ± 0.003 in salinity.

The present position with regard to our knowledge of the relations between the chlorinity, the density and the electrical conductivity of sea-water should be made clear. In 1884, Dittmar reported the results of his chemical analysis of seventy-seven samples of sea-water collected from around the world by the *Challenger* Expedition. These results supported the belief, expressed earlier by Forchhammer (1865), that the ratios, one to another, of the concentrations of the major ions in sea-water are subject to only slight variations. Some subsequent writers have rephrased this to imply that the ratios are exactly constant or have acted as though this were the case. That is to say they have assumed that there is an exact and constant relationship between the chlorinity, density and electrical conductivity of sea-water from any place in any ocean and from any depth. In a critical review of the subject, Carritt and Carpenter (1958) pointed out that this rephrasing is incorrect, and that Dittmar's results themselves indicate the possibility of small variations. Such variations may be of significance, particularly in the study of deep water where the differences in properties are small. After Carritt and Carpenter's review, discussions of these matters made it clear that the time had come for the relations to be reinvestigated, both because of the advances made in analytical techniques since Dittmar's time and because of the improvements made in the techniques for the measurement of electrical conductivity of sea-water and in the greater reproducibility of conductivity measurements compared with titration methods.

An extensive series of measurements of the chlorinity, density relative to pure water, and conductivity of sea-water was carried out under the direction of the late R. A. Cox at the National Institute of Oceanography (now the Institute of Oceanographic Sciences) in England on samples from widely distributed locations in the world oceans. The study revealed that the ionic composition of sea-water does show small variations from place to place and from the surface to deep water. The reasons for these variations in ionic composition are not known. It was found that the relationship between density and conductivity was a little closer than between density and chlorinity (Cox *et al.*, 1970). This is interpreted to mean that the chlorinity, i.e. the chemical composition, shows significant differences from place to place and with depth in the oceans. One of the factors used in arriving at this interpretation is that solutions of the major constituents of sea-water of the same concentration have very similar conductivities and densities. This means that the proportion of one ion to another may change, i.e. the chemical composition may change, but as long as the total weight of dissolved substances is the same the conductivity and the density will be effectively unchanged.

Since one of the main reasons for determining either the salinity or conductivity of sea-water is to deduce the density (which is inconvenient to measure directly), the conclusion is that it is better to do this from conductivity than from chlorinity. For this reason, and because of the practical advantages in use (including *in situ* measurement), the electrical conductivity method is generally used today for the estimation of salinity. (What is actually measured is the *ratio* of the conductivity of the unknown sea-water sample to that of a standard sea-water of known salinity.) It has been demonstrated that, with average observers, conductivity ratio measurements allow density to be estimated with a precision nearly an order of magnitude better than from chlorinity measurements.

On the Practical Salinity Scale 1978 (PSS 78), the *practical salinity* (S) of a sample of sea-water is defined in terms of the ratio K_{15} of the electrical conductivity of the sea-water sample at the temperature of 15°C and pressure of one standard atmosphere to that of a potassium chloride solution in which the mass fraction of KCl is 32.4356×10^{-3} at the same temperature and pressure. A concise review of the earlier definitions of salinity and of the development of the PSS 78 by the Joint Panel on Oceanographic Tables and Standards has been given by Lewis (1980) and includes the formulae for the computation of practical salinity from laboratory and from *in situ* conductivity ratio measurements. (The special issue of the *IEEE Journal of Oceanic Engineering*, Vol. OE-5, No. 1, January 1980, is devoted to (rather technical) details of the PSS 78.)

The International Oceanographic Tables published in 1966 jointly by Unesco and the then National Institute of Oceanography contain the first set of tables for estimating salinity from conductivity ratio values. These tables will be replaced soon by tables based on the PSS 78, but the conversion from conductivity ratio to salinity will usually be carried out by computer using the formulae given by Lewis (1980).

The reasons for the considerable effort by many workers in many laboratories to prepare PSS 78 were that the 1966 Tables only went to 10°C, whereas the bulk of the ocean is at lower temperatures, and the effect of pressure on conductivity needed review, both of these factors affecting *in situ* measurements with CTD instruments with which most salinity data are now collected, and a reproducible primary standard (now the KCl solution) was needed to ensure consistency between laboratories. The PSS 78 is considered valid for the range between $S=2$ to 42, $t=-2°$ to 35°C and pressures equivalent to depths from 0 to 10,000 m.

A useful summary of the history of the measurement of the physical properties of sea-water, of the development of the Practical Salinity Scale (PSS 78) and of formulae for the calculation of *salinity* from electrical

Physical Properties of Sea-Water 17

conductivity, for the *equation of state, density & steric anomalies, specific heat, adiabatic lapse rate, potential temperature* and *Brunt-Väisälä frequency* is given by Fofonoff (1985) together with an extensive list of references to the literature about the determination of these quantities and the relevant formulae. In Unesco (1983) algorithms are given for the calculation of the above quantities as well as for *pressure to depth conversion, freezing point temperature of sea-water* and *sound speed in sea-water*.

3.5 Density

3.51 Units for density

The physical oceanographer is particularly interested in the salinity and temperature of sea-water because they are characteristics which help to identify a particular water body (Chapters 6 and 7) and also because, together with pressure, they determine the *density* (ρ) of sea-water. The latter is important because it determines the depth to which a water mass will settle in equilibrium—the least dense on top and the most dense at the bottom. The distribution of density can also be related to the large-scale circulation of the oceans through the geostrophic relationship (see Chapter 6).

Density is expressed physically in kilograms per cubic metre (kg/m^3) and in the open ocean values range from about 1021.00 kg/m^3 (surface) to about 1070.00 kg/m^3 (at 10,000 m depth). As a matter of convenience, it is usual in oceanography to quote only the last four of these digits in the form of a quantity called $\sigma_{s,t,p}$ defined as:

$$\sigma_{s,t,p} = \text{density} - 1000 \text{ kg/m}^3$$

where s = salinity, t = temperature (Celsius) and p = pressure. This is referred to as the *in situ* value. For many applications in descriptive oceanography the pressure effect on density can be ignored and a quantity $\sigma_{s,t,o}$ is used, commonly abbreviated to σ_t (spoken as "sigma-tee"). This is the density difference of the water sample when the total pressure on it has been reduced to atmospheric (i.e. the water pressure $p = 0$) but the salinity and temperature are as *in situ*. The relationship between σ_t, salinity and temperature is a complicated non-linear one and no simple formula has been devised for it. In practice, values of σ_t are obtained from a nomogram or tables which are entered with the appropriate values of salinity and temperature, or from a polynomial expression in salinity and temperature for computer calculations (e.g. see Millero *et al.*, 1980).

3.52 Effects of temperature and salinity on density

Figure 3.1 shows the σ_t values (curved full lines) for the whole range of salinities and temperatures found anywhere in the oceans. Two points to note

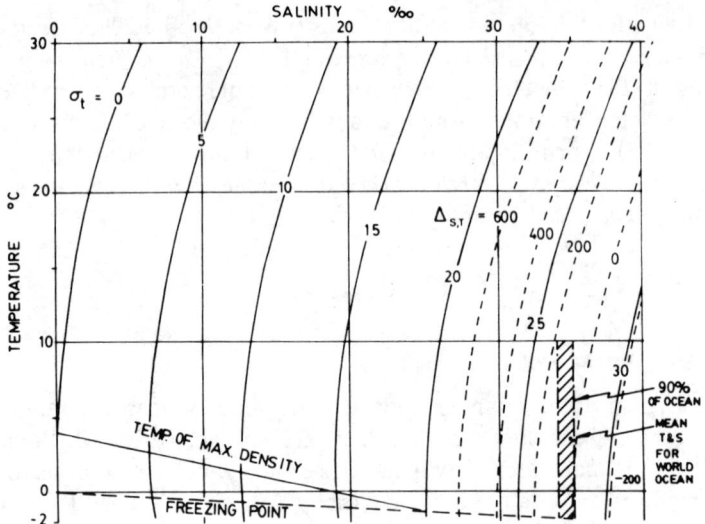

FIG. 3.1. Values of σ_t, $\Delta_{S,T}$, temperature of maximum density and freezing point (at atmospheric pressure) for sea-water as functions of temperature and salinity.

are that the change of σ_t with a change of salinity is almost uniform over the whole range of salinity and temperature but that the variation with temperature change is distinctly non-uniform. To emphasize this point, Table 3.1 shows on the left the change of σ_t ($\Delta\sigma_t$) for a change (Δt) of $+1$K in

TABLE 3.1

Variation of σ_t ($\Delta\sigma_t$) with variations of temperature (ΔT) and of salinity (ΔS) as functions of temperature and salinity

	$\Delta\sigma_t$ for $\Delta t = +1\text{C}°$			$\Delta\sigma_t$ for $\Delta S = +0.5$		
Salinity	0	20	40	0	20	40
Temperature (°C)						
30	−0.30	−0.33	−0.34	0.39	0.38	0.38
20	−0.21	−0.24	−0.27	0.40	0.38	0.38
10	−0.09	−0.14	−0.18	0.41	0.39	0.39
0	+0.07	−0.01	−0.17	0.43	0.40	0.40

temperature and on the right the value of $\Delta\sigma_t$ for a change (ΔS) of $+0.5$ in salinity. It will be seen that at high temperatures, σ_t varies significantly with t at all salinities but as temperature decreases the rate of variation with t decreases, particularly at low salinities (as found at high latitudes or in estuaries). The change of σ_t with ΔS is about the same at all temperatures and salinities. At the same time, it should be noted that the ranges of salinity and temperature found in most of the ocean volume are much less than shown in the whole of Fig. 3.1; the shaded area includes the values for 90% of the volume of the oceans. More extreme values only occur at the surface. (For convenience, values for σ_t are usually quoted without units because it is generally used for descriptive purposes. However, in formulae it must be treated as having units of kg/m^3.)

In Fig. 3.1 the straight full line represents the temperature at which water has its maximum density, starting at about 4°C for fresh water, while the dashed line represents the freezing point as a function of salinity. A point to note about freezing is that at low salinities, water which is cooled reaches its maximum density before freezing and sinks while still fluid, the water "overturning" until it all reaches the temperature of maximum density. On further cooling the surface water becomes lighter and stays near the surface, finally freezing from the surface down, the deeper water remaining unfrozen. However, at salinities greater than 24.7, as sea-water cools the vertical circulation continues until the freezing point, so that the entire water column must be cooled to this temperature, and therefore freezing is delayed.

3.53 Effect of pressure on density and temperature

The reason why one can often omit the pressure component of $\sigma_{s,t,p}$ is because in descriptive oceanography one is usually comparing water masses at the same depth, i.e. the same pressure, or over the same range of depth. However, it should be noted that water is not completely incompressible and the effect of pressure on density is not negligible. For instance, a sample of water of $S = 35.00$ and $t = 0°C$ would have a $\sigma_{s,t,0}$ at the surface (i.e. σ_t) of 28.13 but at a depth of 4000 m its $\sigma_{s,t,p}$ would be increased by compression to 48.49, i.e. the density increased by about 2%.

Associated with the change of density, i.e. volume, with pressure is a change of temperature if the water does not exchange heat with its surroundings (adiabatic change). For instance, if water of salinity 35 and temperature 5.00°C were lowered adiabatically from the surface to a depth of 4000 m its temperature would increase to 5.45°C due to compression. Conversely, if its temperature were 5.00°C at 4000 m depth and it were raised adiabatically to the surface it would cool to 4.56°C due to expansion. This effect has to be considered when water is changing depth significantly. In the second example above the temperature of 5.00°C is called the "*in situ*" temperature (t) while the 4.56°C value is called the "*potential temperature*" (θ). Note that the compressibility of sea-water is a function of both temperature and salinity (as well as pressure) which is why we specified the initial temperature and salinity in the numerical example above. Further references to the effect of temperature on compressibility and hence on the calculation of potential temperature will be given in Sections 4.43 and 7.352.

From thermodynamic considerations, it can be shown that the potential temperature:

$$\theta = t_i + \int_{p_i}^{0} \Gamma(S, t', p) \, dp$$

where
$$t' = t_i + \int_{p_i}^{p} \Gamma(S, t, p) \, dp,$$

S_i, t_i, p_i are the values *in situ* and $\Gamma(S, t, p) =$ adiabatic lapse rate which is the change of temperature per unit pressure change for an adiabatic (no heat exchange) displacement of a water parcel. Fofonoff (1985) presented a general expression for θ as:

$$\theta(S_0, t_o, p_0, p_r) = t_o + \int_{p_o}^{p_r} \Gamma(S, \theta(S_o, t_o, p_o, p), p) \, dp$$

where $\theta(S_o, t_o, p_o, p_r)$ is the temperature which a water parcel of properties (S_o, t_o, p_o) would have if moved adiabatically without change of salinity from an initial pressure p_o to a reference pressure p_r where p_r may be greater or less than p_o. The integration above can be carried out in a single step (Fofonoff, 1977) and an algorithm for the calculation of θ is given in Unesco (1984).

A formula for $\Gamma(S, t, p)$ was given by Fofonoff (1985) as a polynomial with 52 terms. The use of the potential temperature is discussed with an example in Section 4.34. The density of a water sample appropriate to its salinity and potential temperature is called its *potential density* and there is a corresponding *sigma-theta* (σ_θ).

3.54 Specific volume and anomaly

The *specific volume* (α) is the reciprocal of density, with units of m³/kg, and for some purposes is more useful. One application is in the calculation of currents from the distribution of mass by the geostrophic method (Section 6.255). The specific volume *in situ* is written as $\alpha_{s,t,p}$; for convenience a *specific volume anomaly* (δ) defined as:

$$\delta = \alpha_{s,t,p} - \alpha_{35,o,p}.$$

The last quantity is the specific volume of an arbitrary standard sea-water of salinity 35 and temperature 0°C at the pressure p. This standard was chosen so that δ is usually positive. Again the connection between α or δ and salinity, temperature and pressure is complicated. The values of $\alpha_{35,o,p}$ are tabulated for all practical values of p (e.g. Sverdrup *et al.*, 1946, p. 1053). For δ, Bjerknes and Sandström (1910) examined the experimental values as functions of salinity, temperature and pressure to show that δ could be broken down into components as:

$$\delta = \delta_s + \delta_t + \delta_{s,t} + \delta_{s,p} + \delta_{t,p} + \delta_{s,t,p}$$

where δ_s represents the main effect of salinity, δ_t represents that of temperature, $\delta_{s,t}$ represents the interacting effects of salinity and temperature,

Physical Properties of Sea-Water

etc. In practice, the last term is small enough to be ignored, while the two pressure terms, $\delta_{s,p}$ and $\delta_{t,p}$ are smaller than the first three terms. In fact, Montgomery and Wooster (1954) pointed out that in the actual oceans the sum of the first three terms $\delta_s + \delta_t + \delta_{s,t} = \Delta_{S,T}$ is adequate in most practical cases to describe the specific volume of water masses. They proposed calling this term, $\Delta_{S,T}$, the *thermosteric anomaly* and in recent years it has come to be used frequently in place of σ_t to describe the density of ocean waters, at least for the upper layers. The physical units for α, the δ's and $\Delta_{S,T}$ are m^3/kg. Numerical values for $\Delta_{S,T}$ are 50 to 100×10^{-8} m^3/kg; for $\delta_{s,p}$ or $\delta_{t,p}$ they are usually of the order of only 5 to 15×10^{-8} m^3/kg per 1000 m depth. (Previously, a value such as 50×10^{-8} m^3/kg was written as 50 centilitres per tonne (cL/t) to avoid having to write the power of 10.)

Note that although $\Delta_{S,T}$ is the specific volume analogue of σ_t it is not the reciprocal of σ_t. Convenient formulae for calculating equivalent values of σ_t and $\Delta_{S,T}$ in the range $\sigma_t = 23$ to 28 are:

$$\sigma_t = -0.0105\, \Delta_{S,T} + 28.1 \quad \text{(accurate to 0.1 in } \sigma_t\text{)},$$
$$\Delta_{S,T} = -95.1\, \sigma_t + 2675 \quad \text{(accurate to 1 in } \Delta_{S,T}\text{)}.$$

In using these formulae, the power of 10^{-8} in the physical value of $\Delta_{S,T}$ is ignored for calculation, e.g. for $\sigma_t = 25$, the second formula gives $\Delta_{S,T} = 297$, its physical value then being 297×10^{-8} m^3/kg.

In Fig. 3.1, isopleths (lines of constant value) for $\Delta_{S,T} = -200, 0, 200, 400$ and 600×10^{-8} m^3/kg are shown as examples. Isopleths of thermosteric anomaly run substantially parallel to those of σ_t but the values change in the opposite direction. The density of a large proportion of the water of the oceans lies between $\sigma_t = 25.5$ to 28.5 which corresponds approximately to $\Delta_{S,T} = 250$ to -50×10^{-8} m^3/kg.

3.55 Tables and formulae for density and specific volume anomaly

The tables and formulae from which values of σ_t or $\Delta_{S,T}$ are obtained are based on laboratory determinations of relative density at different salinities and temperatures. The classical determinations based on measurements by Forch, Jacobsen, Knudsen and Sörensen and presented in the Hydrographical Tables (M. Knudsen, Ed., 1902) give the relative density of sea-water, referred to pure water, as σ_t as a function of chlorinity, salinity and temperature. In 1970 Cox, McCartney and Culkin reported the results of a new investigation of the relations between salinity and density of sea-water relative to pure water. The new results indicated that the values for σ_o (at $t = 0°C$) in "Knudsen's Tables" are low by about 0.01 on the average in the salinity range from 15 to 40 and up to 0.06 at lower salinities and temperatures. In response to a recommendation by the Unesco Joint Panel on Oceanographic Tables and Standards in 1978 a new equation of state was formulated based on all available precision data (Millero et al., 1980; Fofonoff, 1985) to relate values of α, t, S and p over the ranges of property values from $t = -2$ to $40°C$ and $S = 0$ to 40, and pressures from 0 to 10^5 kPa (equivalent to 0 to 10,000 m depth). It is considered accurate to 5×10^{-6} over oceanic ranges of properties and depths.

The relative densities of sea-water over a range of salinities were determined most recently with a magnetic float densimeter (Millero, 1967). A pyrex glass float containing a permanent magnet floats in a 250 ml cell which contains the sea-water and is surrounded by a solenoid, the whole being in a constant temperature bath. The float is slightly less dense than the most dense sea-water and is loaded with small platinum weights until it just sinks to the bottom of the cell. A current through the solenoid is then slowly increased until the float just lifts off the bottom of the cell. The density of the sea-water is then related to the current through the solenoid. The relation between current and density is determined by carrying out a similar experiment with pure water in the cell. The accuracy of the relative density determined in this way is claimed to be $\pm 2 \times 10^{-6}$ (at atmospheric pressure) but as the absolute density of pure water is known to only $\pm 4 \times 10^{-6}$, this limits the accuracy of knowledge of sea-water density calculated from the formula derived from the work of Millero et al. (1980). Measurements were made on sea-water samples of a range of salinities and over a range of temperatures. The influence of pressure was determined using a high pressure version of the above magnetic float densimeter to provide values for the bulk modulus (K). The latter quantity has also been determined from measurements of the speed of sound (C) in sea-water as $C = \sqrt{K/\rho} = \sqrt{1/\beta\rho}$, where $K = $ bulk modulus $= 1/\beta$ and $\beta = $ compressibility, Section 3.7.)

The relation between density and temperature, salinity and pressure is referred to as the equation of state for sea-water and is:

$$\rho(S, t, p) = \rho(S, t, O)/[1 - p/K(S, t, p)].$$

The polynomial expressions for the equation of state $\rho(S, t, O)$ (EOS 80) and $K(S, t, p)$ derived from the Millero et al. measurements and those of others contain 15 and 27 terms respectively and may be found in Pond and Pickard (1983), Fofonoff (1985) or Unesco (1983). A summary of how the numerical values were assembled from the modern measurements is presented by Fofonoff (1985). Values calculated from the equation of state are considered to be accurate to 9×10^{-6} or better.

For some purposes the possible errors in Knudsen's Tables are not serious. For instance, in the geostrophic method for calculating currents, the oceanographer uses differences between σ_t or $\Delta_{S,T}$ values, not absolute values, and if the salinity and temperature values are not very different, the deficiencies in Knudsen's Tables essentially cancel out and become a minor source of error compared with other sources. This also justifies the oceanographic practice of quoting values of σ_t to four significant figures even though there may be some doubt about the absolute value of the fourth figure.

For plotting σ_t curves, a convenient reference is "Tables for Sigma-T" by Fleming (1939) which give values of salinity and temperature for whole number and fractional values of σ_t from 18 to 30.

The presently available data on density of sea-water are, as mentioned, for the relative density (to pure water). Determinations of the absolute density are needed but this is a very expensive and time-consuming task and only tentative plans are in hand.

One reason why salinity and temperature, and hence density, are important identifying properties of sea-water is because they are "conservative properties" away from the surface. This is to say, below the surface there are no significant processes by which either quantity is changed except by mixing. This absence of *in situ* sources and sinks means that the spreading of water masses in the ocean can be traced from their origin at the sea surface by their characteristic temperature/salinity values as described in Section 6.54. Near the surface, evaporation or precipitation may change salinity, while many surface heat-transfer processes may change the temperature, as discussed in Section 5.3.

3.6 Other Characteristic Properties

Other characteristics of sea-water which are of help in identifying specific water masses are the dissolved oxygen content, concentration of nutrients (phosphate, nitrate, silicate, etc., ions) which will be discussed in Section 4.6, plankton, silt, optical characteristics, etc. These, however, have to be used with care because they are not conservative. Biological processes may change the concentration of oxygen or nutrients without any movement of the water mass, silt may settle out (which may also change the optical properties), etc. These other constituents generally occur in such small concentrations that their variations do not significantly affect the density nor do they affect the relations between chlorinity, salinity and conductivity.

The specific heat of sea-water is an important characteristic of which the classical determinations were reported by Thoulet and Chevallier (1889). In 1959 Cox and Smith of the National Institute of Oceanography reported new measurements estimated to be accurate to 0.05% which gave values 1 to 2% higher than the old ones. A further study (Millero, Perron and Desnoyers, 1973) gave values in close agreement with those of Cox and Smith. Some of the other thermal properties of sea-water and of sea-ice are described in Section 7.54 on "Ice in the Sea".

3.7 Sound in the Sea

In the atmosphere, man receives much of his information about the material world by means of wave energy, either electromagnetic (light) or mechanical (sound). In the atmosphere, light in the visible part of the spectrum is attenuated less than sound, but in the sea the reverse is the case. In clear ocean water, sunlight may be detectable (with instruments) down to 1000 m but the range at which man can see details of objects in the sea is rarely more than 50 m and usually less. Being denied the use of his eyes in the sea, except for close ranges when diving, man has made much use of sound waves to obtain information. With echo-sounders the depth to the sea bottom may be determined to the maximum in the ocean. With SONAR (Sound Navigation And Ranging) the direction and distance to a submarine may be determined to ranges of hundreds of metres and to schools of fish to somewhat lesser ranges. Following Swallow floats (Section 6.251) by sonic means was one of the first techniques used by oceanographers to investigate deep currents directly. Side-scan sonars are used to "take pictures" of limited areas of the ocean bottom to determine its structure and locate shipwrecks. However, the turbulent nature of the ocean and inhomogeneities therein affect the propagation of sound waves and make it impractical to form pictures with them.

As for all wave motions, passage through a medium may be influenced by refraction (change of direction due to change of wave speed), absorption of wave energy by the medium, and scattering by objects in the medium (e.g. silt, gas bubbles, fish, etc.).

The frequencies of sounds which are of interest in the ocean range from 1 Hz (1 hertz = 1 vibration per second) or less to thousands of kHz (kilohertz). As the speed (C), frequency (n) and wavelength (λ) are connected by the wave equation $C = n\lambda$, the wavelengths of sounds in the sea cover a vast range, e.g., from about 1500 m for $n = 1$ Hz to 7 cm for $n = 200$ kHz. However, most underwater sound instruments use a more restricted range from 10 to 100 kHz for which the wavelengths would be 14 to 1.4 cm.

One of man's first applications of sound in the sea was for *echo sounding*, to determine ocean depths more quickly than with a sounding line. In echo-sounding applications, short pulses of sound energy are directed vertically downward and the time taken (t) to the bottom and back to the surface is measured to yield the depth $D = (\bar{C} t)/2$ where $\bar{C}$ represents the mean speed of sound between the surface and the bottom. Usually the same device (transducer) is used as the source to transmit the sound pulse into the water and as receiver to pick it up from the water on its return. Since the size of the transducer of an ordinary echo-sounder is not much greater than the wavelength of the sound, the angular width of the sound beam emitted is large. For a 12 kHz echo sounder, the beam width, in which the energy is no less than one-half of its maximum, will be of the order of 30° to 60°, which makes it difficult to distinguish details of the bottom topography. For special sounding applications, much larger sound sources which form a narrower beam are used. It is also possible to improve the resolution by using higher frequencies to 100 kHz or even 200 kHz but as the absorption of sound energy by sea-water increases roughly as the square of the frequency, this entails loss of range or depth of water which can be penetrated. Echo sounders are also used to detect shoals of fish whose air bladders are good reflectors of sound energy in water.

A second widely used application is in *sonar* equipment which is essentially an echo sounder aligned to send a narrow beam of sound pulses approximately horizontally to determine the range to a reflector, e.g. a submarine or a shoal of fish. The transducer may be rotated through 360° so that the direction to the target is known.

So far we have discussed only the behaviour of sound waves generated by sources introduced for the purpose by man. If one uses a hydrophone to listen to the ambient sound in the sea, a very wide range of frequencies and types of sounds may be heard from low rumbles to high-frequency hisses. Some of the sources which have been identified are microseisms generating low-frequency sounds up to 10–100 Hz, ships producing sounds from 50–500 Hz and up to 1500 Hz, noise generated by wind action at the surface, waves, rain (1–20 kHz), cavitation of air bubbles, animal noises (10–400 Hz), fish and crustaceans (1–10 kHz). Under ice can be heard a variety of noises from 1–10 kHz with highest levels at low frequencies.

The ocean contains a variety of inhomogeneities such as sediment particles, bubbles, plankton, fish, seamounts, etc. which cause scattering of sound waves from a source, and irregular reflections occur from the sea surface and bottom. A result is that a sharp sound pulse from a source is not received at a hydrophone as a simple pulse but is likely to have an irregular tail of later arrival sounds (from the inhomogeneities and irregularities) which is referred to as *reverberation*. One of the sources of reverberation is the so-called *deep scattering layer* which is biological in nature. A feature of this layer is that it exhibits *diel* (day/night) vertical migration of several hundreds of metres, moving toward the sea surface at dusk and back down at dawn. It has been identified as the result of scattering by plankton and (gas-filled) fish bladders and is found in the upper few hundred metres.

The speed of sound waves in the sea is given by the relation $C=\sqrt{1/\beta\rho}$ where β is the adiabatic compressibility of sea-water and ρ is the density. As these quantities depend on temperature and pressure, and to a much lesser extent on salinity, so does the speed of sound. There are a number of formulae for the speed of sound, derived from experimental measurements, and we will present a simple one to illustrate features of the relationship, i.e.:

$$C = 1449 + 4.6t - 0.55t^2 + 1.4(S-35) + 0.017D \quad \text{m/s}.$$

At $t=0°C$, $S=35$, the speed increases by 4 m/s for $\Delta t = +1K$, by 1.4 m/s for $\Delta S = +1$ and 17 m/s for $\Delta D = 1000$ m increase in depth (which represents the pressure effect). In the upper layers, where temperature varies most, sound speed is chiefly determined by this parameter, but in deep water (below about 2000 m) depth (i.e. pressure) is the main determining factor. Only in high latitude waters, where the temperature is close to zero and does not change much, does salinity have any significant effect.

A consequence of these variations of sound speed with water properties and depth, and the typical vertical distributions of these properties is an *in situ* sound speed minimum at depths ranging from near the surface at high latitudes (low temperatures in the upper water) to over 1000 m in mid- and low latitudes. Figure 3.2 shows typical (a) temperature and salinity profiles for a

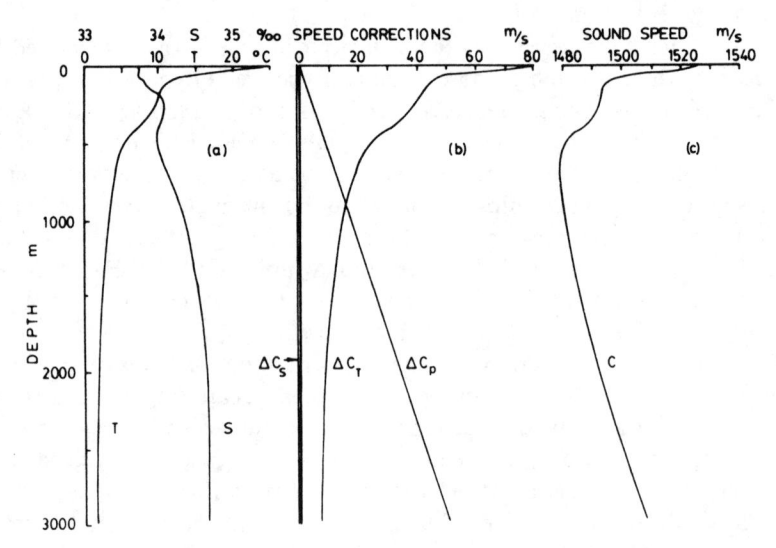

FIG. 3.2. For station in Pacific Ocean at 39° N, 146° W, August 1959: (a) temperature and salinity profiles, (b) corrections to sound speed due to salinity, temperature and pressure, (c) resultant *in situ* sound-speed profile showing sound-speed minimum.

mid-latitude station together with (b) the individual effects of property variations on sound speed. The resultant sound speed profile (c) with a minimum centred at about 700 m depth is also shown. Note that the minimum results chiefly from the reduction of sound speed with decreasing temperature in the upper 700 m followed by the increase with depth (i.e. pressure) below that level. Salinity variations have negligible effects on sound speed in this location. The wider range of temperature variations in the ocean and their strong effect on sound speed make it possible to study acoustic propagation with vertical temperature profiles alone in the upper 1000 m or so.

For a profile in which sound speed increases from the surface to an intermediate depth maximum (e.g. in an upper mixed layer) and then decreases (Fig. 3.3(a)) the acoustic rays will turn toward the surface near the top and toward greater depths below the maximum. The shaded area represents a *shadow zone* into which the rays do not penetrate so that targets could not be detected there. The presence of a sound speed minimum (Fig.

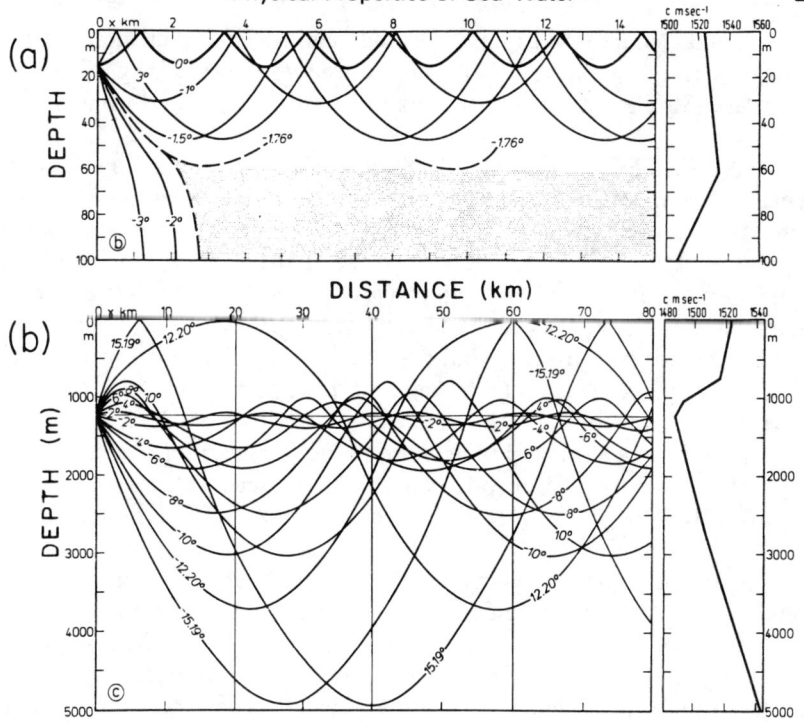

FIG. 3.3. Sound ray diagrams: (a) from a shallow source for a sound speed profile initially increasing with depth in upper mixed layer to a shallow maximum and then decreasing, (b) from a sound source near the speed minimum in the sound channel for a typical open ocean sound speed profile.

3.3(b), typical of mid-ocean regions) gives rise to a *sound channel*. For a source near the depth of the speed minimum, sound waves which are directed at moderate angles above the horizontal are refracted downward, cross the depth of the minimum and are then refracted upward, continuing to oscillate above and below the sound speed minimum depth. Conversely, sound waves initially directed below the horizontal are first refracted upward, cross the minimum and are then refracted downward and also continue to oscillate about the minimum. Because there are many possible starting angles for the waves emitted from the source, there are an equal number of rays following approximately sinoidal paths in the channel. (Note that rays which travel steeply up or down from the source will not be channelled but may travel to the surface or bottom and be reflected there.) This sound channel is referred to as the SOFAR channel (SOund Fixing And Ranging) because sound waves of low frequency (hundreds of hertz) can travel very considerable distances (thousands of kilometres) along it. It permits the detection of submarines at long ranges and has been used for locating lifeboats at sea when the occupants drop overboard an explosive charge designed to explode in the sound channel, the explosion sound being picked up at shore stations established for the purpose. The application to the tracking of drifting subsurface (Swallow) floats to determine deep currents is described in Section 6.251.

The deep SOFAR channel is characteristic of mid- and low-latitudes where the temperature decreases substantially with increase of depth. At high latitudes where the temperature near the surface may be constant (e.g. Fig. 4.6(a), March), or even decrease toward the surface, the increase of sound speed with increase of depth can give rise to a much shallower sound channel which may even be in the surface layer. In the latter case, downward directed sound rays from a shallow source are refracted upward while upward rays from the source are reflected downward from the surface and then refracted upward again (Fig. 3.3(a)). In this situation, detection of deep submarines from a surface ship using sonar equipment mounted in the hull may not be possible and deep-towed sonar equipment may be needed. In shallow water (e.g. < 200 m), reflection can occur both from the surface and from the bottom.

In sonar applications, a signal is sent from a source to a target and the energy which is returned to the receiver (colocated with the source), provides the information on direction (from the direction toward which the source beam is aligned) and range of the target (from the transit time to and from the target). The returned signal will be followed by a "tail" of energy scattered by the inhomogeneities in the sea and from the surface and bottom, i.e. reverberation, which appears as a slowly decaying signal following the main arrival.

However, a feature of the pulse transmission from a source near the sofar channel axis is that at the distant receiver, the signal does not appear as a sharp pulse but as a drawn out signal rising slowly to a peak followed by a sharp cutoff. The peak before the cutoff represents the arrival of the sound energy along the sound channel axis (direct signal), while the earlier arrivals are from sound travelling along the refracted ray routes. It might be thought, from Fig. 3.3(b), that as the refracted rays have to travel a greater distance than the direct ray they would therefore be delayed, but this is an illusion. Figure 3.3(b) is drawn with gross vertical exaggeration, to enable the rays to be shown clearly, but in fact the differences in distances travelled by refracted rays and the direct rays are very small and the greater speed in the refracted ray paths more than makes up for their greater distance, so that the direct ray arrives last.

Some features of very long-range sound channel tracks were considered by Munk et al. (1988), who analysed data for sound transmission from explosives detonated in the sound channel at sea off Perth, Australia, and received in Bermuda. These two locations are almost antipodal, the great circle distance (south-west from Perth, south of Cape Agulhas (South Africa) and north-west up the Atlantic) is 19,840 km but the authors point out that as the earth is ellipsoidal, not spherical, the appropriate minimum distance on the globe is rather along the geodesic and is 19,821 km. Both of these paths pass geometrically clear of Africa and South America but another feature which we have not mentioned yet would cause a likely geodesic to intersect South Africa. This feature is *lateral* refraction in which appropriate variations of sound speed horizontally transverse to the direction of propagation would cause lateral refraction of the wave front in addition to the vertical refraction which we have considered hitherto. As the signals *were* received at Bermuda, several alternatives to overcome this blocking were suggested. These were diffraction around Cape Agulhas, bottom scattering and bottom refraction there, diffraction into the acoustic shadow of Africa by internal wave activity and perturbations in lateral refraction by mesoscale eddy activity off Cape Agulhas. The latter was considered the most likely. Calculations indicated that an eddy refracted ray from Perth to Cape Agulhas, continuing as a refracted ray from there to Bermuda, would have a travel time within a few seconds of the measured time of $13,382 \pm 4$ seconds.

The analysis brought out several points. The effect of the ellipsoidal shape of the earth on the path has been mentioned. The three-dimensional sound-speed field in the Indian and Atlantic Oceans was determined from about 8000 oceanographic stations and showed that the depth of the axis of the sound channel along the likely path varied from about 400 m south-west of Perth to about 1500 m south-east of Bermuda. In addition, the effect of lateral refraction due to horizontal variations in temperature in the sound channel brought out the fact that refraction in this channel can be in both vertical and horizontal directions, not only in the vertical direction as we had tacitly assumed before. An application of sound waves to determine aspects of the structure of the ocean will be described in Section 6.256 on "Ocean Acoustic Tomography".

One last point about sound transmission is that the ratio of the speed of sound in air to that in water is small (about 1 to 4.5) and so only a small amount of sound energy starting in one medium will penetrate through the sea surface into the other, in contrast to the relatively efficiency of passage of light energy through the air/water interface (speed ratio only about 1.33 to 1). This is the reason why a person standing on the shore is unaware of the noises in the sea, and why it is impossible to speak directly from the air to a diver underwater or even to converse underwater (because a speaker's sounds are generated in the air in his throat and little of the sound energy is transmitted into the water). For sound sources to be used for sonar or for depth sounders, the sound energy is generated in solid bodies (transducers), e.g. electromagnetically, in which the speed of sound is similar to that in water and so the two are acoustically "matched" and the energy generated in the transducer is transmitted efficiently into the sea.

A source of more information on sound in the sea is in the text by Urich (1982).

3.8 Light in the Sea

The behaviour of visible light in water is different in degree from that in air, in particular it is absorbed in much shorter distances in the sea than in the atmosphere. We are concerned here with short-wave energy of wavelengths from about 0.4 to 0.8 μm (1 μm = 10^{-6} m), i.e. from the violet to the red of the visible spectrum. When this short-wave energy penetrates into the sea, some is scattered but most is absorbed and causes the temperature of the water to rise. It is the major source of supply of heat to the oceans. The vertical attenuation of the energy is progressive as the radiation passes down through the sea and is different for different wavelengths. The progressive decrease in energy penetrating downward is expressed by an exponential law, $I_z = I_0 \exp(-kz)$, where I_0 is the radiation intensity penetrating through the surface, I_z is the remaining intensity at a depth z metres below the surface and k is the *vertical attenuation coefficient* of the water. The effects of depth and of attenuation coefficient are shown in Table 3.2. The coefficient k depends mainly on the

TABLE 3.2

Amount of light penetrating to specified depths in sea-water as a percentage of that entering through the surface

Depth z (m)	Vertical attenuation coefficient k (m^{-1})			Clearest ocean water	Turbid coastal water
	0.02	0.2	2		
0	I_o = 100%	100%	100%	100%	100%
1	I_z = 98	82	14	45	18
2	96	67	2	39	8
10	82	14	0	22	0
50	37	0	0	5	0
100	14	0	0	0.5	0

absorption of light in the water and to lesser extent on scattering.

The first three columns after the depth column demonstrate the relative influence of the attenuation coefficient k and of the depth z on energy of a particular wavelength; the last two columns represent practical conditions in the sea and will be explained below. The column of the table for $k = 0.02$ demonstrates transmission of the most penetrating component, blue light, passing through the clearest ocean water. Energy penetrates coastal waters less readily than this because of the extra attenuation due to suspended particulate matter and to dissolved materials. The absorbing materials are probably chiefly organic acids, the inorganic salts do not show significant absorption at visible wavelengths. The columns for $k = 0.2$ and 2 are more indicative of the transmission of blue light through less clear ocean water and through very turbid water respectively.

Physical Properties of Sea-Water

For sea-water, the coefficient k varies considerably with wavelength. For clear ocean water (Fig. 3.4(a), curve 1) k has its minimum value at about 0.45 μm wavelength so that blue light is attenuated least (penetrates best) while at shorter wavelengths (toward the ultra-violet) and at longer wavelengths (in the red and infra-red) the attenuation is much greater and penetration correspondingly less. The increased attenuation in the ultra-violet is not important to the heat budget of the oceans because the amount of energy reaching sea level at such short wavelengths is small. The increased absorption is more important in and beyond the red end of the spectrum where more energy is present in the sun's radiation. Virtually all of the energy shorter than the visible is absorbed in the top metre of water, while the energy of wavelength 1.5 μm or greater is absorbed in the top one centimetre or less. In Fig. 3.4(b) (full lines) are shown the relative amounts of energy penetrating clear ocean water to 1, 10 and 50 m as a function of wavelength. The maximum penetration is in the blue while penetration by yellow and red is much less.

In more turbid waters, e.g. near the coast, all wavelengths are attenuated more than in clear water, i.e. k is larger as shown by curve 2 in Fig. 3.4(a), and the least attenuation is in the yellow part of the spectrum. The relative penetration of energy for turbid coastal water is shown by the dashed lines in Fig. 3.4(b). The energy penetrating to 1 m is less and to 10 m much less than through clear water, and the maximum penetration is shifted to the yellow. (The energy reaching 50 m in this turbid water is too small to show on the scale of this graph.)

In clear ocean water the superior penetration of blue and green light is evident both visually when SCUBA diving and also in colour photographs taken underwater by natural light. Red or yellow objects appear darker in colour or even black as they are viewed at increasing depths because the light at the red end of the spectrum has been absorbed in the upper layers and little is left to be reflected by the object. Blue or green objects retain their colour to greater depths. In the clear ocean water there is enough light at 50 to 100 m to permit a diver to work, but in turbid coastal waters almost all of the energy may have been absorbed by 10 m depth.

The above remarks have emphasized the penetration of particular wavelengths. The energy from the sun is composed of a range (spectrum) of wavelengths and, as far as the heat budget is concerned (see Chapter 5), the important quantity is the sum total of energy at all wavelengths penetrating into the water. In any particular body of sea-water the total energy penetration is less than that of the most penetrating component. The last two columns of Table 3.2 indicate the range of penetrations found in actual sea-water.

In addition to its significance for the heat budget, the penetration of radiation into the sea is of interest to biologists in connection with the photosynthetic activity of phytoplankton and algae and with the behaviour of zooplankton and fish. Some of the methods for measuring the attenuation or transmission of radiation through the sea are described in Section 6.26.

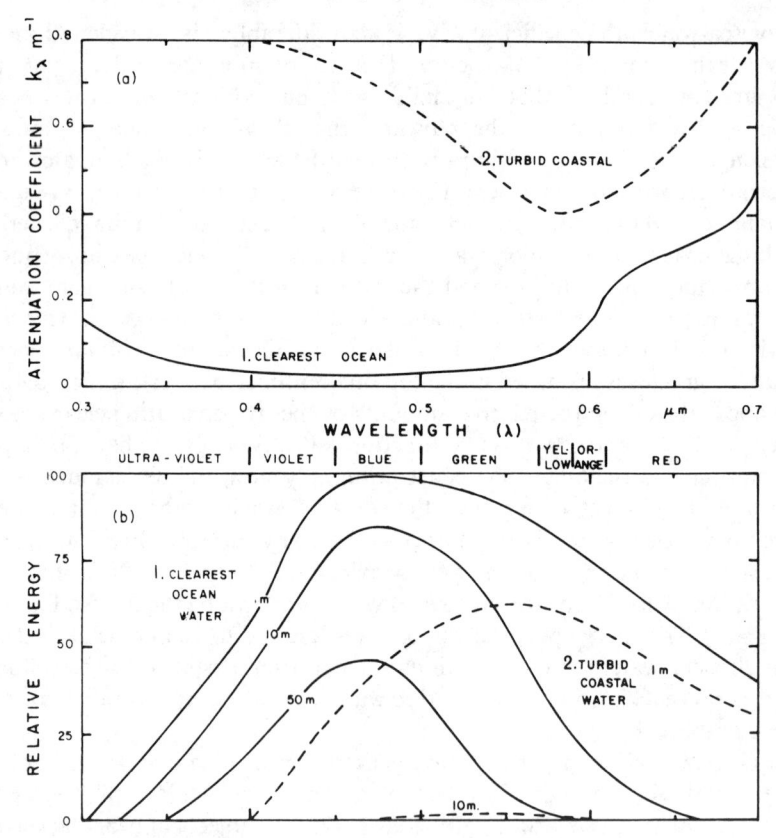

FIG. 3.4. (a) Attenuation coefficient k_λ as a function of wavelength λ for clearest ocean water (full line) and turbid coastal water (dashed line). (b) Relative energy reaching 1, 10 and 50 m depth for clearest ocean water and reaching 1 and 10 m for turbid coastal water.

3.9 Colour of Sea-Water

A number of investigators have considered the reasons for the colour of the sea which ranges from deep blue to green or even greenish-yellow (e.g. Jerlov, 1976). The number of records of sea colour is not great but broadly speaking the deep or indigo blue colour is characteristic of tropical and equatorial seas, particularly where there is little biological production. At higher latitudes, the colour changes through green-blue to green in polar regions. Coastal waters are generally greenish. There are two factors contributing to the blue colour of open ocean waters at low latitudes where there is little particulate matter. In deep water if one looks downward from below the surface, as when snorkelling, the light which one sees is mainly that scattered by the molecules of the water. Because the molecules scatter the short-wave (blue) light much more than the long-wave (red) light the colour seen is selectively blue. In addition, because the red and yellow components of sunlight are rapidly absorbed in the upper few metres, the only light remaining to be scattered from the bulk of the water is the blue light. If one looks at the sea from above the surface, in addition to the blue light scattered from the body of the water one sees some sky light reflected from the surface and the two components add together. If the sky is blue, the sea will still appear deep blue, but if there are clouds the white light reflected from the sea surface will dilute the blue scattered light from the water and the sea will appear less intensely blue. If there are green phytoplankton on the water their chlorophyll content will absorb the blue light and shift the water colour to green. The organic products from plants may also add yellow dyes to the water and these will absorb blue and shift the apparent colour toward the green. This is the situation in the more productive high latitude and coastal waters. In some coastal regions, rivers bring in dissolved organic substances which emphasize the yellowish-green colour. The red colour occurring sporadically in some coastal areas, the so-called "red tide", is caused by blooms of species of phytoplankton of a reddish-brown colour. In other regions, rivers bring in finely divided inorganic material, mud and silt, which may impart their own colour to the water by reflection from the particles. In fjords fed by rivers from glaciers, the surface low-salinity layer may be milky-white from the finely divided "rock flour" produced by abrasion in the glaciers and carried down by the melt water. The material may be kept in suspension by turbulence in the upper layer for a time but when it sinks into the saline water below it flocculates and sinks more rapidly. When diving in such a region one may be able to see only a fraction of a metre in the upper layer but be able to see several metres in the saline water below. The colour of sea-water can be judged most conveniently against the white Secchi disc (Section 6.26) as it is lowered to determine the transparency of the upper water. To judge the colour one may use the "Forel Scale" provided by a set of glass tubes of different shades of blue-coloured water.

CHAPTER 4

Typical Distributions of Water Characteristics in the Oceans

4.1 Introduction

4.11 General

In the previous chapter attention was drawn to temperature and salinity as ocean water characteristics. These quantities vary from place to place in the ocean, and from their distribution we can learn a good deal about the average circulation of the waters. In this chapter, some of the typical distributions will be described so that the reader may gain some feeling for them and be able to recognize normal and abnormal distributions.

A salient feature of the distribution of many water characteristics is that they are horizontally stratified or nearly so. In other words, the sea is made up of substantially horizontal layers as far as these characteristics are concerned and horizontal changes are generally much smaller than vertical ones in the same distance. For instance, near the equator the temperature of the water may drop from 25°C at the surface to 5°C at a depth of 1 km, but it may be necessary to go 5000 km north or south from the equator to reach a latitude where the surface temperature has fallen to 5°C. The average vertical temperature gradient (change of temperature per unit distance) in this case is about 5000 times the horizontal one. However, the horizontal variations do exist and therefore the water properties are distributed in three dimensions. This makes it difficult to display them when we are limited to plotting on paper with only two dimensions. We are usually forced to represent a single real three-dimensional distribution by a number of two-dimensional ones, such as vertical and horizontal sections.

4.12 Collection and analysis of data

To appreciate the way in which our understanding of these distributions is built up, it is helpful to describe briefly how the data are collected. The

Typical Distributions of Water Characteristics in the Oceans 35

oceanographer goes to sea for a cruise in a research ship (Pl. 1) to "occupy a number of oceanographic stations" (sometimes called "hydrographic stations"). This means that at a number of preselected locations (*stations*) the ship is stopped and the oceanographers measure the water properties from surface to deep water. For each station the measurements are then plotted as "vertical profiles", i.e. graphs of temperature, salinity, etc., against depth. The vertical profiles from stations along a line may then be put together to form a "vertical section" of water properties, or the data from a selected depth may be plotted and contoured to show the horizontal distribution of the property at that depth like a map. These are the synopses of the spatial distributions of the data. Originally, the plotting and contouring was done by pencil and paper methods but nowadays the raw data are fed into a computer, checked for consistency, and the plotting done using programmes designed for the purpose. Data reports may also be prepared ready for printing from the checked data.

After the field work and the subsequent plotting of the data, the really interesting part starts. The oceanographer sits down with the plots before him to try to determine the reasons why they are as they are. Among the processes which may be acting are horizontal flow, sinking or upwelling, mixing, diffusion, heat flow through the surface, precipitation etc. There is no set routine method that will be suitable for all situations. The oceanographer does exactly what a detective does in trying to solve a crime. Each assembles all the data (clues) that he can and then tries to deduce from them what really happened and what forces (motives) may be relevant. Each may have all the clues needed, or may lack vital ones, or may be looking at the wrong ones. The oceanographer perhaps has one advantage over the detective. He can sometimes go back to make more observations, whereas the criminal may not be so obliging as to keep repeating his crime.

There is one more dimension to be considered. In addition to the spatial variations of properties (vertical and horizontal) there may be variations with time ("temporal" variations). To observe these the oceanographic ship occupies an *anchor station*, i.e. anchors for a day, a week or longer, or else returns to the same station, time after time, to measure the water properties there. An alternative and more usual approach nowadays is to anchor instrumentation such as current meters, temperature sensors, etc., at specific locations. These instruments automatically record for a period of time (sometimes as long as a year) and are then retrieved for subsequent analysis of their data. These are plotted as *time series* in various ways to see if there are daily ("diurnal") or seasonal variations. Generally the diurnal variations are only appreciable to depths of a few metres and seasonal ones to 100 to 300 m. At greater depths variations take place only over periods of weeks, years or possibly centuries.

Today there are some new opportunities for data collection which depart somewhat from the traditional methods mentioned above. As will be described more fully in the chapter on instrumentation there are now drifting buoys which report regularly, via polar-orbiting weather satellites, on their positions and any sensor data collected (wind, air temperature, sea surface temperature, subsurface temperature profiles, etc.). While the oceanographer may be involved in a cruise to deploy these buoys, the data collection is carried out while the oceanographer sits at home and analyses the data coming in from the satellite. In many cases the buoys are actually deployed by merchant and other "ships of opportunity" that are routinely transiting the region of interest and the oceanographer has no need to go to sea at all to collect his or her data. Still the real challenge remains the same and that is to interpret the buoy data in a physically meaningful way which will shed some light on the processes that cause the data variations observed by the buoy.

Another unique data collection method is the use of such ships of opportunity (also called "volunteer observing ships") to collect upper-layer temperature profile data using the expendable bathythermograph (XBT—see Section 6.24). This device can be deployed from a moving ship and requires only the normal ship power for the recording system. Thus merchant and other ships can be outfitted to collect XBT temperature profiles (down to 500 or 600 m) while they travel along their regular routes. This has become a very effective technique for monitoring the upper-layer thermal variability of the ocean without the high cost of research vessels. The frequent repeat coverage available with these XBT programmes provides a unique challenge for the oceanographer in that now the data must be averaged over time and space to provide adequate space/time coverage of the study region. Although the XBT is less accurate than other research vessel based profiler systems, there are a great many phenomena with signals strong enough to be fully resolved by the XBT data. These phenomena generally exhibit spatial and temporal variability that can only be cost-effectively monitored with such XBT programmes.

4.13 General statistics and area descriptions

Before giving descriptions of some of the typical distributions of water properties, the following statistics on ocean water temperatures and salinities are given for orientation:

(a) 75% of the total volume of the ocean water has properties within the range from $0°$ to $6°C$ in temperature and 34 to 35 in salinity,
(b) 50% of the total volume of the oceans has properties between $1.3°$ and $3.8°C$ and between 34.6 and 34.8,
(c) the mean temperature of the world ocean is $3.5°C$ and the mean salinity is 34.7.

Referring back to Fig. 3.1 it will be seen that the ranges of temperature and salinity found in most of the oceans cover only a very small part of the extreme

Typical Distributions of Water Characteristics in the Oceans 37

ranges that may be found. In Section 6.544 will be given more information on the volumes of ocean water within various ranges of temperature and salinity individually and, what is more important in identifying particular water masses, on the ocean volumes having certain *combinations* of temperature and salinity values.

One feature to be noted about the spatial distribution of water properties is that in the surface and upper waters there is a distinct tendency for the arrangement of some of them to be "zonal", i.e. the value of a property may be much the same across the ocean in the east–west direction but may change rapidly in the north–south direction. As a consequence, when describing ocean water property distributions we frequently wish to refer to the position of an ocean area in terms of its north–south position even when we do not need to specify the latitude explicitly. For such zonal distributions, the adjective "equatorial" refers to the zone near the equator, while the adjective "tropical" refers to zones near the tropics ($23\frac{1}{2}°$ N or S of the equation). The distinction between "equatorial" and "tropical" should be noted as it is often significant; when the two are to be lumped together the term "low latitude" will be used, in contrast to the "high latitudes" which are near the poles, north and south. "Subtropical" refers to zones on the high latitude side of the tropical zones. The term "polar" is properly applied to oceanography only to the Arctic regions but is often used of the ocean close to Antarctica.

4.2 Temperature Distribution

4.21 Surface temperature

The distribution of temperature at the surface of the open ocean is approximately zonal, the lines of constant temperature (*isotherms*) running roughly east–west (Figs. 4.1 and 4.2). (The purpose of the projection in this and some subsequent figures is to show the ocean with areas between parallels of latitude being reasonably correct, rather than being grossly exaggerated at high latitudes as they are on Mercator's projection.) Near the coast where the currents are diverted the isotherms may swing more nearly north and south. Also, along the eastern boundaries of the oceans low surface temperatures often occur due to upwelling of subsurface cool water, e.g. along the west coast of North America in summer, which causes the isotherms to trend equatorward. (Upwelling will be discussed in Section 8.2). The open ocean surface temperature decreases from as high as $28°C$ just north of the equator to nearly $-2°C$ near ice at high latitudes. Values for the surface temperature from south to north for all oceans averaged together are presented in Fig. 4.3. This distribution with highest values at low latitudes and decreasing at higher latitudes corresponds closely with the input of short-wave radiation as will be discussed in Section 5.33. More detailed maps and descriptions of the global distribution of sea temperature etc. were presented by Levitus (1982).

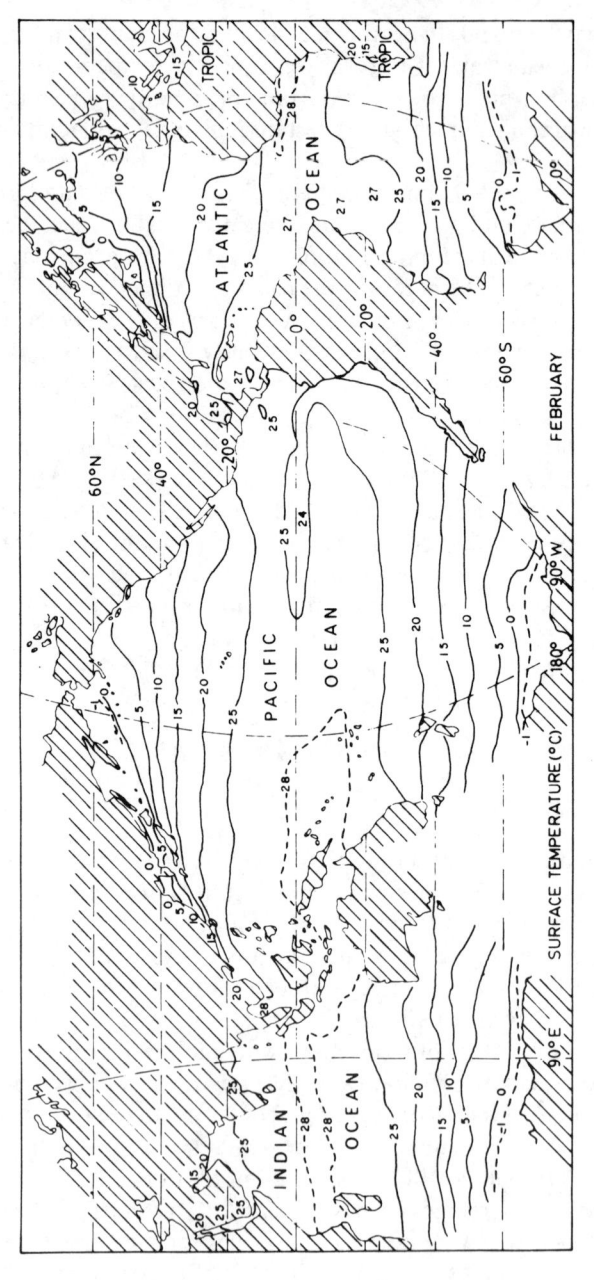

FIG. 4.1. Surface temperature of the oceans in February.

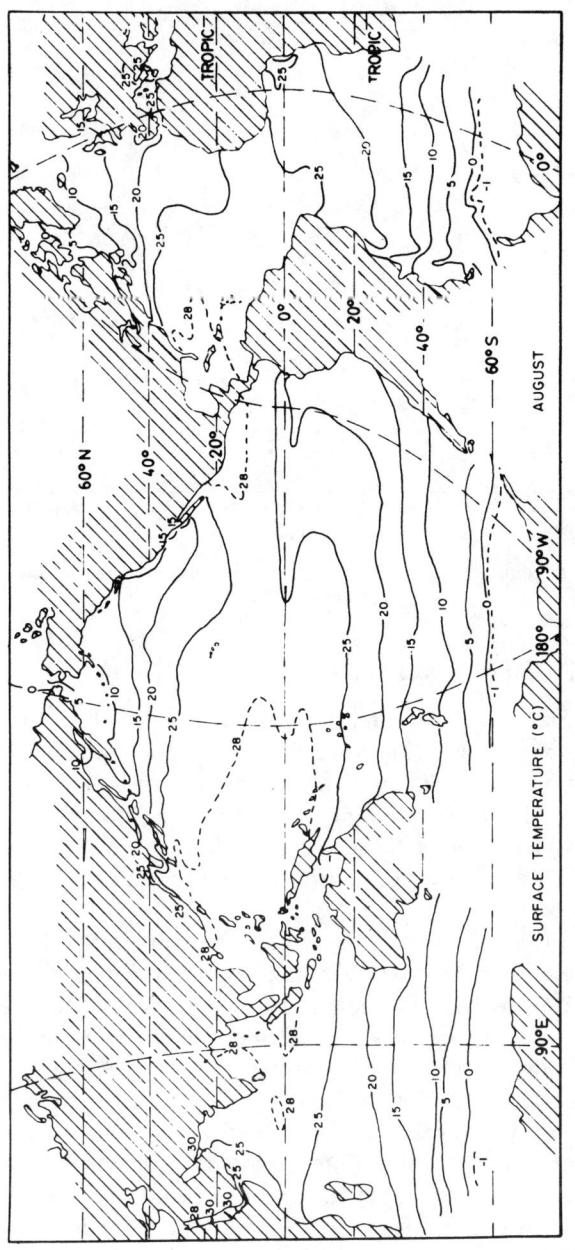

Typical Distributions of Water Characteristics in the Oceans 39

FIG. 4.2. Surface temperature of the oceans in August.

40   Descriptive Physical Oceanography

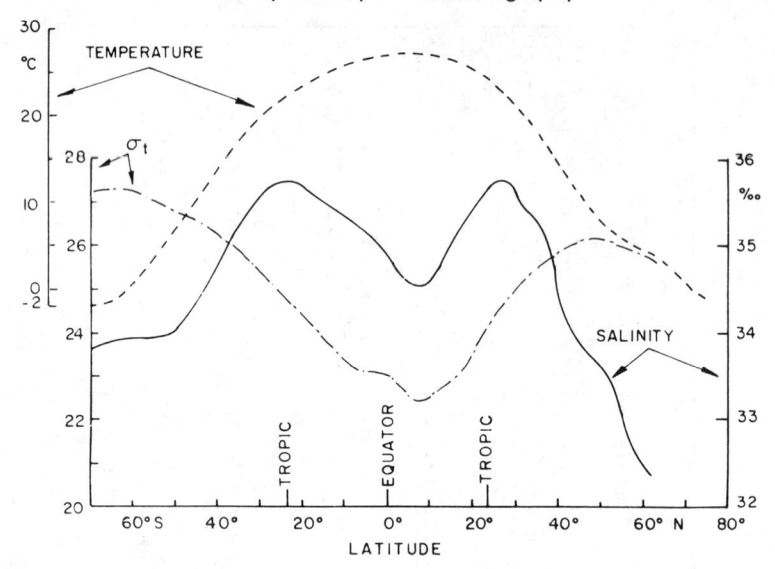

FIG. 4.3. Variation with latitude of surface temperature, salinity and density (σ_t)–average for all oceans.

4.22 Upper layers and the thermocline

Below the surface the water can usually be divided into three zones in terms of its temperature structure (Fig. 4.4). There is an upper zone of 50 to 200 m depth with temperatures similar to those at the surface, a zone below this extending from 200 to 1000 m in which the temperature decreases rapidly, and a deep zone in which the temperature changes slowly. Typical temperatures at low latitudes would be 20°C at the surface, 8°C at 500 m, 5°C at 1000 m and 2°C at 4000 m.

The depth at which the temperature gradient (rate of decrease of temperature with increase of depth) is a maximum is called the *thermocline*. With actual observations of temperature in the sea it is often difficult to determine this depth precisely because of minor irregularities in the temperature/depth profile and it is easier to pick out a "thermocline zone" as a range of depth over which the temperature gradient is large compared with that above and below. Even for this zone, it is often hard to define precisely the depth limits, particularly the lower limit, and one must accept some degree of approximation in stating the depth limits of the thermocline zone. However, in low and middle latitudes it is clear that there is a thermocline present all the time at depths between 200 and 1000 m. This is referred to as the "main" or "permanent" thermocline. In polar waters there is no permanent thermocline.

The continued existence of the thermocline requires explanation. One might expect that as the upper waters are warmest, heat would be transferred downward, despite the inhibiting effect of the stability in the pycnocline (see

Typical Distributions of Water Characteristics in the Oceans

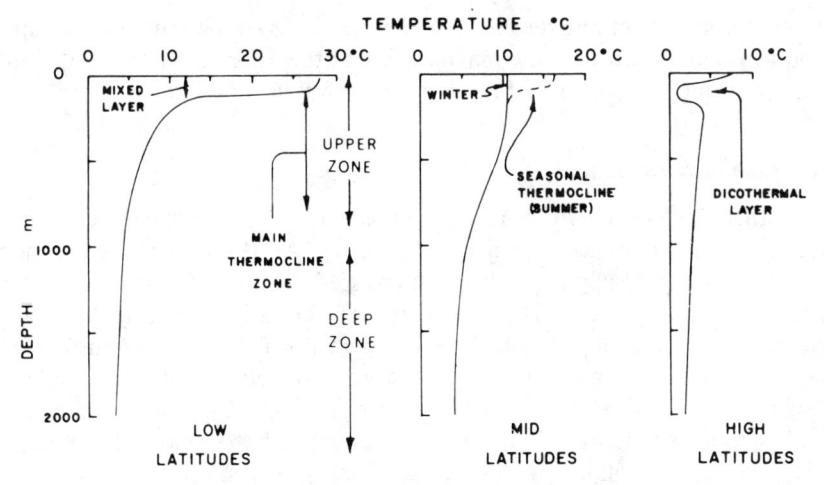

FIG. 4.4. Typical mean temperature/depth profiles for the open ocean.

Sections 4.42, 4.43), and that the temperature difference between the upper and lower layers would eventually disappear. Alternatively one might think that the net input of heat at low latitudes would cause the warm upper wind-mixed layer to deepen steadily. In fact, evidence accumulated over the last 100 years or so indicates that neither of these alternatives is taking place, and that the thermocline between the warm upper layer and the deeper cooler water not only continues to exist but that the depth of the thermocline in any locality appears to be substantially constant. Stommel (1958) suggested that while downward transfer of heat does occur at mid- and low-latitudes (by eddy diffusion), there is a simultaneous upward advection of cool water from below, displaced by the sinking of winter-cooled water in the South and North Atlantic (see Sections 7.11, 7.231, 7.352). Stommel and Arons (1960b) estimated the vertical speeds from temperature/salinity distributions and from deductions from their model of the abyssal circulation and obtained values from 0.5 to 3 cm/day. Unfortunately these speeds are too small to measure with present instruments and so we cannot test the hypothesis directly.

For dynamical reasons (conservation of vorticity, see Pond and Pickard, 1983) the waters which sink in the South and North Atlantic are expected to flow north or south respectively along the western boundaries of the oceans (directly from the source regions in the Atlantic, and from the Antarctic Circumpolar Current in the Indian and Pacific Oceans). From these deep western boundary currents, slow geostrophic currents are proposed flowing out into the main bodies of the deep oceans to displace upward the waters resident there. (Actually, the presence of the cold water on the western side of the South Atlantic, suggesting northward flow there, had been observed by Wüst in 1932 in his Atlas of the water properties in the Atlantic prepared from the data from the *Meteor* expedition (1925–27).)

Another factor limiting the increase of upper-layer temperature in the tropic regions, where the heat input is maximal, is that the waters there turn poleward after transiting the oceans and so carry heat away from the tropic regions.

4.23 Langmuir circulations

Another mechanism which may influence the thickness of the upper mixed-layer is the circulation associated with *Langmuir Cells* (LCs) to which attention was first drawn by Langmuir (1938) who carried out a number of experiments to identify their character. These LCs are made visually evident by the occurrence of numerous long parallel lines or streaks of flotsam at the surface on lakes or sea. They are aligned with the wind direction and typical lateral spacings between these lines are 2–25 m for lakes and 2–300 m at sea. If the wind direction changes suddenly, the lines realign with the wind in 2–3 minutes in lakes and 10–20 minutes in the open sea. They are most commonly observed for wind speeds above 3 m/s but can occur at lower speeds. (It is interesting to note that Langmuir, of the General Electric Company Research Department, identified the main characteristics of LCs using very simple equipment, e.g. leaves, dye, umbrellas for subsurface drift indicators, horizontal aluminum sheets rendered neutrally buoyant with electric light bulbs for estimating vertical velocities, etc. The only electrical instrument which he used was an electrical resistance thermometer for temperature measurements.) An excellent review of the phenomenon and of the theories advanced for the formation of LCs was given by Pollard (1977).

The basic features of LCs are that they take the form of long, helically rotating cells in the upper mixed-layer. Alternate cells rotate in opposite directions, e.g. A, C (clockwise), B, D. (anti-clockwise) (Fig. 4.5) so that convergence and downwelling occurs at the surface (to form streaks of flotsam) between pairs of adjacent cells, e.g. between cells A and B, C and D, while divergence and upwelling occurs between alternate pairs, e.g. B and C. Actually the water in the cells both rotates in the vertical plane and progresses downwind as well, so that its motion is helical. Water in cells A and C then moves in clockwise helices (looking in the wind direction) whose horizontal axes are aligned with the wind direction, while that in cells B and D moves in anticlockwise helices.

The downwelling zones, at speeds of several cm/s, are concentrated in jets occupying one-third or less of the cell width under the streaks while upwelling is more widely distributed at smaller speeds. One set of measurements gave downwelling speeds of about 1 cm/s per 1 m/s wind speed, with considerable scatter about this ratio. The helices are limited to the upper mixed-layer and do not penetrate the thermocline. The horizontal flow speed at the surface in the streaks can be 10 cm/s faster than elsewhere in the surface layer, and the vertical downwelling at the convergences is at about one-third of the surface water speed as driven by the wind. It is stated that LCs only occur for wind

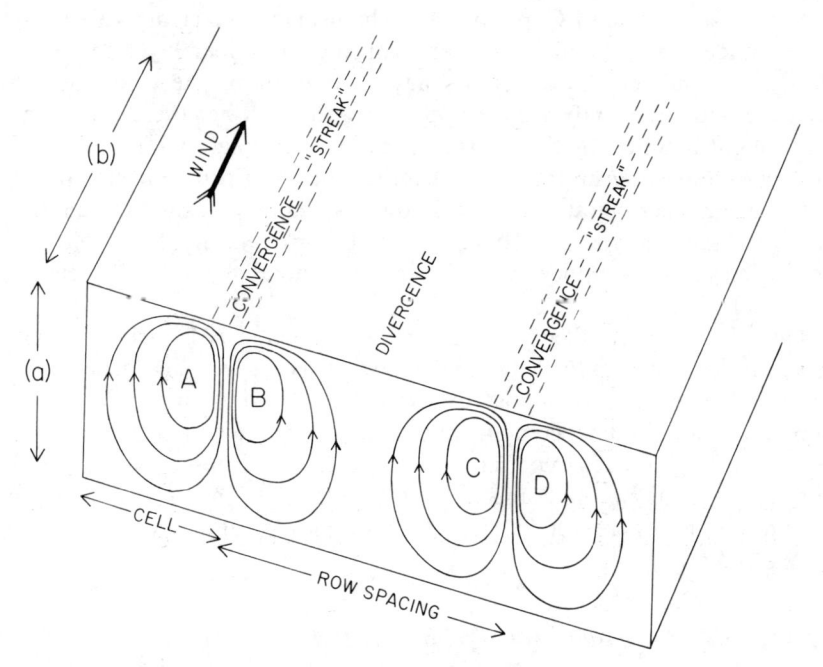

Typical Distributions of Water Characteristics in the Oceans

FIG. 4.5. Schematic diagram of Langmuir Cell circulations: (a) vertical section through helices, (b) streaks on surface formed by convergences between pairs of helices bringing together flotsam. (Note that (a) shows only the component of motion perpendicular to the wind direction—there is also a downwind component making the water motion helical.)

speeds greater than 3 m/s (although Stommel observed them on a small pond at lower speeds) and they appear within a few tens of minutes of wind onset. One characteristic of LCs is that they are only evident when there are wind waves on the water surface. In laboratory experiments it has been shown that surface films which dampen small waves tend to inhibit the formation of LCs.

It is now believed that LCs are due to a non-linear interaction between the vorticity of the wind-driven upper-layer current (whose speed decreases with depth) and cross-wind variation of the Stokes drift of intersecting surface wave trains, which interaction generates the vorticity of the LCs. (In ideal, small-amplitude wave theory, the water particles move in closed circular orbits and there is no net motion of the water mass, only the wave shape moves at the wave speed. However, for real water waves of finite amplitude, the orbits are not closed and there is a small net motion of the water mass in the direction of wave travel. This is called the *Stokes drift*. For a wavelength of 100 m and wave height of 3 m, the shallow water wave speed would be about 12.5 m/s and the Stokes drift speed only about 0.1 m/s.)

Pollard considered Garrett's (1976) theory to be the most plausible but the details of the mathematical theories are beyond the scope of this text.

It is considered that LCs provide a mechanism for converting wave energy to turbulent energy and mixing and causing the upper layer to become deeper. It should be noted that LCs may occur whenever the wind is blowing over the water, they only become apparent to the eye when there is flotsam to be brought together by the convergences (or when material is deliberately strewn on the water in order to reveal the motion). From Garrett's theory, rough estimates suggest that mixing could typically move momentum from the top 1–2 m down to 6–10 m, so that LC mixing might not penetrate through the entire upper mixed-layer, i.e. generally not to the seasonal thermocline.

Note that Ekman's theory of the wind drift (Section 7.12) yields an upper-layer motion *at 45°* to the wind direction whereas the LCs, also generated by the wind, are aligned almost exactly *with* the wind direction and the consequent longitudinal component of motion in the helices is *along* the wind direction. This appears to present a discrepancy but it must be noted that the time scales of the two mechanisms are quite different. The LCs are generated within minutes of the wind starting but the Ekman circulation takes many hours to develop.

4.24 Temporal variations of temperature in the upper layer

The temperature in the upper zone shows seasonal variations, particularly in middle latitudes. The layer between the surface and a depth of 25 to 200 m is usually at much the same temperature as the surface water because of mixing due to wind waves. For this reason it is referred to as the *mixed layer* (Fig. 4.4). In winter the surface temperature is low, waves are large, and the mixed layer is deep and may extend to the main thermocline. In summer the surface temperature rises, the water becomes more stable, and a "seasonal" thermocline often develops in the upper zone (Fig. 4.4). The thermocline zones are of high stability (they are essentially the pycnocline zones), and for this reason they separate the water of the upper from those of the deep zones.

An illustration of the growth and decay of the seasonal thermocline is shown in Fig. 4.6(a). This figure shows monthly temperature profiles from March 1956 to January 1957 taken at Ocean Weather Station "P" in the eastern North Pacific. From March to August the temperature gradually increases due to absorption of solar energy. A mixed layer from the surface down to 30 m is evident all the time. After August there is a net loss of heat energy from the sea while continued wind mixing erodes away the seasonal thermocline until the isothermal condition of March is approached again.

The same data may be presented in alternative forms. In Fig. 4.6(b) is given a time-series plot showing the depth of the isotherms during the year. (The original data include the alternate months which were omitted from Fig. 4.6(a) to avoid crowding.) In Fig. 4.6(c) are plotted the temperatures at selected depths, i.e. *isobaths* of temperature. The different forms in which the

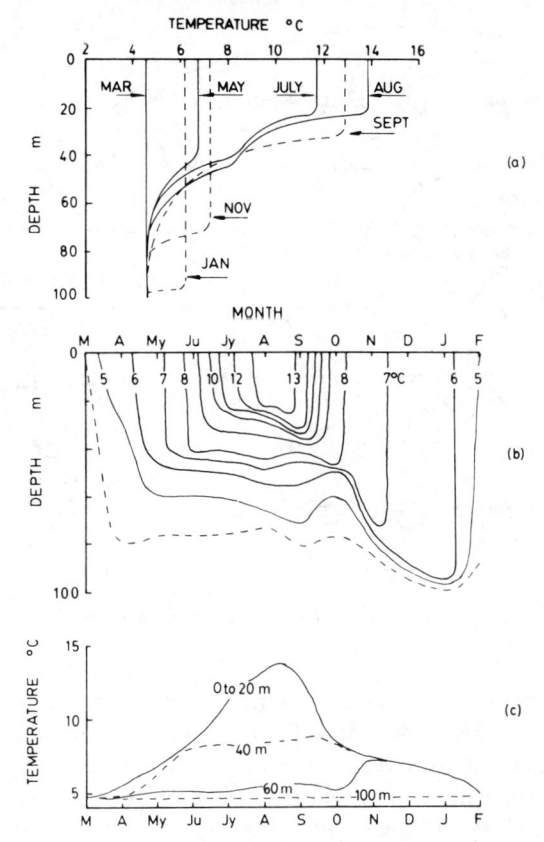

Typical Distributions of Water Characteristics in the Oceans 45

FIG. 4.6. Growth and decay of the seasonal thermocline at 50°N, 145°W in the eastern North Pacific.

thermocline appears in these three presentations should be noted. In Fig. 4.6(a) it appears as a maximum gradient region in the temperature/depth profiles. In Fig. 4.6(b) the thermocline appears as a crowding of the isotherms which rises from about 50 m in May to 30 m in August and then descends to 100 m in January. In Fig. 4.6(c) the thermocline appears as a wide separation of the 20-m and 60-m isobaths between May and October, and between the 60-m and 100-m isobaths after that as the thermocline descends.

At high latitudes the surface temperature are much lower than at lower latitudes, while the deep-water temperatures are little different. In consequence the main thermocline may not be present and only a seasonal thermocline may occur. In high northern latitudes there is often a *dicothermal layer* at 50 to 100 m (Fig. 4.4). This is a layer of cold water, down to −1.6°C, sandwiched between the warmer surface and deeper layers. (Stability is maintained by an increase of salinity with depth through the layer.)

Figure 4.7 shows the annual range of surface temperature over the Pacific as

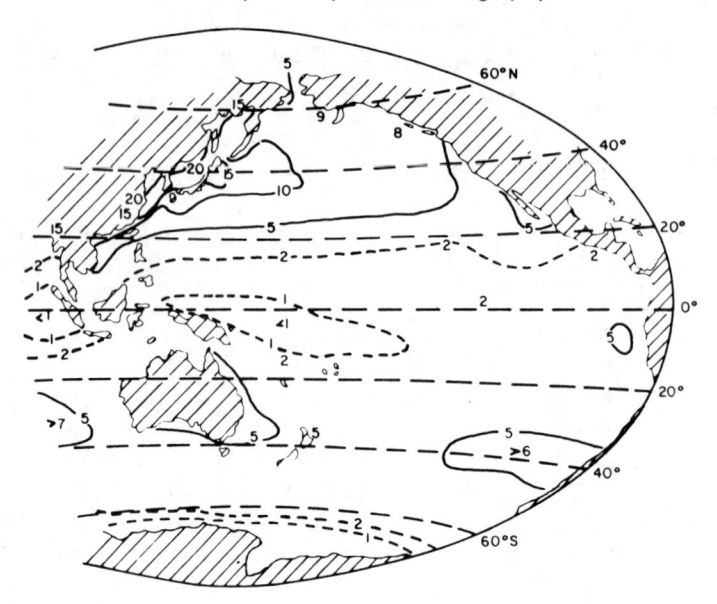

FIG. 4.7. Annual range of sea surface temperature, Pacific Ocean, in kelvins.

an example. Annual variations at the surface rise from 1 to 2K at the equator to between 5 and 10K at 40° latitude in the open ocean and then decrease toward the polar regions (due to the heat required in the melting or freezing processes where sea-ice occurs). Near the coast, larger annual variations (10 to 20K) occur in sheltered areas and particularly in the north-west of the northern oceans for reasons described in Section 5.38. These annual variations in temperature decrease with depth and are rarely perceptible below 100 to 300 m. The maximum temperature at the surface occurs in August/September in the northern hemisphere and the minimum during February/March. Below the surface the time of occurrence of the maxima and minima are delayed by as much as 2 months relative to the times at the surface. Details on the annual variations for the world ocean were given by Levitus (1987).

These statements about annual variations of temperature are made on the basis of observations at fixed stations. Defant (1960) has pointed out that in most oceanic regions the surface water is moving and therefore the statements about annual variations at fixed stations refer not to a particular body of water but to a continually changing one. For a region where the ocean currents are zonal, i.e. east–west, there will be little difference between the annual variation at a fixed station and that in a body of water moving with the current. But in a region where the flow is meridional (north–south) this will not be the case because of the difference in annual mean temperature and in annual variation in the north–south direction. For a planktonic organism which drifts with the water it is the variation in the individual water mass which is important. Little

Typical Distributions of Water Characteristics in the Oceans 47

attention has been paid in the past to determining the annual change of temperature (or other properties) appropriate to an individual water mass. However, modern satellite tracked, freely-drifting buoys (see Section 6.251) measure surface temperature and thus will provide a means for observing temperature changes while moving with the water mass in which they float.

Diurnal variations of sea-surface temperature (SST) had been thought to be small (<0.4K) but recent observations in the open ocean have revealed that larger variations occur frequently although they may be quite localized. It was the continuity of satellite measurements of SST (see Section 6.322) which revealed the frequent occurrence of the larger variations. Such measurements, verified by *in situ* observations from a moored buoy in the Sargasso Sea over a period of two years (Stramma *et al.*, 1986) have shown that diurnal variations to 1K are common and higher values up to 3 to 4K occur occasionally. These values of 1K or more are observed only in conditions of high insolation and low wind speed and are generally limited to the upper few metres of water. Similar values have been observed elsewhere in the North Atlantic and in the Indian Ocean. In sheltered and shallow waters along the coast, values of 2 to 3K are common. It should be noted that much of the daily heat input from solar radiation is used to evaporate water, leaving only part available for raising its temperature. Where there is ice in the water, most of the incoming heat will be used to melt it, so that diurnal variations will be small.

4.25 Deep water; potential temperature

In the deep water below the thermocline zone the temperature generally decreases as depth increases to about 4000 m, the average depth of the world ocean. In the deep trenches, however, the *in situ* temperature often increases slowly with depth beyond 3000 to 4000 m due to the effect of increasing pressure (see Section 3.53). When considering oceanic situations where considerable changes of depth of water masses occur, it is best to plot potential temperature in order to eliminate the effect of change of depth (i.e. pressure) which appears in the *in situ* temperature. (The *potential temperature* (θ) is the temperature which a parcel of water would have if it moved adiabatically (i.e. without gain or loss of heat) from its *in situ* depth to the sea surface. The potential temperature is computed from the *in situ* temperature and pressure (Fofonoff, 1977).

An excellent example of the difference between *in situ* and potential temperature is shown in Table 4.1 and in Fig. 4.8 taken from data of the Dutch *Snellius* Expedition (van Riel, 1934). Figure 4.8(a) shows a sample vertical profile of *in situ* temperature t while Fig. 4.8(b) shows the profile of potential temperature θ. It is seen that while t reaches a minimum at 3500 m and thereafter increases, θ decreases to the bottom. (The salinity changes by only 0.02 between 3500 m and the bottom.) The effect of the correction from *in situ* to potential temperature is more dramatically shown in Fig. 4.8(c) and (d)

Descriptive Physical Oceanography

TABLE 4.1
Comparison of in situ and potential temperatures, etc., in the Mindanao Trench near the Philippine Islands

Depth (m)	Salinity	Temperature in situ $t°C$	Temperature Potential $\theta°C$	Density σ_t	Density σ_θ
1455	34.58	3.20	3.09	27.55	27.56
2470	34.64	1.82	1.65	27.72	27.73
3470	34.67	1.59	1.31	27.76	27.78
4450	34.67	1.65	1.25	27.76	27.78
6450	34.67	1.93	1.25	27.74	27.79
8450	34.69	2.23	1.22	27.72	27.79
10035	34.67	2.48	1.16	27.69	27.79

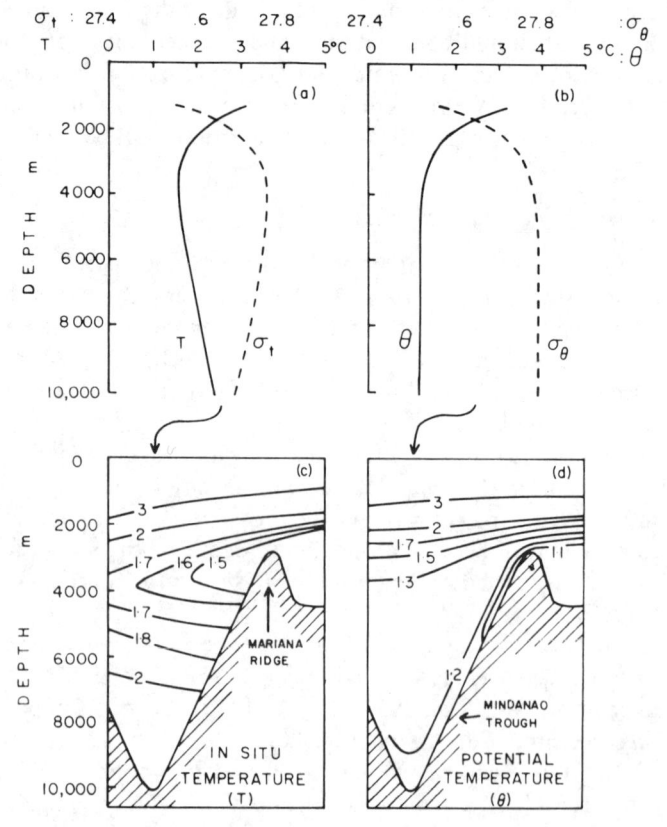

FIG. 4.8. *In situ* and potential temperature and density distributions in the Mindanao Trench: (a, b) vertical profiles, (c, d) vertical sections.

Typical Distributions of Water Characteristics in the Oceans   49

from the Expedition Report. The plot of the *in situ* temperature section (Fig. 4.8(c)) would suggest a flow of cool water over the sill, continuing to descend slowly across the Trench but staying near mid-depth and leaving warmer bottom water undisturbed. The plot of potential temperature (Fig. 4.8(d)) shows a very different pattern with the water which passes over the sill in reality flowing down the slope to the bottom of the Trench.

The profile of σ_t in Fig. 4.8(a) shows a maximum at about 4000 m and then a decrease to the bottom, giving the appearance of instability (more dense water over less dense). However, the potential density (Fig. 4.8(b)) increases to 6450 m and then remains constant showing that when the adiabatic compression with increase of depth is taken into account the water is not unstable but is in neutral equilibrium below 6450 m.

It should be noted that sometimes, in deep water, even a plot of σ_θ may show a decrease with increase of depth, i.e. apparent instability (e.g. Fig. 7.15) because the procedures for calculating σ_θ (and σ_t) neglect the effect of pressure on compressibility, i.e. the values calculated are $\sigma_{S,\theta,0}$ or $\sigma_{S,t,0}$. In very deep water it is better to calculate sigma values for a greater pressure. For example, if calculations are made corresponding to a depth of 4000 m, sigma values are shown as σ_4, potential temperature being used (e.g. Fig. 7.16 which no longer shows the apparent instability in Fig. 7.15 below 3500 m). This matter is discussed in more detail by Lynn and Reid (1968) and Reid and Lynn (1971).

4.3 Salinity Distribution

4.31 Surface salinity

The salinity of the surface waters is basically zonal in distribution (Fig. 4.9) although not as clearly so as the temperature. The average surface salinity distribution (Fig. 4.3) is different from that for temperature in that it has a minimum just north of the equator and maximum values in the sub-tropics at about 25°N and S of the equator. The minimum and maxima are evident in the individual oceans in Fig. 4.9. Values decrease toward high latitudes. Observations make it clear that surface salinity is determined by the opposing effects of evaporation increasing it and precipitation decreasing it as is shown in Fig. 4.10. The salinity maxima of Figs. 4.3 and 4.10 are in the trade wind regions where the annual evaporation (E) exceeds precipitation (P), so that (E–P) is positive, while the temperature maximum is near the equator because the balance of energy into the sea has a single maximum there (e.g. see Fig. 5.7.2(f)).

Examples of the areal distribution of (E–P), as distinct from the global average of Fig. 4.10, were given as annual and seasonal averages for the North Atlantic Ocean by Schmitt *et al.* (1989) together with references to other sources of such data.

It will be noted from Fig. 4.3 that the density of the surface water has a single

50   Descriptive Physical Oceanography

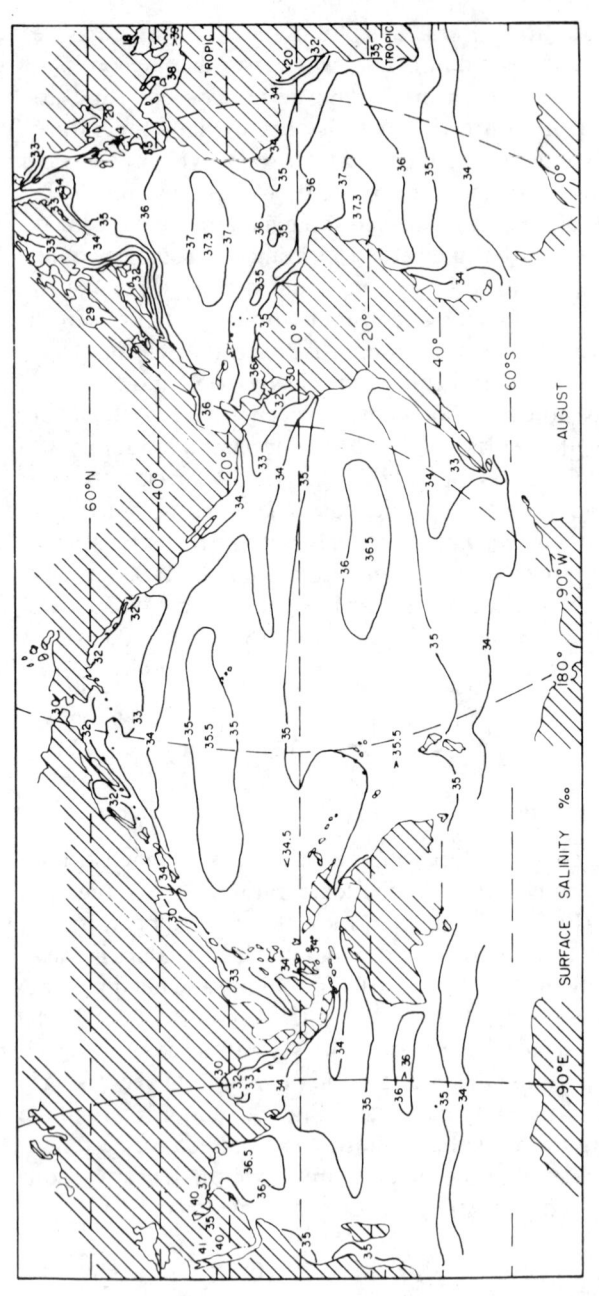

FIG. 4.9. Surface salinity of the oceans in August.

Typical Distributions of Water Characteristics in the Oceans 51

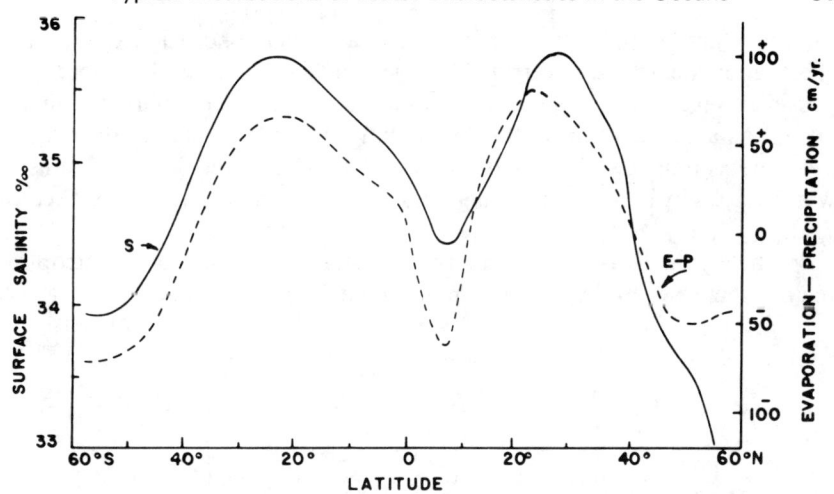

FIG. 4.10. Surface salinity (S, average for all oceans) and difference between evaporation and precipitation (E–P) versus latitude.

minimum at low latitudes, corresponding to the single temperature maximum. The salinity does exert some influence on density but not sufficient for the tropical maxima to appear on the σ_t curve as density maxima.

The range of surface salinity values in the open ocean is from 33 to 37. Lower values occur locally near coasts where large rivers empty and in the polar regions where the ice melts. Higher values occur in regions of high evaporation such as the eastern Mediterranean (39) and the Red Sea (41). On the average the North Atlantic is the most saline ocean at the surface (35.5), the South Atlantic and South Pacific less so (about 35.2) and the North Pacific the least saline (34.2).

More detailed descriptions of the global salinity distribution and seasonal changes were presented by Levitus (1982, 1986). Incidentally, the maps in the latter paper which show the locations from which observations were available reveal the great disparity in numbers of observations between the two hemispheres, there being far more ($\sim 90\%$) in the northern than in the southern ($\sim 10\%$). (This is true of temperature observations also.)

4.32 Upper-layer salinity

The vertical salinity distribution cannot be summarized quite as simply as the temperature distribution. In the upper water the reason for this is that density, which is the factor responsible for determining the stable position of a water body in the vertical direction, is determined chiefly by temperature in the open ocean (except in the polar seas). Therefore, water of higher temperature (lower density) is generally found in the upper layers and water of lower temperature (higher density) in the deeper layers. The variations in salinity which occur in open ocean waters are usually not sufficient in their effect upon density to override the effect of temperature. Therefore it is possible to have either high or low salinity in the warmer surface and upper layers. As a consequence of this less important role in dictating the density structure, salinity acts as a "passive tracer" indicating the flow directions of water masses by their salinity signatures (minima or maxima), e.g. see Section 7.522.

In the vertical distribution in the equatorial, tropical and subtropical regions there is a marked salinity minimum at 600 to 1000 m (Fig. 4.11) with

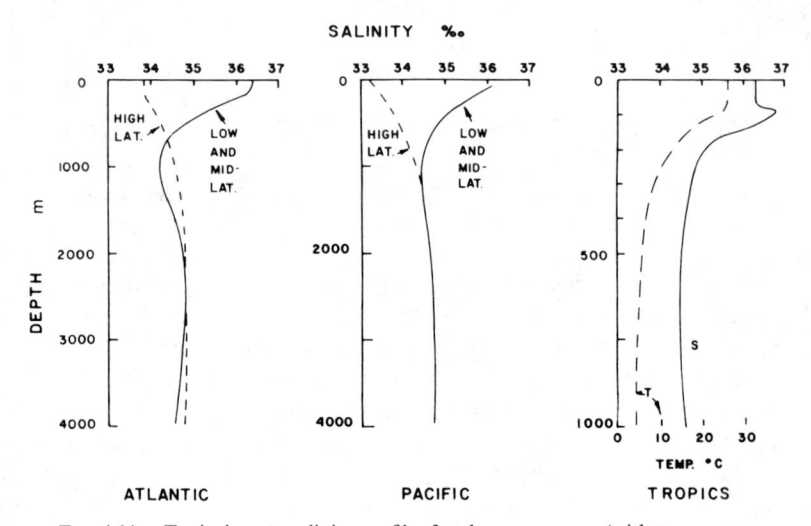

FIG. 4.11. Typical mean salinity profiles for the open ocean (with temperature profile for the tropics).

the salinity then increasing to 2000 m. In the Atlantic it decreases slightly below this. In the tropics there is often a sharp salinity maximum at 100 to 200 m depth close to the top of the thermocline (see also Fig. 7.33). This results from water sinking at the tropic salinity maxima (Fig. 4.3) and flowing equatorward. In high latitudes, where the surface value is low, the salinity generally increases with depth to about 2000 m, with no subsurface minimum. In coastal regions, where there is much river runoff, there is generally a zone of rapid increase of salinity, the *halocline*, between the upper, low-salinity water and the deeper more saline water. Here the pycnocline is determined by the salinity distribution rather than by temperature.

4.33 Deep-water salinity

In the deep waters, 4000 m or deeper, the salinity is relatively uniform at 34.6 to 34.9 throughout the world ocean. Remembering that the deep-water temperature also has a small range ($-0.9°$ to $2°C$) this means that the deep-water environment is very uniform in character.

4.34 Temporal variations of salinity

Information on temporal variations of salinity is much less than for temperature which is more easily measured. Annual variations of surface salinity in the open ocean are probably less than 0.5. In regions of marked annual variation in precipitation, such as the eastern North Pacific and the Bay of Bengal, and near ice, there are large annual variations. These variations are confined to the surface layers because in such regions the effect of reduced salinity may override the effect of temperature in reducing the sea-water density. This keeps the low salinity water in the surface layer. Diurnal variations of salinity appear to be very small.

Temporal salinity variations may be large at water mass boundaries where advective and diffusive changes occur. These water mass boundaries can be found both at the surface and at depth. Again, these boundaries are usually marked by fluctuations in salinity maxima or minima. Examples will be presented later when temperature-salinity diagrams (see Section 6.54) are discussed.

4.4 Density Distribution

4.41 Density at the surface

The distribution of the density of sea-water at the ocean surface can be described roughly by stating that the value of σ_t increases from about 22 near the equator to 26 to 27 at $50°$ to $60°$ latitude, and beyond this it decreases slightly (Fig. 4.3).

4.42 Subsurface density and the pycnocline

More important, however, is the distribution of density in the vertical direction. A guiding principle here is that the density normally increases as depth increases. This is simply a consequence of the general tendency in Nature for a system to settle down to a state of minimum energy. This is the case in still water when the least dense water is at the surface and the most dense at the bottom. The density in the sea does not increase uniformly with depth, however. In equatorial and tropical regions there is usually a shallow upper layer of nearly uniform density, then a layer where the density increases rapidly with depth, called the *pycnocline*, and below this the deep zone where the density increases more slowly with depth (Fig. 4.12). There is little variation with latitude of the deep water σ_t which is about 27.9. As a

Descriptive Physical Oceanography

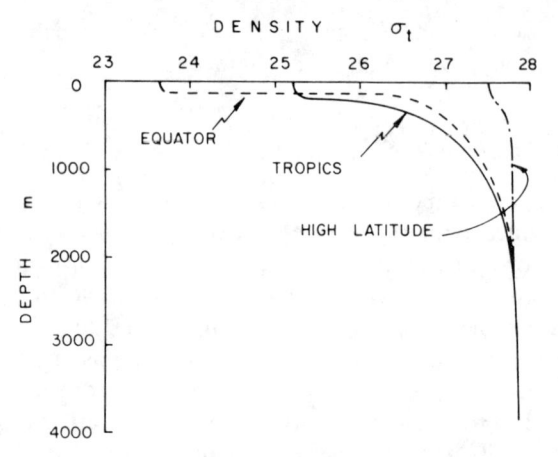

FIG. 4.12. Typical density/depth profiles for low and high latitudes.

consequence, in high latitudes where the surface σ_t rises to 27 or more there is a much smaller increase of density with depth than in the low latitudes and the pycnocline is less evident.

4.43 Static stability

The rate of change of density with depth determines the water's *static stability* or unwillingness to be moved vertically. That is, if a parcel of water is moved adiabatically down/up a short distance and finds itself surrounded by water of greater/lesser density it will tend to rise/sink toward its original position, then the water in that location is said to be statically stable. The rate of return to the original position will depend on the difference in density between the parcel and its surroundings at the displaced position. A further point to be taken into account is that the density of the parcel itself will increase/decrease as it is moved down/up because the pressure on it will increase/decrease respectively.

The mathematical derivation of expressions for the stability of a water column in terms of its density distribution is presented in detail in Pond and Pickard, (1983), but we will give a summary here. The magnitude of the tendency for the parcel to return to its original position is a quantitative measure of the stability (E) of the water column. The full expression for E is rather complex (see Pond and Pickard, *loc. cit.*) but a definition of this quantity for small displacements is $E = -(1/\rho)(\partial\rho/\partial z)$ where the water is stable, neutral, or unstable according as the numerical value of $E > 0$, $= 0$ or < 0 respectively.

For small displacements (metres to tens of metres), it can be shown that $E = -(1/\rho)(\partial\sigma_t/\partial z)$ in the upper layers, i.e. a first approximation to stability is that σ_t shall increase with depth. (Note that as z is taken to be positive upward,

Typical Distributions of Water Characteristics in the Oceans 55

an increase of depth ∂z is represented by a negative number, e.g. for a move from $z = -100$ m to $z + \partial z = -110$ m, $\partial z = -10$ m, so that if σ_t increases from where $z = -100$ m to where $z = -110$ m, then E is positive, i.e. the water is stable.

For larger vertical displacements, a better approximation is $E = -(1/\rho) (\partial \sigma_\theta / \partial z)$. However, remember that the potential density anomaly σ_θ represents the value of that quantity if the water sample were taken adiabatically all the way to the sea surface where $p = 0$. If one is considering the stability of abyssal waters, terms omitted in the approximation for E given immediately above become important, chiefly because the compressibility of water is a function of temperature, cold water being more compressible than warm water. Therefore, for abyssal waters it is better to estimate stability by considering displacements over small depth ranges, e.g. to 4000 m, for which the potential density anomaly is written as $\sigma_4 (= \sigma_{s,\theta,4})$ as is discussed in Section 4.25.

A more exact expression for stability is $E = -(1/\rho)(\partial \rho/\partial z) - (g/C^2)$ where $\rho = in\ situ$ density (from the equation of state), $g =$ acceleration due to gravity and $C =$ speed of sound at the same depth. The addition of the term (g/C^2) allows for the effect of temperature on compressibility because sound waves are compression waves.

In the upper 1000 m in the open ocean, values of E range from 1000×10^{-8} m^{-1} to 100×10^{-8} m^{-1}, while larger values occur in the pycnocline. Below 1000 m depth, values of E decrease below the last value above, and in abyssal trenches it may be as low as 1×10^{-8} m^{-1}. In the upper 50 m of the sea, values of $E = -25$ to -50×10^{-8} m^{-1}, i.e. indicating static instability, are frequently observed over extensive areas in sub-tropical regions (possibly resulting from an increase of salinity caused by evaporation) suggesting that the vertical overturn which would be expected in the presence of static instability only occurs when E reaches a sufficiently negative value. For instance, the above discussion of stability has been of the static case and has neglected effects of heat conduction, friction and eddy diffusion which may be significant for the dynamic case of actual overturn. (It has been shown in laboratory experiments with thin layers of fluids heated from below, that vertical convection (e.g. Bénard cells) only starts when E has values of -16 to -64×10^{-8} m^{-1}.)

Unstable conditions are uncommon below the surface layer; they may be found near the interface between bodies of water of different density distributions in the process of mixing, e.g. at the northern edge of the Gulf Stream. Again this is a dynamic case. (A more detailed discussion of stability will be found in Pond and Pickard, *loc. cit.*)

In the initial discussion above we imagined a parcel of water being moved up or down, and in a stable fluid it would experience a return force, toward its original (equilibrium) position, which would increase with stability. For example, if a parcel were actually displaced upward, the return force would cause it to accelerate down toward its equilibrium position and, having mass and therefore momentum, it would overshoot, slow down to a stop in the denser fluid below, accelerate up and continue to oscillate up and down (for a time). The frequency of such oscillations (N), is called the *buoyancy frequency* or *Brunt-Väisälä frequency* where $N^2 = gE \approx g[-(1/\rho)(\partial \sigma_t/\partial z)]$. Obviously, the greater the stability, the higher the frequency of vertical oscillation. The frequency f in cycles/sec (hertz) is $f = N/2\pi$ or the period $\tau = 2\pi/N$. Values in the upper ocean for $E = 1000 \times 10^{-8}$ to 100×10^{-8} m^{-1} are $\tau = 10$ to 33 min, while for the deep ocean with $E = 1 \times 10^{-8}$ m^{-1}, $\tau \approx 6$ h. The frequency $N/2\pi$ is the maximum possible frequency for internal waves in water of stability E.

The water in the pycnocline is very stable. That is to say, it takes much more energy to displace a particle of water up or down in the pycnocline than in a region of lesser stability. A result is that turbulence, which causes most of the mixing between different water bodies, is less able to penetrate through this layer than through the less stable water. The pycnocline then, although it is too slight to offer any barrier to the sinking of bodies which are much denser than water, offers a real barrier to the passage of water and water properties in the vertical direction, either up or down. Due to the presence of the pycnocline, the vertical density distribution can be approximately represented as a thin, upper low-density layer of 100 to 500 m thickness lying over the remainder of the ocean with a near-uniform higher density. The so-called "two-layer" ocean is a convenient simplified description often used in analytical and numerical ocean models. Results of these models are expressed in terms of variations in the position of the interface between the two layers which represents the pycnocline. The use of two-layer models is more completely discussed by Pond and Pickard (1983).

4.44 Geographic distribution of density

Perhaps the most illuminating graphical presentation of the distribution of density is a north–south vertical section through the ocean, such as Figs. 7.15 and 7.44, showing isopycnals of σ_θ. Those in the upper layers tend to be concave upward, showing the increase from equator to pole. Below about 2000 m, however, the total range of values is only from about 27.6 to 27.9 for σ_θ.

A further point to bear in mind when considering the ocean circulations later is that there is a strong tendency for flow to be along surfaces of constant potential density. In the upper layers we can regard this as to be along surfaces of constant σ_θ. For instance, the processes which give ocean waters their particular properties act almost exclusively at the surface, and one can trace the origin of even the deepest water back to a region of formation at the surface somewhere. Since deep ocean water is of high density this implies that it must have formed at high latitudes because only there is high density water found at

Typical Distributions of Water Characteristics in the Oceans 57

the surface. After formation it sinks along constant density surfaces. When the word "sink" is used here it does not necessarily imply that the water goes straight down like a stone. The sinking is often combined with horizontal motion so that the water actually moves in a direction only slightly inclined below the horizontal. The slopes of the constant density surfaces in Figs. 7.15 and 7.44 are exaggerated because of the vertical scale exaggeration used in plotting the data.

In the low- and mid-latitude open ocean, most of the variations of density in the upper 1000 m are due to variations of temperature; at greater depths, salinity variations may play a significant part. The effect of salinity on the deep water is more evident in the Atlantic (Fig. 7.15), where there is an obvious salinity structure, than in the Pacific (Fig. 7.44) where the deep waters are more uniform. Only in certain areas, such as the north-east Pacific and in the polar regions, does the variation in salinity play a part in the upper layers. In coastal waters, fjords and estuaries, salinity is often the controlling factor in determining density at all depths, while the temperature variations are of secondary importance. The information in Table 3.1 is useful in estimating the relative effects of temperature and salinity changes on density in a particular situation.

4.5 Dissolved Oxygen Distribution

In addition to the solids dissolved in sea-water there are also gases. One which has been widely used as a water characteristic is oxygen, expressed as the number of millilitres of oxygen at NTP dissolved in one litre of sea-water (mL/L). The SI unit of μmol/kg is coming into use but as mL/L are most common in the literature to date, this unit will be used in this text. (Equivalent values for the units are given in the Appendix.)

The range of values found in the sea is from 0 to 8 mL/L, but a large proportion of values fall within the more limited range from 1 to 6 mL/L. The atmosphere is the main source of oxygen dissolved in sea-water and at the surface the water is usually very close to being saturated. Sometimes, in the upper 10 to 20 m, the water is supersaturated with the oxygen which is a by-product of the photosynthesis by marine plants. Below the surface layers the water is usually less than saturated because oxygen is consumed by living organisms and by the oxidation of detritus. Low values of dissolved oxygen in the sea may often be taken to indicate that the water has been away from the surface for a long time, the oxygen having been depleted by the biological and detrital demands.

Figure 4.13 shows typical dissolved oxygen profiles for the Atlantic and the Pacific for three latitude zones. Common features are: (1) the high values close to the surface, (2) the oxygen minimum in the upper 1000 m between the tropics, (3) the relatively high values below 1000 m in the Atlantic (North Atlantic Deep Water), (4) low values in the North Pacific and (5) similar

58   Descriptive Physical Oceanography

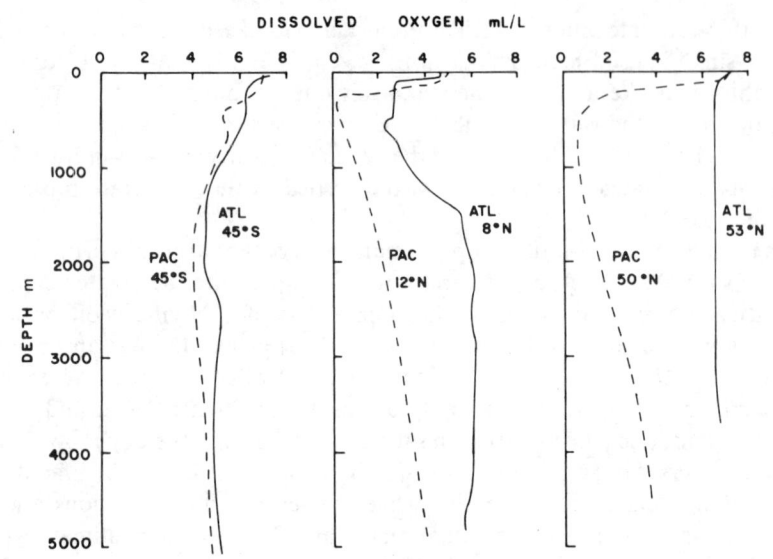

FIG. 4.13. Profiles of dissolved oxygen for Pacific Ocean (data from Reid, 1965) and for Atlantic Ocean (data from GEOSECS Atlas, 1976).

distributions in the southern latitudes in both oceans. Distributions in the Indian Ocean are similar to those in the Pacific (south and tropics). The lower values in the deep water of the Pacific compared with the Atlantic indicate that this water has been away from the surface for a longer period of time and that the deep-water circulation may be slower in the Pacific. In certain regions, such as the Black Sea and the Cariaco Trench (off Venezula in the Caribbean), there is no oxygen but hydrogen sulphide is present instead (from the reduction of sulphate ion by bacteria). This indicates that the water has been stagnant there for a long time.

The conspicuous oxygen minimum in the vertical profiles from between the tropics is apparent in the western Atlantic (Fig. 7.15), in the mid-Pacific (Fig. 7.44) and is particularly evident in the eastern Pacific (Fig. 7.40 and 7.43) and in the northern Indian Ocean. Suggestions for the cause for this minimum are basically either that this intermediate depth water is in regions of minimal motion so that there is little circulation or mixing to refresh the water and replace the oxygen consumed, or that biological detritus accumulates in this region because of the increase of density with depth (stability) and uses up the oxygen. Neither suggestion is accepted as satisfactory in itself and the oxygen minimum still requires a full explanation. As the production and utilization of oxygen in the sea are essentially biochemical matters they will not be pursued further here but it must be remembered that whenever oxygen is considered as a water property it must be used with caution since it is non-conservative.

Typical Distributions of Water Characteristics in the Oceans   59

4.6 Other Water-Motion Tracers

4.61 Introduction

Other water properties which may be used as flow tracers or for identifying water masses are the so-called nutrients, e.g. phosphate, nitrate and silicate ions (discussed in the next section), dissolved gases other than oxygen, and plankton which are small organisms which drift with the water (as distinct from nekton, the free swimming fishes and mammals, which move about of their own volition). These characteristics must be used with caution because, like oxygen, they are non-conservative, i.e. they may be produced or consumed within the water mass, and generally their rates of production and consumption are not well known. In addition, radioactive elements are used as tracers, and the fact that they do decay at known rates is used to estimate the "age" of ocean water as described in Section 6.4. An extensive review of chemical and radioactive tracers is presented by Broecker and Peng (1982) in their *Tracers in the Sea*—here we will limit ourselves to the main features of the distributions of the major so-called nutrients which are of interest to marine biologists.

4.62 Nutrients

The distributions of phosphate, nitrate and silicate, as the more important nutrients are shown as south–north vertical sections in Figures 4.14 and 4.15 for the Pacific and Atlantic Oceans, respectively. The section for the Pacific is down the 180° ($\pm 10°$) meridian from 50°N to the equator and then down the 170°W meridian to about 68°S, while the Atlantic section is down the western basin. (The non-linearity of the latitude scales is because the sections are plotted along the actual cruise tracks which did not follow meridians.)

For all three nutrients, the values are low in the upper few hundred metres with higher values in the deeper water. In the Pacific, these deeper distributions are in the form of mid-depth maximum value tongues extending from north to south with cores at 1000/2000 m for phosphate and nitrate and at 2000/3000 m for silicate, and values decreasing from north to south. In addition there are maxima at the south in the Antarctic Bottom Water (Section 7.231).

In the Atlantic, the mid-depth tongues extending from north to south are *minimum* value cores associated with the North Atlantic Deep Water (Section 7.352) and again there are maximum values in the south in the Antarctic Bottom Water.

The actual values in the mid-depth cores are higher by a factor of about two for phosphate and nitrate in the Pacific than in the Atlantic and by a factor of three to ten for silicate. This is attributed to the slower mid-depth and deep circulations in the Pacific than in the North Atlantic. The lower dissolved

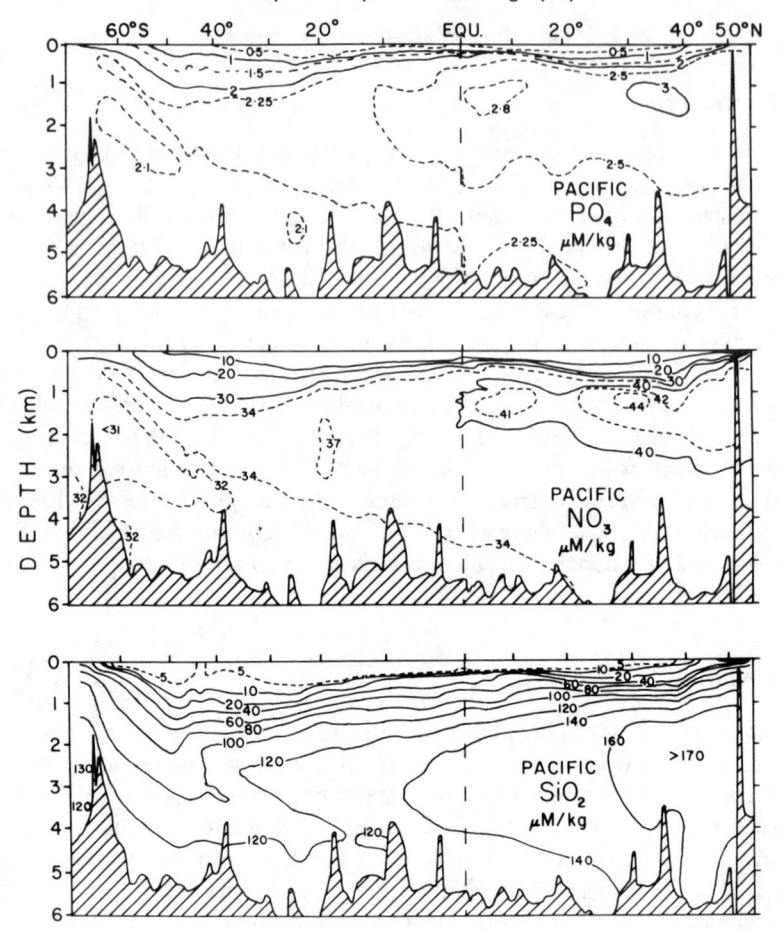

FIG. 4.14. Pacific Ocean—south–north vertical section of nutrients along the 180° (±10°) meridian (data from GEOSECS Atlas, Craig et al., 1981).

oxygen values in the Pacific than in the Atlantic are attributed to the same cause (cf. Fig. 7.44 and 7.15). A conspicuous feature of the phosphate and nitrate concentrations is that their ratio is everywhere close to 1 phosphorus atom to 15 nitrogen atoms, although the reason for this is not known.

The low values of the nutrients in the upper layers is because of their utilization by phytoplankton in the euphotic zone, while the increase in deeper waters is because of their release back to solution by biological processes during the decay of detrital material sinking from the upper layers. It is assumed that replenishment in the surface layers is chiefly by physical processes of vertical diffusion, overturn and upwelling. The upwelling processes active along the east sides of the oceans (Section 8.2) are the reasons for the high biological productivity observed there. In addition, in the upper

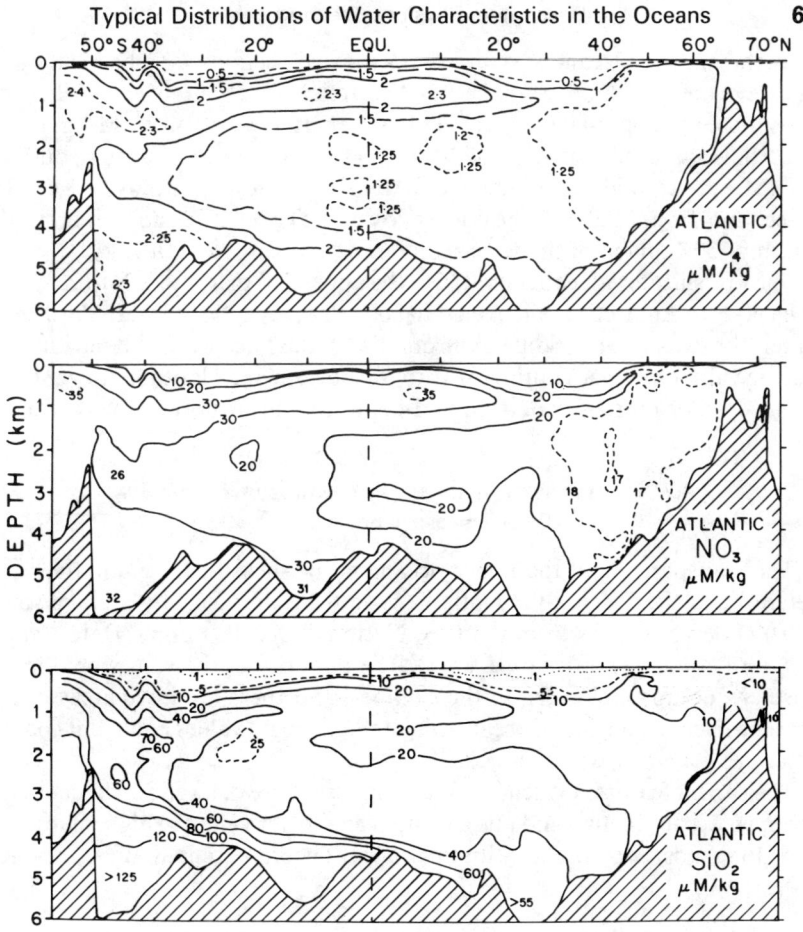

Typical Distributions of Water Characteristics in the Oceans

FIG. 4.15. Atlantic Ocean—south–north vertical section of nutrients along the western trough (data from GEOSECS Atlas, Bainbridge, 1976).

200–300 m, physical processes of up- and downwelling associated with upper-layer divergences and convergences modify nutrient distributions locally. For example, in the equatorial Pacific (Fig. 7.34) higher surface values of phosphate (and silicate) occur in the upwelling regions at and north of the equator associated with the divergence of the South Equatorial Current at the equator and between the North Equatorial Counter Current and the North Equatorial Current at about 10°N (see Section 7.614) and these are regions of high biological productivity.

The nutrient distributions in the Atlantic show considerable similarity to the salinity distribution (Fig. 7.15) with a relatively low value minimum tongue corresponding to the North Atlantic Deep Water extending south and higher values in the Antarctic Intermediate and Bottom Waters extending north.

In the Indian Ocean, Wyrtki (1971) showed sections for phosphate and silicate from 10°N off Somalia to Antarctica at about 65°S, 95°E. For phosphate the upper-layer values are low (generally less than 1 μM/kg) in the upper few hundred metres from 10°N to 40°S but increase further south to 1.5 μM/kg. At mid-depth there is a maximum tongue at over 2 μM/kg descending from 1000 m in the north to over 2000 m at 25°S and further south. South of 50°S, most of the water column has values over 2 μM/kg.

The low silicate values are limited to the upper 200 m from 10°N to 25°S and from 50°S to Antarctica, but extend to 800 m between 25°–45°S. In the deeper water, the distribution is different from that in the Pacific and Atlantic in that silicate values increase continuously to the bottom at all latitudes to values of 120 μM/kg or more. There is no mid-depth tongue as in the other oceans.

4.7 Transparency of the upper ocean from Secchi disc measurements

The transparency of the upper ocean can be measured quantitatively by determining the Secchi disc depth (Section 6.26) and recently a file of some 120,000 Secchi depths from the U.S. National Oceanographic Data Centre has been studied (Lewis et al., 1988). The majority of the values are for the northern oceans and taken in the summer, and there are large areas of the southern hemisphere open oceans where there are no values at all, but coastal areas are generally well sampled.

The salient features evident from the analysis were of large Secchi depths in the open oceans in low- and mid-latitudes and lower values in higher latitudes and along most coasts. The latitudinal variation is apparent in Fig. 4.16 which

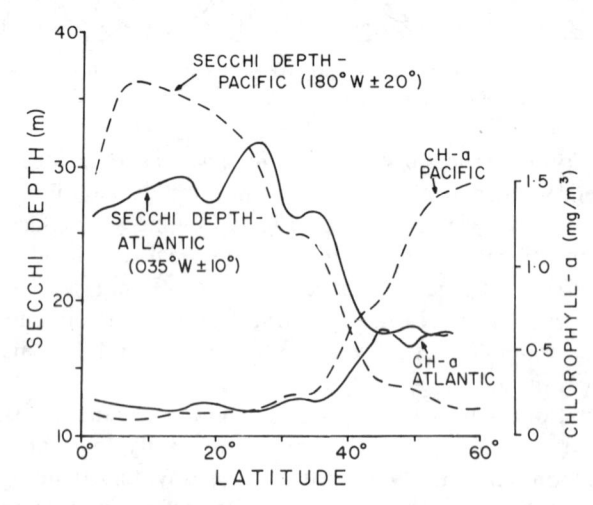

FIG. 4.16. Mean Secchi disc depths as functions of latitude in the Pacific and Atlantic Oceans.

Typical Distributions of Water Characteristics in the Oceans 63

shows averages of Secchi depths along $180° \pm 20°$W for the Pacific and along $35° \pm 10°$W for the Atlantic. Lewis *et al.* (1988) discussed sources of error and concluded that they were small relative to the latitudinal variations shown and that the prime source of variability in the open was attenuating material in the water (i.e. the value of $(\alpha + K)$ in Section 6.26). The lower Secchi depths corresponded to higher chlorophyll-a values. The most marked feature (Fig. 4.16) is the sharp decrease in Secchi depths beyond about 30° latitude, corresponding to higher productivity in the higher latitudes. In the North Pacific, there is a distinct seasonal variation in the position of the transparency front which migrates about 1000 km from a northerly position at $\sim 28°$N in September/November to $\sim 18°$N in March/May, attributed to increased upwelling and consequent increased biological production in higher latitudes.

The large Secchi depths in the Atlantic are in the Sargasso Sea, a region of notably low biological productivity. In a polynya in the Weddell Sea in 1986, a Secchi disc was visible to four observers at 79 m and disappeared at 80 m. This was claimed as a record and it may well be because the Secchi depth calculated for distilled water was 80 m. In coastal waters, values of 10 to 2 m are common, and in silty waters near to rivers and in estuaries, values of less than 1 m are to be expected.

CHAPTER 5

Water, Salt and Heat Budgets of the Oceans

IN THE basic sciences much use is made of a number of conservation principles such as conservation of energy, of momentum, etc., and these rather simple principles have very far reaching results and valuable applications. Conservation of heat energy as applied to the oceans will be discussed later in this chapter. First we will discuss two other principles, the second of which is peculiar to oceanography. They are the conservation of volume and the conservation of salt.

5.1 Conservation of Volume

The principle of conservation of volume (or the equation of continuity as it is sometimes called) follows from the fact that the compressibility of water is small. It says that if water is flowing into a closed, full container at a certain rate it must be flowing out somewhere else at the same rate. "Containers" such as bays, fjords, etc., in the oceans are not closed in the sense that they have lids on (except when frozen over), but if one observes that the mean sea level in a bay remains constant (i.e. after averaging out the tides) then there is no flow through the upper surface and the bay is equivalent to a closed container. One might say that this principle is just common sense. This may be true but nonetheless it is science too. It may lead to interesting results. For example, many of the fjords of Norway, western Canada and Chile have large rivers flowing into their inland ends, but on the average the mean sea level in them remains constant. We conclude from the principle of continuity of volume that there must be a simultaneous outflow elsewhere. The only likely place is at the seaward end, and if we measure the currents there we find that in fact there is a net outflow of the surface layer. The direction is correct to balance the inflow from the river but when we check we find that there is a much greater volume flowing out to the sea in this surface layer than in from the river. If conservation of volume is to apply there must be another inflow; the current measurements show that this is from the sea below the outflowing surface

Water, Salt and Heat Budgets of the Oceans

layer. The reason for this situation is that the river water, being fresh and therefore less dense than the sea-water of the fjord, stays in the surface layers as it flows toward the sea. However, it picks up sea-water from below en route and the outflowing surface layer includes not only the river water but also the extra salt water picked up. The latter is often in much greater volume than the river water, and the surface outflow to the sea is therefore correspondingly greater than the inflow from the river. In addition, the salt water which has been picked up and flushed out of the fjord must be replaced; this is the cause of the sub-surface inflow from the sea. (This type of circulation is nowadays referred to as an "estuarine" one; it will be discussed in more detail in Section 8.42.)

If we represent these flows schematically as in Fig. 5.1 and add precipitation (P) on to and evaporation (E) from the water surface, the conservation of volume principle may be stated symbolically as:

$$V_i + R + P = V_o + E,$$

or rearranged slightly as

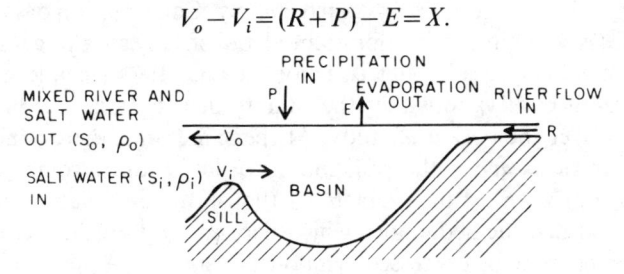

$$V_o - V_i = (R + P) - E = X.$$

FIG. 5.1. Schematic diagram of basin inflows and outflows for Conservation of Volume discussion.

Here V stands for *volume transport*, a phrase which is used when we express flow in terms of volume per second (usually m^3/s) rather than as a linear speed (m/s). The second equation simply says that the net volume flow of salt water balances the net volume flow of fresh water (when averaged over a suitable time period). This is an example of a *steady-state* situation in which some or all parts of a system may be moving but at no point is there any *change* of motion (or of property) with time.

5.2 Conservation of Salt

5.21 Principle

The principle of conservation of salt asserts that the total amount of dissolved salts in the ocean is constant. When one first learns that the rivers of the world contribute to the sea a total of about 3×10^{12} kg of dissolved solids per year, the conservation of salt seems to be contradicted. In principle it is, but in practice it is contradicted only to a negligible extent. The total amount of salt dissolved in the ocean waters is about 5×10^{19} kg, and therefore the amount brought in each year by the rivers increases the average ocean salinity by about one part in 17 million per year. But we can only measure the salinity of sea water to an accuracy of about ± 0.003, or about 1500 parts in 17 million if we take the mean ocean salinity as 35. In other words, the oceans are increasing in salinity each year by an amount which is only one fifteen-hundredth of our best accuracy of measurement. So for all practical purposes we can assume that the average salinity of the oceans is constant, at least over periods of tens or even hundreds of years. Furthermore, when we apply the principle of conservation of salt to a limited volume where there is no significant input of salt by rivers, the principle applies with even more rigor.

The principle of conservation of salt has been demonstrated above for the world ocean as a whole but in practice it is usually applied to smaller bodies of water. It turns out to be most useful when applied to bodies of water which have only limited connection with the main ocean, e.g. the Mediterranean Sea, a bay or a fjord, as will be demonstrated shortly. Conservation of salt in such water bodies is sometimes taken for granted, but strictly speaking it should be verified before being used. That is, before we use the principle we should determine from observations that the salinity distribution does not change significantly over the period of study. At the same time, there is nothing to prevent us from *assuming* the principle in order to draw some (tentative) deductions. But we must then remember that until conservation of salt has been demonstrated the deductions which depend on it are subject to doubt.

The principle may be expressed symbolically as:

$$V_i \cdot \rho_i \cdot S_i = V_o \cdot \rho_o \cdot S_o$$

where S_i and S_o are the salinities respectively of the inflowing and the outflowing sea-water, and ρ_i and ρ_o the respective densities (Fig. 5.1). Since the two densities will be the same within 3% at the most (the difference between ocean and fresh water) the ρs can in practice be cancelled leaving:

$$V_i \cdot S_i = V_o \cdot S_o.$$

This equation can be combined with the second equation for conservation of volume (Section 5.1) to give Knudsen's relations (Knudsen, 1900):

$$V_i = X \cdot S_o/(S_i - S_o) \text{ and } V_o = X \cdot S_i/(S_i - S_0).$$

One can draw some qualitative conclusions from these relations. In the first case, if both S_o and S_i are large they must be similar (because there is an upper limit to S in the ocean), therefore $(S_i - S_o)$ must be small and both $S_o/(S_i - S_o)$ and $S_i/(S_i - S_o)$ must be large. Therefore V_i and V_o must be large compared with X, the excess of fresh water inflow over evaporation. In the second case, if S_o is much less than S_i, then V_i must be small compared with X while V_o will be

Water, Salt and Heat Budgets of the Oceans

only slightly greater than x. For the same value of X for both, the exchange of water with the outside in the first case will be large, while in the second case the exchange will be small. One may therefore expect that the body of water in the first case will be less likely to be stagnant than that in the second case.

5.22 Two examples of applications of the two conservation principles

5.221 The Mediterranean Sea

The Mediterranean Sea is one from which evaporation exceeds precipitation plus river runoff (i.e. for the volume transport equation in Section 5.1, $E > (R+P)$ and X is negative) so that there is a net loss of volume as fresh water which must be made up by inflow of salt water from the Atlantic. As shown in Fig. 5.2(a) the inflow of less saline water through the Strait of Gibraltar is in

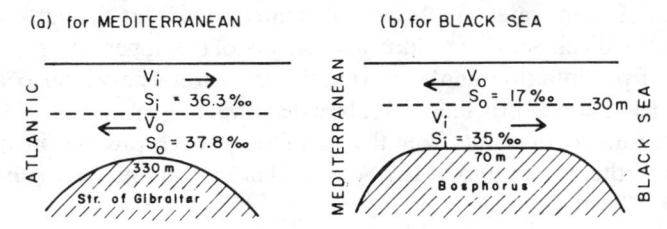

FIG. 5.2. Schematic diagram of inflow and outflow characteristics for: (a) Mediterranean Sea, (b) Black Sea.

the upper layer and the outflow is more saline and is deeper because this denser water has sunk from the surface where it was rendered more saline by net evaporation. The two salinity ratios in the equations above for V_i and V_o have values of about 25 which imply that the salt-water flows are both greater by this factor than the fresh-water balance X. Direct measurements of the upper-layer currents give an average value for $V_i = 1.75 \times 10^6$ m³/s. Then from the equations in Section 5.21, $V_o = 1.68 \times 10^6$ m³/s and $X = (R+P) - E = -7 \times 10^4$ m³/s, i.e. evaporation exceeds fresh-water input by 7×10^4 m³/s. The value above for V_i implies an inflow of 5.5×10^4 km³/year and at this rate it would take about 70 years to fill the Mediterranean (3.8×10^6 km³ volume). This may be taken very roughly as a measure of the *residence time*, i.e. the time required for replacement of all the Mediterranean water (sometimes called *flushing time*). The saline outflow at depth, represented by V_o is an important source of salinity for the mid-depth waters of the North Atlantic.

5.222 The Black Sea

For the Black Sea (Fig. 5.2(b)) the salinity ratios above are 1 and 2 respectively, indicating that here the salt-water flows V_i and V_o are of the same order as the fresh-water balance X. Measured values are approximately $V_i = 6 \times 10^3$ m³/s and $V_o = 13 \times 10^3$ m³/s, giving $X = (R+P) - E = 6.5 \times 10^3$ m³/s, i.e. there is a net inflow of fresh water to the Sea. In this case, the value of V_i implies an inflow of saline water of 0.02×10^4 km³/year. Compared with the Black Sea volume of 0.6×10^6 km³, this suggests a residence time of about 3000 years.

These residence or flushing-time calculations are very rough but the contrast between 70 years for the Mediterranean and 3000 years for the Black Sea is notable for two marginal seas which are neighbours and are connected.

Oceanographic studies support the contrast as the bulk of the Mediterranean water has an oxygen content of over 4 mL/L whereas the Black Sea water below 200 m has no dissolved oxygen but much hydrogen sulphide (over 6 mL/L). The Mediterranean is described as "well flushed" or "well ventilated" whereas the Black Sea is stagnant below 95 m. As will be described in Section 7.41, the physical reason for the ventilation of the Mediterranean is that quantities of deep water are formed by winter cooling at the surface in the north; in the Black Sea, the salinity and density of the upper water is too low, because of precipitation and river runoff, for even severe winter cooling to make it dense enough to sink to replace deep water.

Other examples of results from the use of these conservation principles will be given in the description of the South Atlantic (Section 7.322) and of the Arctic Sea (Section 7.53).

5.3 Conservation of Heat Energy; the Heat Budget

5.31 Heat-budget terms

It has already been stated that the temperature of the ocean waters varies from place to place and from time to time. Such variations are indications of heat transfer by currents, absorption of solar energy, loss by evaporation, etc. The size and character of the variations in temperature depend on the net rate of heat flow into or out of a water body, and calculations of this quantity are referred to as *heat-budget* studies. In what follows, the symbol Q will be used to represent the rate of heat flow measured in joules per second (watts) per square metre, i.e. W/m², generally averaged over 24 hours or over 1 year. (The relationship of these SI units to others which have been used is given in the Appendix.) A subscript will be used to distinguish the different component of the heat budget. These components are:

Q_s = rate of inflow of solar energy through the sea surface,
Q_b = net rate of heat loss by the sea as long-wave radiation to the atmosphere and space,
Q_h = rate of heat loss/gain through the sea surface by conduction,
Q_e = rate of heat loss/gain by evaporation/condensation,

Q_v = rate of heat loss/gain by a water body due to currents which are usually in the horizontal direction so that Q_v is measured through a vertical area of one square metre. The transfer of properties by current flow is called *advection* to distinguish it from the transfer due to diffusion; Q_v is called the *advective term*.

Other sources of heat inflow, such as that from the earth's interior, change of kinetic energy of waves into heat in the surf, heat from chemical or nuclear reactions, etc., are small and can be neglected. The heat budget for a particular body of water can then be stated by the equation:

$$+Q_s+Q_b+Q_h+Q_e+Q_v=Q_T$$

where Q_T is the total rate of gain or loss of heat of the body of water as in Fig. 5.3 which also gives some average values for the Q terms.

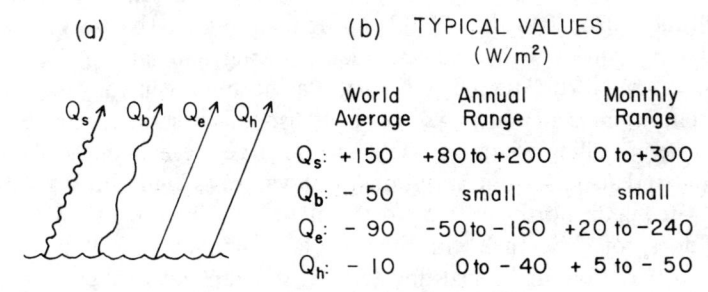

(a) (b) TYPICAL VALUES (W/m²)

	World Average	Annual Range	Monthly Range
Q_s:	+150	+80 to +200	0 to +300
Q_b:	−50	small	small
Q_e:	−90	−50 to −160	+20 to −240
Q_h:	−10	0 to −40	+5 to −50

FIG. 5.3. (a) Schematic diagram showing the heat-budget components, (b) typical global annual average and range values for surface terms.

The above is a symbolic equation. When it is used for heat-budget calculations it is necessary to enter numerical values with a positive sign if they represent gain of heat by the water or with a negative sign if they represent loss from the sea. In practice, Q_s values are always positive, Q_b values are always negative, Q_h and Q_e values are generally negative but may be positive in limited areas at times. Q_v may be positive (inflow of warm water or outflow of cold) or may be negative (inflow of cold water or outflow of warm). Furthermore, as water is substantially incompressible, there will generally be balanced inflow and outflow volumes from a particular sea region (continuity of volume) and the advection of heat by both must be taken into account. (Note also that Q_s, Q_b, Q_h and Q_e are in W/m² and must be multiplied by the sea surface area (m²) of the body of water being considered and Q_v must be multiplied by the vertical area through which advection occurs in order to obtain the total heat flow rate Q_T in watts (W) into (+) or out of (−) the water body to cause its temperature to change.)

If the temperature of a body of water is not changing, this does not mean that there is no heat exchange. It simply means that the algebraic sum of the terms on the left of the heat-budget equation is zero—net inflow equals net outflow, an example of a steady-state condition.

If we apply the heat-budget equation to the world ocean as a whole, Q_v will be zero because then all the advection is internal and must add up to zero. Also if we average over a whole year or number of years the seasonal changes average out and Q_T becomes zero. The equation for the oceans as a whole then simplifies to:

$$Q_s + Q_b + Q_h + Q_e = Q_{sfc} = 0.$$

Note that the typical values in Fig. 5.3(b) for these four terms are only intended as an indication of the general range of annual average values (see Budyko, 1974) and must not be used for specific calculations. It should also be noted that the amount of data on climate for the oceans is limited and so only rounded-off values are given. Note that *monthly* averages show larger ranges. For instance, monthly averages of Q_s vary widely from winter to summer in high latitudes (from 0 to about 300 W/m² in polar regions) but vary less at low latitudes. Q_h varies with time and place, having maximum values in the north-western North Atlantic and North Pacific, but is generally the smallest term. It may represent a small gain of heat seasonally in some coastal localities (e.g. where upwelling occurs). Q_e is the second largest term in the heat-balance equation and also has large variations, with values as high as 240 W/m² loss being noted in the north-western North Atlantic in winter. Q_b is the only term which does not vary much with time or place. The reason will be apparent later. The variations in the heat-flow terms in different localities give rise to the temperature characteristics of the regions, and the terms will be discussed individually below.

5.32 Short- and long-wave radiation; elements of radiation theory

Before discussing the radiation terms, Q_s and Q_b, some aspects of electromagnetic radiation theory will be reviewed briefly. First, Stefan's Law states that all bodies radiate energy at a rate proportional to the fourth power of their absolute temperature (°K = °C + 273°). This energy is in the form of electromagnetic radiation with a range or spectrum of wavelengths. Second, the concentration of energy is not the same at all wavelengths but has a marked peak at a wavelength λ_m given by Wien's Law; $\lambda_m \cdot T = 2897$ μm °K, where T is the absolute temperature (°K) of the radiating body. For a body at a high temperature the radiant energy is concentrated at short wavelengths and vice versa.

The sun has a surface temperature of some 6000°K and radiates energy in all directions at a rate proportional to 6000^4. According to Wien's Law this energy is concentrated round a wavelength of 0.5 μm (1 μm = 10^{-6} metre); 50% of this energy is in the visible part of the electromagnetic spectrum (about 0.35 to 0.7 μm) while 99% is of wavelength shorter than 4 μm. This energy is

Water, Salt and Heat Budgets of the Oceans 71

referred to as *short-wave* radiation and is the source of the Q_s term in the heat budget. The *long-wave* radiation term Q_b represents the electromagnetic energy which is radiated outward by the earth (land and sea) at a rate depending on the absolute temperature of the earth. Taking an average temperature of $17°C = 290°K$ for the sea, it is radiating energy at a rate proportional to 290^4. This is a very much smaller rate than that for the sun, and as the temperature is lower the wavelength is longer. The wavelength at which the sea radiation reaches its maximum is about 10 μm (i.e. in the infrared); 90% of the sea radiation is in the wavelength range from 3 to 80 μm and this is referred to as *long-wave* radiation in contrast to that from the sun which is chiefly less than 4 μm.

5.33 Short-wave radiation (Q_s)

5.331 Incoming solar radiation

A small fraction of the sun's total radiated energy reaches the earth's atmosphere. In Fig. 5.4 this is represented at the top left as 100 parts of

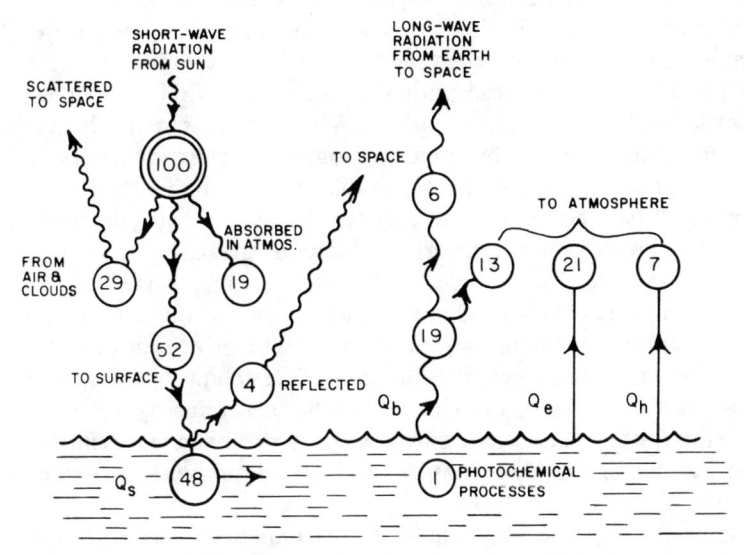

Fig. 5.4. Distribution of 100 units of incoming short-wave radiation from the sun to the earth's atmosphere and surface–long-term world averages.

incoming short-wave radiation. Of this about 29 parts are lost to space by scattering from the atmosphere and clouds, 19 parts are absorbed in the atmosphere and clouds, and about 4 parts are reflected from the sea surface. The remaining 48 parts enter the sea. A small part is scattered upward and the remainder constitutes the Q_s term of the heat budget. Of this 48 parts, about 29 parts reach the sea as *direct* radiation from the sun and 19 parts as *indirect* scattered radiation from the atmosphere (*sky* radiation). Note again that this distribution represents a long-term world-area average; instantaneous values vary diurnally, seasonally and with locality and cloud cover.

The rate at which energy reaches the outside of the atmosphere from the sun is called the *solar constant* and, as obtained from satellite measurements above most of the earth's atmosphere, is from 1365–1372 W/m² perpendicular to the sun's rays. On account of absorption and scattering in the atmosphere, only 50% or less reaches the earth's surface.

5.332 Effects of atmospheric absorption, solar elevation, clouds, etc.

The rate at which short-wave solar energy enters the sea, Q_s, depends upon a number of factors discussed in the following paragraphs.

The first factor is the length of the day, i.e. the time during which the sun is above the horizon, which varies with the season and the geographic latitude. In the following discussion this factor has been taken into account wherever possible for figures quoted for the heat-budget terms, and values given are to be understood to refer to the average over a 24-hour period.

The second factor affecting Q_s is absorption in the atmosphere. This depends on the absorption coefficient for short-wave radiation and on the elevation of the sun. The absorption is the combined effect of that due to gas molecules, to dust, to water vapour, etc. When the sun is vertically overhead, i.e. at an elevation of 90° above the horizontal, the radiation passes through the atmosphere by the shortest possible path and the absorption is a minimum. When the sun is at an elevation of less than 90°, the path of the radiation is greater and the absorption therefore greater.

The elevation of the sun has a second effect. If one considers a beam of radiation from the sun of one square metre cross-section this will cover an area of one square metre of calm sea surface when the sun is vertically overhead. At lower elevations, the beam strikes the sea surface obliquely and is distributed over a larger area than one square metre. The energy density, or amount per square metre of sea surface, therefore decreases as the sun moves further from the vertical. The energy density on the sea surface is proportional to the sine of the angle of elevation of the sun.

Figure 5.5 shows the daily inflow of solar radiation at the earth's surface, assuming an average atmospheric transmission and no clouds, as a function of latitude and time of year. The main features are: (1) the highest values occur at about 30°N and S latitude in the respective hemisphere summers, (2) there is no short-wave input at high latitudes during the polar winter and (3) the amount of energy input is greater in the southern hemisphere than in the northern. The reason for the last feature is simply that the earth, in its elliptical orbit, is nearer to the sun during the southern summer than during the northern one.

So far in this discussion it has been tacitly assumed that there was no cloud

Water, Salt and Heat Budgets of the Oceans   73

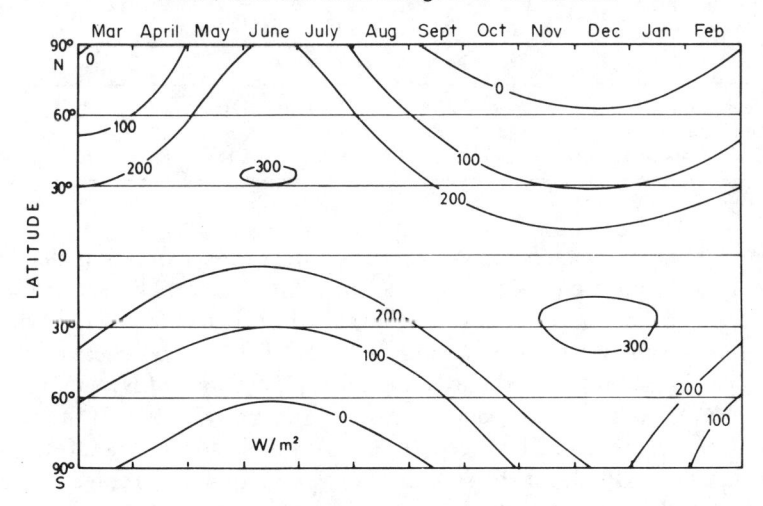

FIG. 5.5. Daily short-wave radiation Q_s in watts/m² received at the sea surface in the absence of cloud.

in the sky. The effect of cloud is to reduce the average amount of energy reaching the sea surface below it because of the absorption and scattering by the cloud. The effect of the cloud may be taken into account by multiplying the mean energy which would arrive in the absence of cloud by a factor based on the cloud amount C where C is the proportion in eighths (oktas) of sky covered by cloud as seen in plan view.

In addition to direct sunlight, the sea also receives a significant amount of energy from the sky, i.e. sunlight scattered by the atmosphere, clouds, etc. The skylight component increases in importance at high latitudes. For instance, at Stockholm (59°N) for a clear sky in July about 80% of Q_s will be direct sunlight and only 20% skylight. In December, only 13% will be direct sunlight and 87% skylight. It must be remembered, however, that the total amount of energy reaching the ground will be less in December than in July, and the 87% of skylight in December will represent a smaller energy flow than the 20% in July.

The next factor affecting the incoming short-wave radiation is reflection at the sea surface. This depends on the elevation of the sun and the state of the sea (calm or waves). It is necessary to calculate the effect separately for direct sunlight which strikes the sea surface at a specific angle of incidence and for skylight which comes from all directions. For a flat sea, the amount of reflection depends on the sun's elevation as in Table 5.1. For skylight it is calculated that the average amount reflected is about 8%, leaving 92% to enter the water. These figures are all affected by waves but no very good figures for reflection in their presence are available, and the figures of Table 5.1 are used as the best available. A few per cent of the radiation entering the sea may be scattered back to the atmosphere.

74 Descriptive Physical Oceanography

TABLE 5.1
Reflection coefficient for sea water

Sun's elevation:	90°	60°	30°	20°	10°	5°
Amount reflected (%):	2	3	6	12	35	40
Amount transmitted into water (%):	98	97	94	88	65	60

The rate Q_s at which short-wave energy enters the sea and is available to raise its temperature depends upon all the above factors. Direct measurements of the energy arriving at the sea surface can be made with a pyranometer as described in Section 6.26 but it is not practical to do so over large areas or for prediction, and for such purposes it is usual to refer to tables or to use a formula. One source is Budyko (1963) and other formulae are discussed by Reed (1985). A simple formula adapted here from Laevastu (1963) for the short-wave radiation input (averaged over 24 hours) in the absence of clouds (Q_{so}) but allowing for the average atmospheric loss is $Q_{so} = 0.4\ A_n t_d$ W/m² where A_n = noon altitude of the sun in degrees, t_d = length of day (sunrise to sunset) in hours (both obtained from a nautical almanac). Note that this expression is given in W/m², i.e. energy per second per square metre, for convenience in comparison with other heat flow components, but in practice it should only be used for calculating whole day values, i.e. it should be multiplied by (24 × 3600) to yield the daily energy input. (This formula should only be used for approximate calculations; for instance it does not take into account the difference between the northern and southern hemisphere input rates. For serious work, tables such as those of Budyko mentioned above should be used.)

To allow for the reduction by cloud cover (C in oktas), a multiplier $(1 - 0.0012\ C^3)$ should be used, i.e. $Q'_s = Q_{so}(1 - 0.0012\ C^3)$ where Q'_s is the rate at which radiant energy arrives at the sea surface. For example, for half of the sky covered $C = 4$ and the multiplier = 0.92, for complete overcast ($C = 8$) it would = 0.39.

Some of the energy arriving at the sea surface is reflected back toward the sky and Laevastu (1963) gave the following relation, derived from field observations, for the amount of short-wave radiation reflected from the sea surface: $Q_r = 0.15\ Q'_s - (0.01\ Q'_s)^2$. Then the complete expression for the short-wave radiation actually penetrating into the sea is:

$$Q_s = Q'_s - Q_r = 0.85\ Q'_s - 10^{-4}\ Q'_s\ \text{W/m}^2. \qquad (5.1)$$

(Actually, we find that this expression, which we present as an example because it is simple to use, although derived from field measurements gives rather high values and we recommend that values calculated using it be multiplied by 0.7.)

5.333 Spatial and temporal variations

Average values for Q_s allowing for atmospheric absorption, mean cloud amount, scattering, etc., are given by Budyko (1974) with much information on other components of the heat and water budgets also. The annual average value for the short-wave radiation input to the oceans ranges from about 90 W/m² at 80°N through a broad maximum of 220 to 240 W/m² between 25°N and 20°S, a minimum of 90 W/m² at 60°S and rising to 110 W/m² at 80°S. Seasonally the ranges of values are greater. In the summer hemisphere, monthly values (averaged over June or December, respectively) are from 240 to 300 W/m² at 80° latitude, decrease through a minimum of 130 to 150 W/m² at 55° latitude, increase to a maximum of 260 to 280 W/m² at 30° to 25° latitude, and then decrease to the equator. In the winter hemisphere, at the same time, from the equator values decrease with increasing latitude to about 15 W/m² at 60° latitude and to zero poleward of about 70° latitude. The high values in the polar regions are due to the length of the polar summer day and to the low content of water vapour (an absorber) in the atmosphere there.

5.34 Long-wave radiation (Q_b)

5.341 Determining factors

The back radiation term, Q_b, in the heat budget takes account of the net amount of energy lost by the sea as long-wave radiation. The value of this term is actually the difference between the energy radiated outward from the sea surface in proportion to the fourth power of its absolute temperature, and that received by the sea from the atmosphere which also radiates at a rate proportional to the fourth power of its absolute temperature. The outward radiation from the sea is always greater than the inward radiation from the atmosphere and so Q_b always presents a loss of energy from the sea.

The long-wave back radiation Q_b is determined by calculating the rate of loss of long-wave energy outward from the sea from Stefan's Law and subtracting from this the long-wave radiation coming in from the atmosphere. This is measured with a radiometer as described in Chapter 6.

If direct measurements of Q_b are not available it is possible to estimate the heat loss by means of data published by Ångström (1920). He showed that the net rate of loss depends upon the absolute temperature of the sea surface itself and upon the water-vapour content of the atmosphere immediately above it. The temperature of the sea determines the rate of outward flow of energy. The water-vapour content effectively determines the inward flow from the atmosphere because the water vapour in the atmosphere is the main source of its long-wave radiation. Ångström's data were published in the form of a table of values of Q_b as a function of water temperature and of water-vapour pressure. The latter quantity is not measured directly as a rule but instead it is determined from the air temperature and the relative humidity. The latter is easily measured with a psychrometer (wet and dry bulb thermometers). The data were adapted by Lönnquist (Laevastu, 1963) to the formula:

$$Q_b = (143 - 0.9\, t_w - 0.46\, e_a)(1 - 0.1\, C) \text{ W/m}^2 \qquad (5.2)$$

where t_w = water temperature (°C), e_w = relative humidity above the water surface and C = cloud cover in oktas. Figure 5.6 shows values calculated from

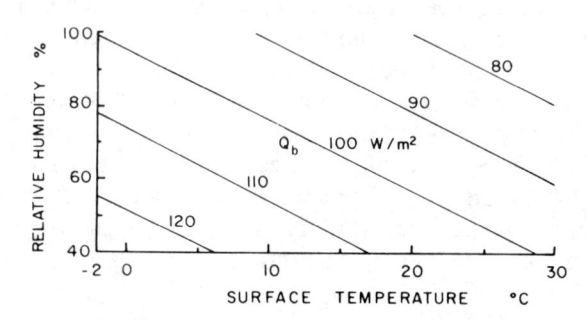

FIG. 5.6. Back radiation Q_b from a water surface as a function of surface temperature and relative humidity above the surface in absence of cloud.

equation (5.2) for Q_{bo}, i.e. for the cloudless condition $C=0$, for a range of values of t_w and e_a likely to be found at sea. Values of Q_{bo} for oceanic conditions range from 120 to 70 W/m² with the higher rates occurring at low temperatures and low humidities and vice versa.

The value of Q_b decreases as the sea-surface temperature increases for the following reason. A rise of sea-surface temperature causes an increase of the outward radiation from the sea but is accompanied by an increase of humidity in the atmosphere immediately above it. The temperature of this lower atmosphere follows that of the sea. However, the amount of water vapour increases exponentially, i.e. more rapidly than the temperature, with the result that the atmosphere's radiation into the sea increases more rapidly than the sea's outward radiation. The net result is a decrease of Q_b (i.e. a reduced loss from the sea) as the sea temperature increases. If only the atmospheric humidity increases, the inward radiation from the atmosphere to the sea increases and therefore Q_b decreases.

5.342 Temporal and spatial variations

The back radiation terms does not change much either daily or seasonally or with location because it depends on the absolute temperature, not the Celsius temperature, and because the relative humidity does not change much over the sea. For instance, a seasonal change of sea temperature from 10° to 20°C would give rise to a change of outward radiation proportional to

$293^4/283^4$ or about 1.15, i.e. only a 15% increase. At the same time the atmospheric radiation inward would increase and reduce the net rate of loss below this figure. The small seasonal and geographic changes of Q_b are in contrast to the large changes of Q_s.

5.343 Effect of clouds

The effect of cloud cover is allowed for in Equation (5.2) by the factor $(1-0.1C)$, or if the clear sky values of Fig. 5.6 are used they must be multiplied by that factor. The value for C refers to substantial cloud, i.e. thin cirrus cloud is less effective than thicker cloud such as heavy stratus or cumulus. It is seen that with the sky completely covered with substantial cloud, i.e. ($C=8$) the value of the factor is 0.2, i.e. the loss of energy as long-wave radiation is sharply reduced by cloud cover. This effect of cloud is well known on land where the frost which results from radiation cooling (i.e. the Q_b term for land) is more frequent on clear nights than on cloudy ones. The reason for the big difference between clear and cloudy conditions is that the atmosphere, particularly its water-vapour content, is relatively transparent to radiation in the range from about 8 to 13 μm which includes the peak of the radiation spectrum for a body of the temperature of the sea. In clear weather, energy between 8 and 13 μm wavelength radiated by the sea (and the land) passes through the wavelength "window" in the atmosphere and out into space where it is lost from the earth system.

It should be noted here that for energy in this long-wave part of the spectrum, water has a very high absorption coefficient. The incoming long-wave radiation from the atmosphere is all absorbed, not in the top metres of the sea but in the top millimetres. Similarly the outward radiation is determined by the temperature of the literal surface or skin of the sea. In practice the "sea-surface" temperature as measured is that of a bucket of water dipped from the upper half-metre or so. If the sea surface is disturbed by wind and waves the *bucket temperature* is assumed to represent the skin temperature but very little work has been done in this skin layer. It may seem rather trivial at first to go out to study the top millimetre of the sea, but apart from the heat-budget aspect it is probable that other processes such as the early stages of wind generated waves are determined by stresses in this surface skin. Also, as the upper layer of the ocean is often well mixed vertically this skin temperature, which may be sensed remotely from aircraft or setellites, may be representative of that entire layer.

5.344 Effect of ice and snow cover

When the sea surface becomes covered with a layer of ice, and especially if snow covers the ice, there is a marked change in the heat-radiation budget. For a water surface, the average proportion of short-wave radiation (sun+sky=

Q_s) reflected is relatively small (10 to 15%) and the proportion absorbed is therefore large. For ice or snow, the proportion of short-wave radiation reflected is much larger (50 to 80%) leaving a smaller proportion to be absorbed. However, the size of the Q_b loss term is much the same for ice as for water, and the result is a smaller net gain $(Q_s - Q_b)$ by ice and snow surfaces than by water. In consequence, once ice forms it tends to be maintained. It has been estimated that the balance in the Arctic Sea is relatively fine and that if the sea-ice were once melted the increased net heat gain $(Q_s - Q_b)$ might maintain the Arctic Sea free of ice (Donn and Shaw, 1966). This, however, would increase the amount of evaporation and there might be marked increases in precipitation on the high north latitude lands which at present receive a relatively small precipitation (mostly as snow in the winter). It should be emphasized that this latter idea about possible changes in the Arctic is very speculative as it is difficult to be sure that the factors which might be changed have been correctly assessed, or indeed that all the factors that would affect the situation have been included. However, it is often by making predictions and then observing their success or otherwise that we test our understanding of natural processes.

5.35 Evaporation (Q_e)

The evaporation term is an important one but not easy to determine directly. The reason why evaporation enters into the heat budget is that for it to occur it is necessary to supply heat from an outside source or for heat to be taken from the remaining liquid. The second is the more usual case for the sea. (It is the reason why one often feels cold when one stands with one's wet body exposed to the wind after swimming.) Therefore evaporation, besides implying loss of water volume, also implies loss of heat. The rate of heat loss is $Q_e = F_e \cdot L_t$, where F_e is the rate of evaporation of water in kilograms/sec per square metre of sea surface and L_t is the latent heat of evaporation in kilojoules. For pure water this depends on the temperature of the water t°C as $L_t = (2494 - 2.2\ t)$ kJ/kg. At 10°C, the latent heat is about 2472 kJ/kg, much greater than the value of 2274 kJ/kg (540 cal/gm) at the boiling point. Basically, then, to determine Q_e we must determine F_e.

A direct method for this is to determine the rate of water loss from a pan of water but there are serious practical difficulties in doing this at sea. For instance, a pan of water on deck is in a different wind and humidity regime than at the sea surface, spray may get into the pan, screens to prevent this would alter the wind flow, a shallow pan may get warmer or cooler than the sea surface, etc. Even on land, when studying evaporation from reservoirs, it has been difficult to get consistent results with pans of different construction. From such measurements as have been made at sea, it appears that the evaporation rate from a pan at deck level is likely to average about twice the rate from the sea surface. The average amount of evaporation per year from the sea surface is about 120 cm, i.e. the equivalent of the sea surface sinking by that amount due to evaporation. Local values range from an annual minimum of as little as 30/40 cm/yr in high latitudes to maxima of 200 cm/yr in the tropics associated with the trade winds, and decreasing to about 130 cm/yr at the equator where the mean wind speeds are lower and the heat input is reduced seasonally by increased cloud cover. The high values for evaporation in the tropics are the reason for the high surface salinity values there (e.g. Figs. 4.3, 4.9). For large area estimates and for prediction, a formula using more easily measured parameters is desirable.

In principle, F_e can be estimated by the application of a formula of the type $F_e = -K_e \cdot de/dz$, where K_e is a diffusion coefficient for water vapour and de/dz is the gradient of water vapour concentration in the air above the sea surface. In small volumes, where the vapour as a whole is stationary, the process of diffusion is due to the random movement of molecules and K_e is the molecular diffusion coefficient. However, in nature, and in particular over the sea, the air is usually in motion (wind) and the motion is turbulent. A result of this is that the air eddies, consisting of bulk movements of the air, tend to carry air properties with them (*eddy transfer*) down the property gradient, humidity in this case, at a much greater rate than for pure molecular motion. Where there is turbulent motion it is possible to define empirically a quantity called the *eddy diffusivity* (A_e) for water vapour and to describe the rate of water vapour transfer as $F_e = -A_e \cdot de/dz$. The eddy transfer process also applies to heat in the atmosphere and to diffusion of salt and to conduction of heat in the sea where the motion is usually turbulent. The eddy transfer rate is generally so much greater than the molecular rate that the latter may be neglected.

The reason why the introduction of eddy diffusion (and eddy conduction) introduces some difficulty is that the eddy coefficients are not constant quantities but depend on the character of the turbulence in the air. This depends on various factors such as wind speed and the size of ripples or waves on the sea surface. We do not know enough about turbulence in nature to say with any certainty what the value of an eddy coefficient will be in every situation. All that we can really say is that if we happen to have measured an eddy coefficient on a previous occasion when the wind conditions, etc., were much the same, then the eddy coefficient will probably be much the same. As we cannot judge the character of turbulence by eye and it is not easy to measure, even with sophisticated instruments, this leaves us in an unhappy situation. It would not be so bad if the eddy coefficients only varied by a small amount with wind, etc., conditions; unfortunately they may vary over a range of ten- or a hundredfold. The whole problem of eddy transfer of properties in fluids is a part of the more general problem of fluid turbulence which is one of the more pressing problems of physics requiring solution at the present time. Research is under way and we are gradually acquiring some understanding. (The eddy transfer process is discussed in more detail in Pond and Pickard (1983) in terms of the eddy transfer of momentum, i.e. eddy viscosity.)

In practice, a semi-empirical flow formula is frequently used for evaporation as: $F_e = 1.4(e_s - e_a) \cdot W$ kg per day per square metre of sea surface, where W = wind speed (m/s) at 10 m height, so that:

$$Q_e = F_e L_t = 1.4(e_s - e_a) \cdot W \cdot (2494 - 2.2\, t_s) \cdot 10^{-3} \text{ W/m}^2. \quad (5.3)$$

In this formula e_s is the saturated vapour pressure over the sea-water and e_a is the actual vapour pressure in the air at a height of 10 m above sea level, both of these pressures being expressed in kilopascals (101.35 kPa = 760 mm of mercury), while W is the wind speed in metres per second at 10 m height. The saturated vapour pressure over sea-water (e_s) is a little less than that over distilled water (e_s'). For a salinity of 35, $e_s = 0.98 e_d$ at the same temperature. The saturated vapour pressure over distilled water may be obtained from tables of physical or meteorological constants. If the water-vapour content in the air is given as relative humidity then the value e_a is equal to the saturated vapour pressure over distilled water at the temperature of the air multiplied by the relative humidity expressed as a fraction, not as a percentage. For example, at an air temperature of 15°C the saturated vapour pressure is 1.71 kPa (= 12.8 mm Hg). If the relative humidity is 85% then the actual vapour pressure in the air is 1.71 × 1.45 kPa.

This practical formula is basically a simplified version of the theoretical eddy diffusion-flow formula above. In that formula de/dz is the vertical humidity gradient, de being the change in humidity over a vertical distance dz. In the practical formula $(e_s - e_a)$ is the change of humidity over a vertical distance dz of 10 m between the sea surface and the height where e_a is measured. (In the practical formula the factor of 10 does not appear explicitly, having been absorbed in the numerical constant.) The W of the practical formula represents the variation of A_e in an elementary fashion. The values of A_e do not necessarily vary directly as the wind speed but we expect turbulence to increase in some way as wind speed increases and therefore eddy diffusion should increase as wind speed increases. Hence the use of W in the formula does at least give some variation of eddy diffusion in what is certainly the right direction. The actual numerical value of A_e is, of course, not the same as the wind speed; again the factor of proportionality between them is hidden in the numerical constant in the practical formula.

In most regions of the ocean, it turns out that e_s is greater than e_a and therefore as all the other terms in the practical formula are positive, the value of F_e is positive and so is Q_e in these regions. This is entered in the heat-budget equation numerically as a negative quantity as it represents a loss of heat from the sea due to evaporation in such cases. In fact, as long as the sea temperature is more than about 0.3K greater than the air temperature, there will be a loss of heat from the sea due to evaporation. Only in a few regions is the reverse the case, when the air temperature is greater than the sea temperature and the humidity is sufficient to cause condensation of water-vapour from the air into the sea. This results in a loss of heat from the air into the sea. The Grand Banks

Water, Salt and Heat Budgets of the Oceans 81

off Newfoundland, and the coastal seas off northern California are examples of regions where the heat flow Q_e is into the sea (numerically positive). The fogs that occur in these regions are the result of the cooling of the atmosphere.

5.36 Heat conduction (Q_h)

5.361 Eddy conduction

Another reason that heat may be gained or lost from the sea surface is that there is often a temperature gradient in the air above the sea. If the temperature decreases upward from the sea surface, heat will be conducted away from the sea and Q_h will be a loss term. If the air temperature decreases downward toward the sea, heat will be conducted into the sea and Q_h will be a gain term. The rate of loss or gain of heat is proportional to the temperature gradient, heat conductivity (for which we use an eddy conductivity, A_h) and the specific heat of air at constant pressure (C_p) as: $Q_h = -C_p \cdot A_h \cdot dt/dz$. Wyrtki (1965) used $A_h = \rho_a \cdot C_d \cdot W$ for calculations of the heat budget of the Pacific Ocean, where ρ_a = density of air = 1.2 kg/m^3. C_d = drag coefficient at the sea surface = 1.55×10^{-3}, W = wind speed in m/sec and with $C_p = 1008$ J/(kg K), then:

$$Q_h = -1.88 \cdot W \cdot (t_s - t_a) \text{ W/m}^2. \tag{5.4}$$

5.362 Convection

In situations where the sea is warmer than the air above it, there will be a loss of heat from the sea because of the direction of the temperature gradient. However, the phenomenon of convection will also play a part in assisting the transfer of heat away from the sea surface. Convection occurs because the air near to the warm sea gets heated, expands, and rises carrying heat away rapidly. In the opposite case where the sea is cooler than the air, the latter is cooled where it is in contact with the sea, becomes denser, and therefore tends to stay where it is and convection does not occur. The consequences is that for the same temperature difference between sea and air, the rate of loss of heat when the sea is the warmer is greater than the rate of gain when the sea is the cooler. In the tropics the sea is generally warmer than the air, on the average by about 0.8K, and the result is that Q_h is a loss term. In middle and high latitudes the temperature difference is more variable but on the whole the sea is warmer than the air and consequently here also Q_h is generally a loss term.

5.363 Bowen's Ratio

If, in the full heat-budget equation of Section 5.31, we consider the situation when $Q_v = 0$ (no advection) and when $Q_T = 0$ (steady state), and we introduce the quantity $R = Q_h/Q_e$ called *Bowen's Ratio* (Bowen, 1926), then $Q_e = (Q_s - Q_b)/(1 + R)$. From this equation, if we have values for Q_s and Q_b and can determine R, we can obtain a value for Q_e. The two radiation terms have already been discussed and it remains to discuss R, the ratio of the heat-conduction term to the evaporation term. Earlier it was explained that the molecular transfer rates for heat and for water vapour are considered negligible by comparison with the eddy transfer rates associated with turbulence. The expressions for $Q_h = -C_p \cdot A_h \cdot dt/dz$ and $Q_e = -L_t \cdot A_e \cdot de/dz$ are similar because the transfer mechanism due to turbulent eddy movements of the air above the sea is of the same physical nature for both. If one goes further and assumes that the numerical values of A_h and A_e are the same, then these two terms will cancel out in Bowen's Ratio, leaving only the ratio of the temperature and the humidity gradients. Each of these gradients may be expressed approximately by the difference of the respective quantity (temperature or humidity) between the sea surface and a level above the sea surface. If the temperature and humidity gradients are both measured over the same range of height dz between the sea surface and some height above sea level (e.g. on the ship's mast or bridge), then the dzs will also cancel in Bowen's Ratio. This then reduces to the simple form: $R = 0.062(t_s - t_a)/(e_s - e_a)$. Here t_s(°C) and e_s (kPa) represent measurements at sea level while t_a and e_a represent the measurements at 10 m height. Both temperature and water-vapour pressure may be measured relatively easily. Hence a value for R is obtained, and this may be used in the earlier equation together with measurements of Q_s and Q_b to obtain a value for Q_e and hence $Q_h = R \cdot Qe$.

Before going on it should be pointed out that the above simplified practical formula for determining R from the meteorological observations depends on the assumption that A_h and A_e are numerically the same. This is based on the simple argument that the transfers of heat and water vapour are both due to the turbulent motion of the air above the sea. The process of turbulence consists in the transfer of properties from larger to successively smaller eddies and eventually, at the end of the scale, molecular transfer must play a part. While the eddy transfer coefficients for heat, water vapour and other properties tend to have much the same values for high turbulence, they are not numerically equal for low turbulence. The assumption that they are equal when simplifying the Bowen's Ratio formula is therefore an approximation but it seems to be a reasonably good one, judged on the basis of consistency of deductions obtained by its use compared with other evidence.

Average values for R based on many sets of meteorological data are of the order of $+0.1$ in equatorial and tropical regions and increase to about 0.45 at 70°N. Remembering that $R = Q_h/Q_e$ this indicates that the heat flow term is usually smaller than the evaporation term. It should be noted that the average values for R are positive because t_s and e_s are usually greater than t_a and e_a respectively. Therefore both of the difference terms in the expression for R are positive. However, it is possible for R to be negative in the regions where the

Water, Salt and Heat Budgets of the Oceans 83

sea temperature t_s is less than the air temperature t_a. A negative value for R indicates the unusual condition of Q_h representing heat flow into the sea.

If a value for R were available, then Q_e could be obtained from the relation $Q_e = (Q_s - Q_b)/(1 + R)$ derived above. This relation, however, depends on the two assumptions that Q_T and Q_v are both zero, and is of limited application. It is not too difficult to check on the first assumption but the second requires detailed current measurements and these are among the most time and effort consuming aspects of experimental physical oceanography if they are to be done well. The heat-budget method for determining Q_e is most useful for large areas for checking heat-budget estimates obtained by other methods. In practice, Q_e is usually obtained from the semi-empirical flow formula, and then with a value for R we can calculate $Q_h = R \cdot Q_e$.

5.37 Some comments on the calculation of the heat budget terms

Equations 5.1–5.4 which we have presented are by no means the only formulae for the heat budget terms, and variations on these may be found in two of the more recent calculations of heat budget distributions for ocean areas—those by Wyrtki (1965) for the Pacific Ocean and by Bunker (1976a, b) for the North Atlantic. For Q_s, Wyrtki used tables compiled by Berliand (1960) for Q_{so} and corrected these for reflection and cloud cover, and used a more complicated formula for Q_b than our Equn. 5.2. His formulae for Q_e and Q_h were of the same form as our Equns. 5.3 and 5.4. Bunker (1976b) discussed the four formulae, and in particular, values which might be used for the exchange coefficients A_e and A_h in the expressions for Q_e and Q_h. For his study of the North Atlantic heat budget (Bunker, 1976a) he used "formulas and tables of Budyko (1963)" for Q_s and Q_b. The reader is referred to the papers by Bunker for more detail on heat budget calculation formulae.

In a recent paper, Smith (1988) reviewed the values for the coefficients for wind stress and heat flux over the sea. Using the formulae (converted to our notation):

Evaporation: $Q_e = L_t A_e \rho (q_s - q_a) W$,
Heat Conduction: $Q_h = A_h \rho C_p (t_s - t_a) W$,

where A_e and A_h are the eddy coefficients for water vapour and heat transfer respectively, q_s and q_a are the mixing ratios for water vapour at the sea surface and at 10 m height, t_s and t_a are the temperatures of the sea surface and at 10 m height, and W is the wind speed at 10 m height. (Note: the mixing ratio $q = (0.622\,e)/(p-e) \doteq (0.622\,e)/p$ where e = pressure of water vapour in the atmosphere and p = atmospheric pressure.) Smith gave graphs and tables for A_h as functions of $(t_s - t_a)$ from $-20K$ to $+20K$ and $W = 0 - 25$ m/s. We abstract a portion of his table as Table 5.2:

Note that as wind speed increases, values for A_h tend toward unity. The reason for the large differences at low wind speeds is related to the stability of

TABLE 5.2

Some values for A_h as functions of $(t_s - t_a)$ and wind speed W

	Wind speed W m/s			
$(t_s - t_a)$ (K)	2	5	10	20
−10			0.75	0.96
−3		0.62	0.93	0.99
−1	0.34	0.87	0.98	1.00
+1	1.30	1.10	1.02	1.00
+3	1.50	1.19	1.06	1.01
+10	1.87	1.35	1.13	1.03

the air over the sea. For example, for $(t_s - t_a) = -1\,\text{K}$, i.e. the sea is cooler than the air ($t_s < t_a$), and the stability in the air is positive so that heat conduction is inhibited, but when the sea is warmer than the air, e.g. $(t_s - t_a) = +1\,\text{K}$, the air will be unstable and heat conduction away from the sea is promoted. The blank areas in the table are for highly stable conditions (unusual) where Smith's analysis broke down.

For evaporation, Smith commented that "measurements in open sea conditions are relatively rare, particularly for high wind speeds . . ." but after reviewing the available data he recommended $A_e = 1.20\,A_h$ as the best available figure, the variation with stability being expected to be similar for evaporation as for turbulent heat conduction.

Smith also presented graphs/tables for values for the wind-stress drag coefficient C_d which has a significantly different behaviour in that values increase markedly with wind speed (W) rather than converging to a constant value. For example, for neutral stability ($t_s - t_a = 0$):

$W =$	2	5	10	20	25 m/s
$C_d =$	0.98	1.03	1.30	1.80	2.04

For stable/unstable conditions, values are lower/higher respectively but the difference from neutral stability decreases markedly as wind speed increases.

5.38 Geographic distribution of the heat-budget terms

5.381 Pacific Ocean

Annual average distributions of the various Q terms for the Pacific, based on calculations by Wyrtki (1965), are presented in Fig. 5.7(b) to (f). Q_s has a range from +70 to +220 W/m² (gain) with a strong latitude dependence, related to solar elevations. The low values in the north are due to large cloud amounts there (Fig. 5.7.1(a), while high values around 5°S 145°W and 20°N 160°E result from smaller cloud amounts. Q_b (Fig. 5.7.1(c) is relatively

Water, Salt and Heat Budgets of the Oceans 85

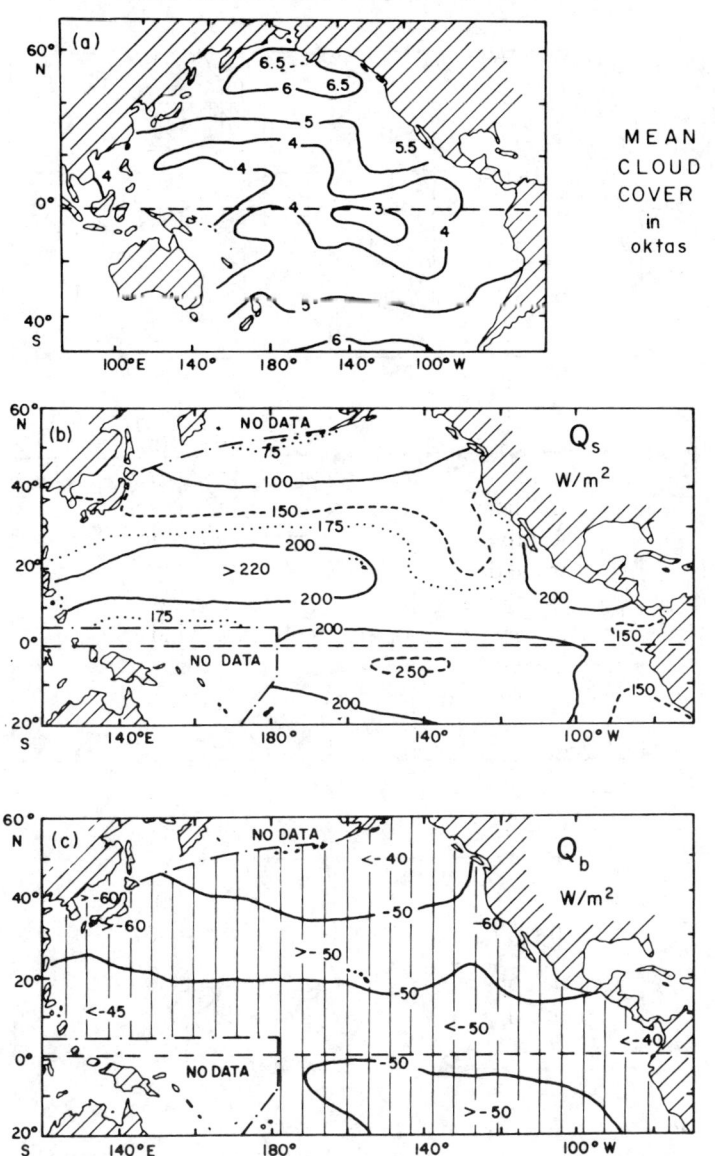

FIG. 5.7.1. Pacific Ocean: (a) annual mean cloud cover in oktas; annual mean values in watts/m² for (b) Q_s, (c) Q_b. (Note: + values = gain by sea, − value = loss by sea; shading emphasizes loss areas.)

uniform with a range only from -40 to -65 W/m² (loss). The low values in the north are due to large cloud amounts there, and higher values in the northwest are due to conditions to be explained in the next paragraph.

Q_e, with a range from -25 to over -150 W/m² (Fig. 5.7.2(d)), shows large

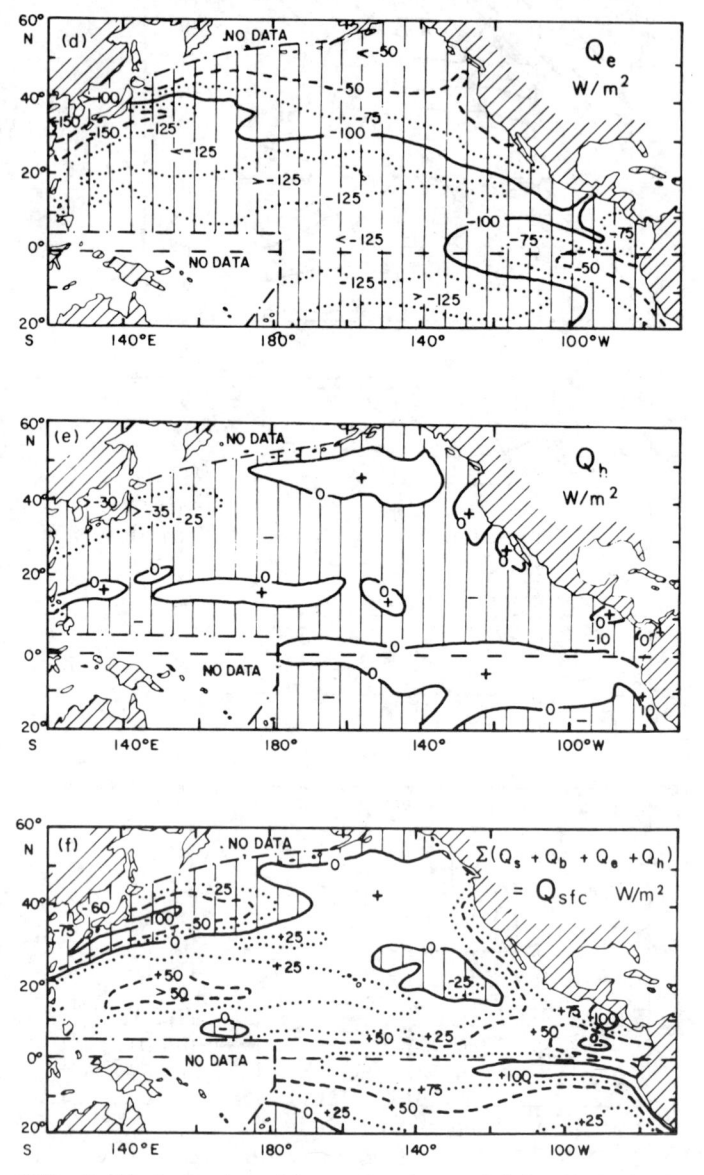

FIG. 5.7.2. Pacific Ocean: Annual mean values in watts/m² for: (d) Q_e, (e) Q_h, (f) $Q_{sfc} = (Q_s + Q_b + Q_e + Q_h)$. (+ values = gain by sea, − values = loss by sea; shading emphasizes loss areas.)

values in the tropics because $(e_s - e_a)$ is high and there are the steady trade winds to assist evaporation. Even higher values are found in the north-west over the Kuroshio Current off Japan in the winter. The reason for this is that the westward-flowing North Equatorial Current (Fig. 7.31) turns north along the western boundary of the ocean and carries warm water, with relatively high vapour pressure, to higher latitudes where the air temperatures and vapour pressures are lower. This gives rise to a strong humidity gradient resulting in a large upward flow of water vapour from the sea. The maximum values for Q_e occur at the western side of the ocean in winter. The latter fact appears surprising at first, but is simply because the water temperature, and therefore the vapour pressure, of the northward-flowing water does not decrease much in winter, but the temperature and humidity of the air do decrease (cold, dry continental arctic air from the north-west). Consequently, $(e_s - e_a)$ is greater in winter than in summer in this north-western part of the ocean making Q_e larger in winter than in summer. Q_b is largest in winter because of the low humidity of the air. Lowest values of Q_e occur over the upwelling regions (cold water) along the western costs of the America (eastern sides of the ocean).

Q_h (Fig. 5.7.2(e)), with a range from -10 to $+40$ W/m^2, is generally small over most of the ocean and negative (loss) over large areas. It becomes strongly negative in the north-west off Japan. It reflects the generally smaller values of $(t_s - t_a)$ over much of the ocean except in the north-west in winter, for the reasons given above.

The sun of the above flows through the surface, $Q_{sfc} = (Q_s + Q_b + Q_h + Q_e)$, has high gain values of $+100$ W/m^2 in low latitudes, particularly in the east, south of the equator, due to high values for Q_s and low values for Q_e. The sum has high loss values of -100 W/m^2 in the north-west due to the large negative values for Q_b, Q_e and Q_h in winter.

It is interesting to note that the above distribution of Q_{sfc} (Fig. 5.7.2(f)) leads to an annual average of 12×10^{14} W gain to the North Pacific through the surface. As the mean temperature of that ocean does not appear to be increasing, conservation of heat energy indicates that there must be a balancing loss. This must be as a Q_v term and must therefore be an outflow of warm, i.e. upper layer, water. This means that there must also be an inflow of subsurface (cool) water. It is calculated that the outflow volume would have to be about 8 Sv (8×10^6 m^3/s), and the vertical speed as the subsurface inflow replaces the upper water would be of the order of 1 cm/day averaged over the whole North Pacific. The location of the outflow is uncertain. Heat may also be leaving the surface in high latitudes where water, in the North Pacific, sinks to intermediate depths.

For the South Pacific, marked differences in the land boundaries and the absence of a strong western boundary current, such as the Kuroshio, may be expected to lead to substantial zonal heat-exchange differences from the North Pacific.

5.382 Atlantic Ocean

Detailed maps of the heat budget terms, wind characteristics, etc. were presented by Isemer and Hasse (1987) in *The Bunker Climate Atlas of the North*

Descriptive Physical Oceanography

Atlantic Ocean where it is evident that the general distributions of the heat budget terms for that ocean are very similar to those for the North Pacific. For the South Atlantic, a shorter description with maps of $(Q_s - Q_b)$, Q_h, Q_e and resultant values for Q_{sfc} was given by Bunker (1988).

5.383 World ocean; northern hemisphere

In Fig. 5.8(a) are shown the average annual values for the heat-budget terms

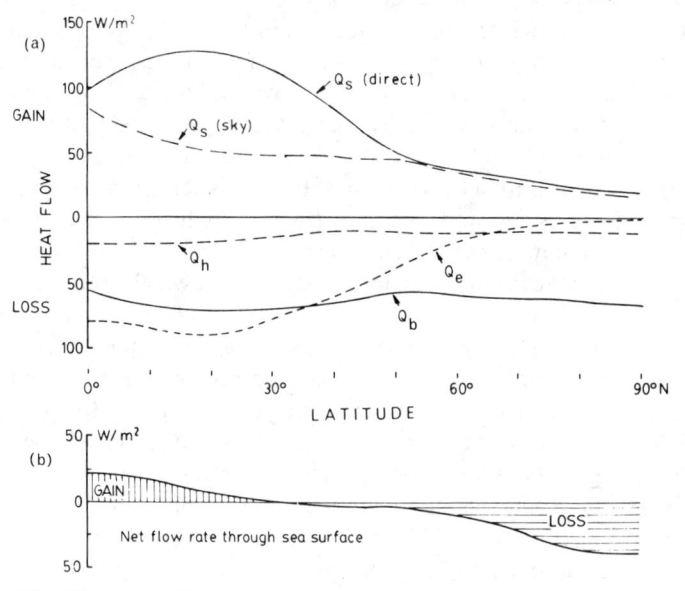

Fig. 5.8. Values (world ocean average, northern hemisphere) for heat-flow terms through the sea surface as a function of latitude.

in the northern hemisphere. Some of the features revealed by this figure are that the direct sun's radiation (corrected for cloud) dominates to about 50°N but that beyond this the skylight component is equally important, and that while the evaporation-loss term decreases markedly toward the pole, the back radiation term is much the same at all latitudes. The balance between the gain and loss terms (Fig. 5.8(b)) shows a net gain from the equator to 30°N and a net loss beyond this. At first sight there appears to be a much greater loss than gain but this is not really the case. The quantities shown in Fig. 5.8 are the annual average rates of flow *per square metre* of sea surface (W/m²). To obtain the total flow in or out for any latitude zone one must multiply by the total sea area in that zone. This area is less at high than at low latitudes and with this correction the gain and loss are more nearly, but not exactly, balanced. As there is no indication that the oceans as a whole are getting warmer or cooler we would expect an exact balance. We must conclude that the fault lies in our having insufficiently accurate knowledge of the individual heat-flow terms to permit the budget to be balanced exactly.

For the northern hemisphere as a whole (land and sea), and averaged over the year, Q_s is greater than Q_b between the equator and about 40°N, and there is a net gain of heat by radiation at these lower latitudes. At higher latitudes, Q_s is less than Q_b and therefore there is a net loss of heat by radiation. Since the average temperatures over the earth remain substantially constant we conclude that there must be a net advective flow of heat to the north, from the lower latitudes of net radiation gain to the higher ones of net loss. This heat flow toward the pole is effected both by the ocean and by the atmosphere. These transport warm water or air toward the pole and cooler water or air toward the equator. Recent estimates, based on 9 years of radiation measurements from satellites, indicate that in the northern hemisphere the contribution to the heat transport by the ocean rises to a maximum of 60% of the total at 20°N, is 25% at 40°N and 9% at 60°N (Oort and Vander Haar, 1976). These estimates compute the ocean's contribution to poleward heat transport as the difference between the satellite radiation measurements and observations of atmospheric heat transport from meteorological data. Errors in the atmospheric data may help to explain the large differences between the above residual computation of ocean heat transport and traditional estimates of poleward oceanic transport based on oceanographic observations.

Recent interest in the possible effects of the ocean on the world's climate has prompted a new series of studies into the advective heat fluxes in the ocean. One surprising result has been the suggestion that the oceanic heat flux in the mid-latitude South Atlantic is toward the equator rather than toward the pole. This contradicts coventional thought which requires the South Atlantic to carry heat from the warm equatorial zone to the colder polar latitudes. A summary of the heat and fresh-water fluxes for the world's oceans, based on work by Stommel, is presented in Fig. 5.9. Since the advection (flow) carrying heat in the ocean also carries the particular salinity character of the water, the fluxes of heat and salt are often studied together to provide insight into the accompanying mass or volume flux. The variety in the results of these studies emphasizes both our imperfect understanding of the advective process and the lack of comprehensive data for evaluating the process.

5.384 Large-scale programmes on the heat budget

Apart from the satellite measurements referred to in the previous section, the statements on the heat budget are based on information available before 1964 which is sparse for the northern hemisphere and less than adequate for the southern. The inadequacies in the meteorological data have long been recognized and plans are in hand for acquiring much more comprehensive data. The major developments are expected from the Global Atmospheric Research Programme (GARP) which includes international projects to

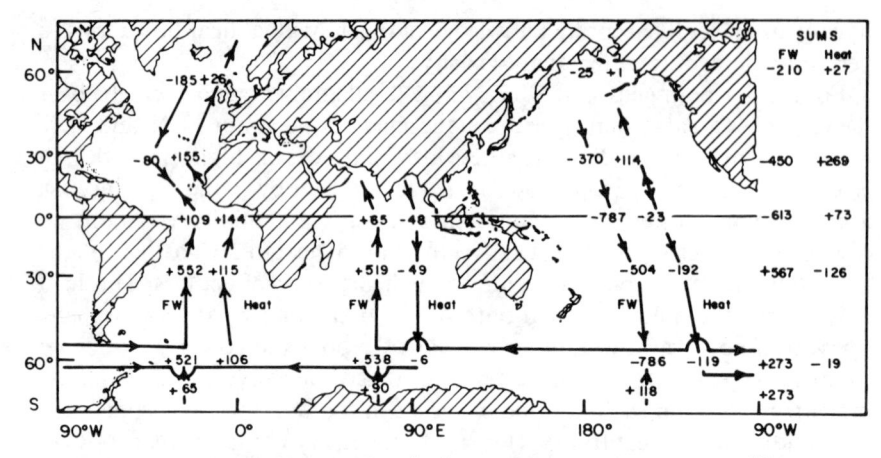

FIG. 5.9. Fresh-water flux (FW in 10^3 m^3/s) on left, and heat flux (heat in 10^{13}W) on right, at various latitudes, positive northward.

improve our understanding of the behaviour of the atmosphere. Because so much of the energy which drives the atmospheric circulation comes from solar heat collected by the oceans and returned to the atmosphere as sensible and latent heat, this programme will contain a significant oceanographic measurement component. One of the first major projects under GARP was the GARP Tropical Experiment (GATE) in the eastern tropical Atlantic in 1974 to improve our knowledge of tropical atmospheric interactions, including those with the ocean. A total of thirty-eight ships and twelve aircraft took part in the Experiment over a period of 6 months. A second major effort, called the First GARP Global Experiment (FGGE), took place in early spring and summer of 1979. A fleet of ships from many countries collected a wide variety of oceanographic and atmospheric data in the equatorial regions while many free-floating, satellite-tracked buoys were deployed in the Southern Ocean in an effort to improve significantly the measurement of atmospheric pressure around Antarctica.

A major oceanographic/meteorological experiment to study global climate change and its effects on ocean circulation and ocean/atmosphere interaction, the World Ocean Circulation Experiment (WOCE), is now developing, with planning and theoretical studies (numerical modelling) having been under way for some years and field observations planned to start in 1990, e.g. see Mayes, 1989.

Much of the modern interest in the interaction between the ocean and atmosphere comes from the realization that weather cannot be predicted with any skill for periods longer than a few days. The atmosphere is characterized by short term changes and, knowing that most of the energy which drives the atmospheric motions comes from the sea (as latent heat of evaporation), it is hoped that longer-period weather fluctuations, known as climate changes, may somehow be related to the longer thermal memory of the ocean and its interactions with the atmosphere. Some aspects of studies of such possible interactions between sea-surface temperature (SST) and the climate of North America will be discussed in Section 7.622.

CHAPTER 6

Instruments and Methods

6.1 Introduction

Before describing the techniques and methods of physical oceanography something must be said about the aims and limitations of field observations.

A fundamental goal of many physical oceanographers has been to determine the three-dimensional circulation of the oceans as a function of time. The obvious way would seem to be to go to sea with current meters and to measure it directly. Unfortunately, current meters only give information on the velocity (speed and direction) of the water at the location of the instrument itself, and experience indicates that large variations in current velocity can occur over small distances as well as over small time intervals. Also, in the present state of current meter development it is costly, even with moored instruments, to measure currents even at only a few points at any one time. The limited number of oceanographic ships and of oceanographers available to deploy and recover these current meter moorings also limits the number of spot current measurements that may be made. In consequence, direct measurements of currents have to be restricted to key localities of limited area and for purposes such as testing specific theories. The total of direct current measurements of subsurface currents has provided only a small proportion of our observed knowledge of the ocean circulation.

Failing a sufficiency of direct measurements, the synoptic oceanographer has been forced to use indirect methods. The chief indirect method has been to observe the distributions of water properties, which can be done more expeditiously than observing currents, and to deduce the flow from these distributions. In addition to the fact that measuring properties is a lot easier than directly measuring currents there is the underlying assumption that the ocean processes dictating the property distribution (currents, mixing, etc.) have time-scales that are longer than the time taken to collect the measurements from a research vessel. Thus the suite of observations collected during an oceanographic cruise are often treated as though they were all collected at the same time. In the majority of cases, this method only reveals the path followed by the water and gives little information on speed. The path is better than nothing but the synoptic oceanographer is always on the look out for any characteristic of the property distributions which will give him an idea of the speed as well as direction. The rate of oxygen consumption has been used in a tentative manner but the built-in clock of radioactive decay offers more promise as it is independent of the physical and biological character of the environment. Carbon-14 (^{14}C), deuterium and tritium, for example, have been used although in the sea their use as clocks is by no means straightforward. They will be discussed in section 6.4.

The other indirect procedure is to use the geostrophic or dynamic method. Geostrophic currents represent that part of the ocean current field that is due to the horizontal distribution of mass (i.e. density) as expressed by the horizontal gradient of pressure in the ocean. While it takes some time before the geostrophic balance between the pressure gradient and the Coriolis force (due to the earth's rotation) is established it is surprising just how many ocean currents have been found to be in geostrophic balance. Not only the main currents but also the smaller mesoscale eddies and meanders are in geostrophic adjustment.

The real advantage of the geostrophic method is that it allows one to compute a part of the circulation from measurements of temperature and salinity (which are used to compute the density and hence the pressure gradient). The computation of geostrophic currents is a topic more properly treated in a text on dynamical oceanography but since the results of such computations have become a primary tool for the descriptive oceanographer to estimate the current patterns the subject will be briefly treated in Section 6.255. Usually the results of the geostrophic computation are expressed in terms of a field known as "dynamic height" which represents the pressure gradient field driving the geostrophic currents. Geostrophic currents flow along lines of equal dynamic topography oriented so that the high topographies are on the right in the northern hemisphere and on the left in the southern.

Even when using the indirect methods, the time factor for data collection is significant. The ships available for oceanographic research on the high seas (e.g. Pl. 1) have speeds of only 10 to 15 knots (say 15 to 25 km/h), the distances to be covered are large (thousands of kilometres), and the time taken at each station to sample the water at a sufficient number of depths may be measured in hours. The German research vessel *Meteor* spent 2 years in one study of the South Atlantic alone (e.g. Emery, 1980). For even a small area it may take weeks or months for one ship to complete a survey and if the variations with time are to be studied, years may be required. Some multiple-ship studies have been made of limited areas (e.g. GATE, MODE) and of whole ocean areas (e.g. IIOE) but the organization of such expeditions involves tremendous effort and expense. For a whole ocean it is impracticable to obtain a truly simultaneous picture, although the new satellite observational techniques will help, at least for surface features. The synoptic oceanographer therefore has to make the assumption that, when he analyses them, the data from his cruise or cruises may be considered as simultaneous (sometimes referred to as "synoptic"). It is certainly fortunate that such checks as are available suggest that many of the main features of the open ocean are in a reasonably steady state and therefore the oceanographer's assumption is frequently justified. In fact, it is when he comes into shallow coastal waters that difficulties arise because the variations in properties with position are often greater and the period of change shorter than in the open sea. Also, as a result of MODE and other measurement programmes, oceanographers have realized that the ocean is populated by many small to medium scale (mesoscale) circulation features (*eddies*) analogous to the weather systems in the atmosphere. Proper evaluation of these features requires rapid sampling and the advent of profiling temperature instruments, deployed from aircraft, has helped to provide more nearly "synoptic" pictures of the upper ocean thermal structure over limited areas.

In an effort to overcome these sampling limitations, oceanographers have developed innovative methods which reduce the need for ship time and expand the areas of coverage. As mentioned earlier in Section 4.12, these methods include satellite-tracked and interrogated drifting buoys, XBT measurements from merchant and other ships of opportunity, aircraft observations and satellite remote sensing. In all of these methods new technologies make it possible to sample faster, over a wider area, and require less input from the oceanographer in terms of effort in collecting the data. Only the continuation of this trend will make it possible to observe the large geographic regions that are the domain of the physical oceanographer.

6.2 Instruments

In the following sections some of the basic instruments used in physical oceanography will be described, emphasizing the principles rather than trying to give detailed descriptions.

6.21 *Winches, wire, etc.*

One of the most essential pieces of equipment on an oceanographic vessel is a *winch* with a drum holding *wire rope* on which instruments are lowered into the sea. For lowering bathythermographs and small instruments, a light-duty winch with some 500 m of 2- to 3-mm diameter steel wire rope and a motor of 1 to 2 kW is used. For water sampling and temperature measurements, a medium-duty winch (Pl. 2) with 2000 to 5000 m of 4-mm diameter wire rope and a 7- to 15-kW motor may be used, while for heavier work, such as dredging, coring, etc., winches with up to 15,000 m of 10- to 20-mm wire and 75 to 150 kW have been used. The wire rope used is multi-strand for flexibility, and made of galvanized or stainless steel (more expensive) to resist corrosion. (Sea-water is one of the most corrosive substances known, given time to act.) The winches must be capable of reeling the wire in or out at speeds up to

100 m/min but must also be controllable in speed so that an instrument can be brought accurately to a position for operation or to where it can be reached for recovery. For instruments which telemeter their information to the surface, steel cable with one or more insulated electrical conductors incorporated is used and the winch must have slip rings to transmit the electrical signals from the wire to the deck instruments while the winch drum is turning.

6.22 Depth measurement

The determination of the depth to which an instrument has been lowered is not always easy. The wire is passed over a *meter wheel* (Pl. 3) which is simply a pulley of known circumference with a counter attached to the pulley to count the number of turns, thus giving a direct indication of the length of wire passed out over it. This length gives the maximum depth to which the instrument on the wire could reach. In calm conditions with negligible currents this will be the actual depth. More often the ship is drifting with the wind or surface currents and the wire is then neither straight nor vertical so that the actual depth will be less, sometimes much less, than the length of wire paid out.

The depth of an instrument can be estimated by measuring the hydrostatic pressure at its level, as this is proportional to depth (and to the density distribution in the vertical direction). One pressure-measuring device is a bourdon tube moving the slider of an electrical potentiometer, but this needs an electrical cable to transmit the depth information to the ship. It may be accurate to ± 0.5 to 1%. Another device is the electrical strain-gauge pressure transducer which uses the change of electrical resistance of metals with mechanical tension. It consists of a resistance wire firmly cemented to a flexible diaphragm, to one side of which the *in situ* hydrostatic pressure is applied. As the diaphragm flexes with change of pressure, the tension in the wire changes and so does its resistance which is measured to provide a value for the pressure and hence for the depth. Accuracies to $\pm 0.1\%$ or better of full-scale depth range are claimed, with resolution to $\pm 0.01\%$ or better. The "Vibratron" pressure gauge applies the water pressure to vary the tension in a stretched wire which is caused to vibrate electromagnetically. The frequency of vibration depends on the wire tension and hence on the depth. The vibration frequency is determined to give a measurement of depth to about $\pm 0.25\%$ accuracy. The use of the protected/unprotected reversing thermometer combination for the estimation of depth is described in the section on temperature measurement.

6.23 Water properties

6.231 Water-sampling bottles

In order to determine the properties of a sample of sea-water it is necessary first to obtain the sample. For a "surface" sample, a bucket on a rope often suffices to obtain water for temperature and salinity measurement. A plastic bucket is best, as less likely to pollute the sample than a metal one, and the experienced oceanographer will be found using a small one containing a litre or so rather than a full-sized bucket.

For subsurface samples a variety of *water-sampling "bottles"* (e.g. Pl. 5) are available. These are generally metal or plastic tubes with either plug valves at each end (Nansen bottle) or spring-loaded end-caps with rubber washers. The bottle with the ends open is attached to the wire and lowered to the desired depth. There it is closed by the tripping action of a *messenger* (a small metal weight which is slid down the wire). Generally a number of bottles (12 to 24) are attached in series at predetermined intervals along the wire (a *bottle cast*) and closed in succession. (Each in turn releases a messenger to close the next below it.) When the bottles have been brought back on deck the water samples are drawn through a tap, following a routine designed to obtain a pure sample. In some designs, the bottle when tripped is released at its upper end and rotates through 180° about a hinge at its lower end where it is clamped to the wire. This is for the purpose of operating the *reversing thermometers* described later, and leads to the bottles being referred to as "reversing water bottles". In other designs, the bottle remains stationary while a frame carrying the reversing thermometers rotates. A capacity of 1.25 litres is common for these bottles but for special purposes, such as C^{14} analysis, larger bottles are used up to several hundred litres capacity.

Another arrangement of water bottles is in the form of a so-called *rosette sampler* (Pl. 7). In this, twelve to twenty water bottles are mounted in a single frame which is attached to the end of the oceanographic wire. This has an electrical conductor incorporated and the bottles can be closed when desired by electrical command from on deck. This rosette arrangement is generally used in conjunction with a CTD sensor head with deck read-out so that water samples can be obtained to check the CTD or to obtain confirmation of interesting features in the water property profiles.

6.232 Temperature measurement

For measuring the temperature of a surface bucket-sample, an ordinary mercury-in-glass thermometer is generally used, taking care not to expose the bucket to the sun (heating) or to the evaporating influence of the wind (cooling).

Another special method for determining the sea-surface temperature makes use of Stefan's Law that the rate of emission of heat radiation from an object, in this case the sea surface, is proportional to the fourth power of its absolute temperature (see Section 5.32). The radiation is measured by a radiation bolometer which uses a small thermistor as the detecting element. The electrical resistance of the thermistor depends on its temperature which depends on the amount of heat radiation falling on it from the sea. In practice the temperature of the sea is not measured absolutely but is compared with that of a constant temperature enclosure by placing the thermistor at the focus of a parabolic mirror which is wobbled rapidly so as to look alternately at the sea and at the enclosure. This gives rise to an alternating current proportional to the difference between the two temperatures. This *radiation thermometer* has its chief value in determining the sea temperature from an aircraft. In this application it can be used to examine a considerable area of sea in a short time in order to get a nearly simultaneous picture. Strictly speaking, since it operates at long wavelengths (see Section 5.32), it measures the temperature of the surface skin, a fraction of a millimetre thick, of the sea. However, in the presence of wind mixing, it is probable that this does not differ very much from the bulk temperature of the upper mixed layer. A serious source of error with the airborne radiation thermometer is the variability in signal due to the absorption by water drops in the atmosphere between the sea surface and aircraft. This requires the aircraft to fly at as low an altitude as practicable, i.e. hundreds of metres rather than thousands.

For measuring subsurface temperatures the basic instrument has been the *protected reversing thermometer* (Pl. 6) developed especially for oceanographic use. It is mercury-in-glass thermometer which is attached to a water sampling bottle. When the latter is closed to collect the sample the thermometer is inverted and, as a result of its construction, the mercury "breaks" at a particular point and runs down to the other end of the capillary to record the temperature *in situ* at the depth of reversal. The break occurs in the capillary stem above the bulb at a point where a short side-arm is placed. It is really rather surprising that the mercury should break as consistently as it does—to better than ± 0.01K in a good thermometer in laboratory tests. After the thermometer has been reversed it becomes almost insensitive to subsequent changes of temperature and it is read when it is brought back on deck. This insensitivity subsequent to reversal is necessary because the surface temperature is usually higher than the deep-water temperature and, as it was brought back to the surface, an ordinary thermometer would warm up and "forget" the deep-water temperature. After corrections for scale errors and for the small change in reading due to any difference between the *in situ* temperature and that on deck, the reversing thermometer yields the water temperature to an accuracy of about ± 0.02K in routine use.

The "protected" part of its name arises because the thermometer is enclosed in a glass outer case to protect it from the pressure of the water.

One way to determine the depth of a sampling bottle is to use an *unprotected reversing thermometer* (Pl. 6) together with a protected one. The unprotected thermometer has a hole in its glass outer case; as a result the water pressure compresses the glass of the bulb and causes this thermometer to indicate a higher apparent temperature than the protected one. The difference in reading between the two thermometers is a measure of the compression of the glass, which depends on its known compressibility and upon pressure, i.e. upon depth. A pair of thermometers, one protected and one unprotected, therefore serves to measure both the temperature *in situ* and the depth, the latter to about $\pm 0.5\%$ or to ± 5 m whichever is the greater.

Another widely used instrument was the *bathythermograph* in which a liquid-in-metal thermometer causes a metal point to move in one direction over a smoked or gold plated glass slide which is itself moved at right angles to this direction by a pressure sensitive bellows. The instrument is lowered to its permitted limit in the water (60, 140 or 270 m) and then brought back. Since pressure is directly related to depth, the line scratched on the slide forms a graph of temperature against depth. It is read against a calibration grid to an accuracy of ± 0.2K and ± 2m if well calibrated. The great advantage of the bathythermograph is that, although it is less accurate that the reversing thermometer, it gives a continuous trace of temperature against depth instead of only the values at spot depths given by those thermometers. Although still in use, the mechanical bathythermograph has, to a large extent, been superseded by an electronic expendable instrument described in the next section.

In the CTD instruments described later there are generally two thermometers. One is used to determine the temperature as a function of depth as a characteristic of ocean water masses while the other is used in the circuits for the calculation of salinity from conductivity. Platinum or copper resistance thermometers are generally used for both purposes.

Thermistor "chains" consisting of a cable with a number of thermistor elements at intervals are sometimes moored along with current meters to record the temperature at a number of points in the water column. A "data logger" samples each thermistor sequentially at intervals and records temperatures as a function of time.

6.233 Salinity measurement

The classical (Knudsen) method of measurement is to determine the chlorinity by titration with standard silver nitrate solution (e.g. Strickland and Parsons, 1972) and then to calculate the salinity from the formula given in Section 3.4. In routine use, an accuracy of ± 0.02 is considered reasonable, with rather better accuracy if special care is taken and replicate tritrations made. A careful operator may titrate fifty samples per day. It must be remembered that this method is a volumetric one, whereas salinity is defined gravimetrically (i.e. by mass). In consequence it is necessary either to correct for deviations of the temperature of the solutions from the standard, or preferably to carry out the titrations in a temperature-controlled room. This titration method is practical but not very convenient to use on board ship.

The estimation of salinity through the electrical conductivity measured by means of an A.C. bridge was in use by the U.S. Coast Guard for the International Ice Patrol in the western North Atlantic since about 1930. The

Instruments and Methods 99

method was not more widely used for many years because of the bulk and expense of the equipment required. This is because the conductivity is as much a function of temperature as of salinity, which necessitates thermostating the samples to $\pm 0.001 K$ during measurement. However, improvements in circuits and equipment encouraged a number of laboratories to bring this method into wider use from about 1956 and an accuracy of ± 0.003 is obtained in routine use. This is substantially better than the titration method and makes it possible to distinguish water masses which were previously not distinguishable. One of the great advantages of the electrical salinometer is that it uses a null-balance method which is much less tiring for the operator to use than the end-point method of chemical titration. However, the variability in use of the characteristics of the platinum electrodes posed problems with the earlier electrode-type salinometers.

In 1957 Esterson of the Chesapeake Bay Institute described an electrical salinometer which avoided the electrode problem by using an inductive (electrodeless) method. Then Brown and Hamon (1961) in Australia described an inductive salinometer design which has now come into wide use. In this instrument the temperature effect is taken care of not by thermostating the sample but by measuring the temperature while the conductivity is being measured and correcting for its effect automatically in the electrical circuit. The salinity may be measured to a precision of ± 0.003 over the range from 32 to 39, and with a little practice an operator can measure the salinity of up to forty-five samples per hour.

A new laboratory salinometer, the Canadian 'Autosal" by Guildline (Pl. 8) based on a design by Dauphinee (Dauphinee and Klein, 1977), has come into wide use. This uses a four-electrode conductance cell of small dimensions in a thermostat bath (to $\pm 0.001 K$/day) with a precision of ± 0.001 or better. The sea-water flows continuously from the sample bottle through a heat exchanger in the thermostat, to bring it to a specified temperature, and then through the cell. The conductance bridge is balanced semi-automatically and the *conductivity ratio* of the sample relative to that of Standard Sea-Water (see below) is displayed digitally. Salinity is then obtained from the conductivity ratio and the temperature using the Unesco/N.I.O International Oceanographic Tables (until recently) or the Practical Salinity Scale 1978 Formula or Tables referred to in Section 3.4. The circuits are such that variations of electrode surface conditions do not affect the measurement. The size of the instrument is about $60 \times 50 \times 55$ cm and it may be used on shipboard as well as in a shore laboratory.

The refractive index of sea-water is also related to salinity (and to temperature) and the interference type of refractometer has been used in the past with a claimed accuracy of ± 0.02. Recently a refractometer has been developed which can be installed in a profiling instrument to measure salinity *in situ* rather than in a laboratory setting.

One feature of all the above methods must be noted—they are all comparative rather than absolute. The so-called *Standard Sea-Water* is prepared (now at the Institute for Ocean Sciences in England) to an accurately known salinity and conductivity ratio by comparison with a standard potassium chloride solution as described previously (Section 3.4; ref. Lewis, 1980) so that the Standard Sea-Water has a salinity of 35.000. Samples of this Standard, sealed in glass ampoules (Pl. 8), are used by oceanographic laboratories throughout the world to standardize the silver nitrate used for titration or the electrical conductivity salinometers which are more often used now. One advantage of this procedure is that all oceanographic laboratories use a common standard for salinity, reducing the possibility of systematic errors occurring and hence making it possible to combine data from different expeditions or surveys in the same area or world-wide.

The above methods are all laboratory methods, but *in situ* measurement of water properties has always been something to aim for. About 1948 one of the first *in situ* salinometers was developed. A variety of substances and organisms in ocean waters cause fouling of the electrodes and consequent change of calibration, and in practical instruments it is necessary to design the conductivity sensor so that the electrodes may be cleaned routinely. In the mid-1950s inductive salinometers were developed for *in situ* use (Hamon, 1955; Hamon and Brown, 1958) and a number of such instruments are now available from several manufacturers (see Baker, 1981). Because they measure Conductivity, Temperature and Depth (actually pressure) they are referred to as CTD instruments. In a CTD, a unit consisting of conductivity, temperature and pressure sensors is lowered through the water on the end of an electrical conductor cable which transmits the information to indicating and recording units on board ship. The digital transmitting units have claimed accuracies of ± 0.005 (conductivity accuracy expressed as equivalent salinity accuracy), ± 0.005K and $\pm 0.15\%$ of full-scale depth, with resolutions 5 to 10 times better than these figures. The sensor head of the Guildline instrument is shown in Plate 9. It uses an electrode cell for conductivity measurement, rendering calibration simpler than for the inductive type conductivity sensor because it can be carried out in a smaller calibration space. The Neil Brown instrument built in the United States uses the same principles and has become a standard of high accuracy for *in situ* measurements. There are also somewhat less sensitive self-contained instruments (STDs) in which salinity is calculated and plotted internally as a graph of salinity versus depth as well as a temperature/depth graph. These STDs are lowered on the end of the ordinary steel cable used for bottle casts. Some instruments can also measure other water properties such as dissolved oxygen content and turbidity.

In an STD, salinity is calculated from simultaneous temperature and conductivity measurements; the slower response time of the temperature sensor than of the conductivity sensor often results in large "spikes" in the salinity record. For CTDs with which salinity is computed subsequently from the temperature and conductivity records it is possible to apply some compensation for the different response times and reduce the spiking.

The above instruments are all designed for repeated use and are quite expensive. Another class of instruments, to measure temperature, salinity or other properties, which are less precise but also less expensive and are considered expendable will be described in Section 6.24.

6.234 Density measurement

The standard laboratory method, using a weighing bottle, to determine density is not practical at sea because of the motion of the ship, and it is too slow for routine use on shore. The method of weighing a quartz sinker immersed in the water sample has been used by some laboratories on shore. The simple hydrometer is not to be despised for coastal or inshore work where large variations occur, particularly in the surface layers, and high accuracy is not required. Sets of three hydrometers are available to cover the range from 1000 to 1031 kg/m^3, corresponding to 0 to 41 in salinity which can be estimated to about 0.2.

A suggestion made by Richardson for the direct measurement of density was later applied by Kremling (1972). The water sample is placed in a glass tube which is supported at one end only and to the free end of which is attached a piece of iron. The tube is caused to vibrate by passing an alternating electric current through a coil near the iron and the natural frequency of vibration of the tube is determined. This depends on the mass of the tube and contents and therefore on the density of the sea-water content which is thereby determined. In the report by Kremling the accuracy was estimated at ± 0.005 in σ_t and the values obtained agreed with those of Cox et al. (1970) in indicating that Knudsen's Tables are low by about 0.013 in σ_t. This instrument has the merit of permitting continuous measurement of density if the tube is a U-tube through which the sea-water flows. It is also possible that the method might be adapted to permit the *in situ* measurement of density for which no other practical method is yet available.

More recent measurements by Millero et al. (1976, 1980), using a magnetic float densimeter (Millero, 1967), estimated accurate to ± 3 in 10^6 (i.e. to ± 0.003 in σ_t), indicate that Knudsen's values for relative density are low by about 0.009 in σ_t at 35 salinity.

A variety of calculations in dynamic oceanography require also the pressure effect on density. (The full relation between ρ, T, S and p is referred to as the *equation of state* of sea-water.) Previously the measurements of relative density reported by Ekman (1908) were used but recently a new equation of state was announced by Millero et al. (1980, see Section 3.55) with a precision of ± 5 in 10^6 (standard deviation) over the ranges of values found in the ocean, i.e. about 0.005 kg/m^3 in σ_t. The new measurements indicate that Ekman's values are up to 60 in 10^6 in error, i.e. up to 0.06 in σ_t.

Failing adequate means to measure sea-water density quickly in the field with the accuracy needed for geostrophic calculations it is usual to estimate it indirectly from salinity and temperature, using the equation of state.

6.24 Expendable instruments

In wide use now is the *expendable bathythermograph* (XBT) which uses a thermistor as temperature-sensitive element (Pl. 10). The thermistor is in a small streamlined weighted casing which is simply dropped over the ship's side. It is connected by a fine wire, on special free-unwinding spools, to a recorder on the ship which traces the temperature of the water in a graphical plot against depth. The latter is not sensed directly but is estimated from the time elapsed since release, using the known rate of sink of the freely falling thermistor casing. This casing is relatively inexpensive and is not recovered. These XBTs are available for depth ranges from 200 to 1800 m, and can be used from ships underway and have even been used from circling aircraft. They can also be dropped from aircraft (AXBT) in a small buoy which contains a radio transmitter to send the temperature/depth information (to 300 to 800 m) to the aircraft which is continuing its flight.

Expendable instruments have provided the oceanographer with simple tools for rapid sampling. This has proved important for synoptic sampling from multi-ship or aircraft surveys and has led to wider use of *ships of opportunity* as described in Section 4.12. In an effort to extend such technology to other important parameters, recent developments have produced an expendable velocimeter (speed of sound) and an expendable current profiler, using the electro-magnetic principle. An expendable CTD (XCTD) is also available but needs further development to be truly reliable. Plans also call for the construction of an expendable microstructure velocity profiler to study turbulence on the centimetre scale. These more exotic expendables are all considerably more expensive than the XBT and therefore have been less widely used.

6.25 Current measurement

There are two basic ways to describe fluid flow, the *Eulerian* method in which the velocity (i.e. speed and direction) is stated at every point in the fluid, and the *Lagrangian* method in which the path followed by each fluid particle is stated as a function of time. In both cases the statements are usually made with respect to axes which are stationary relative to the solid earth. In theoretical studies the Eulerian method is the easier to use, but in describing the circulation of the oceans, as in Chapter 7, the Lagrangian method is used more frequently.

Typical horizontal current speeds in the ocean range from about 200 cm/s (about 200 km/day) in the swift western boundary currents (Gulf Stream, Kuroshio), through 10 to 100 cm/s in the equatorial currents, to a fraction of

1 cm/s in much of the surface layer and in the deep waters. Vertical speeds are estimated to be very much less, of the order of 10^{-5} cm/s or 1 cm/day.

6.251 Lagrangian methods

The simplest Lagrangian current indicator is an object floating in the water with a minimum of surface exposed to the wind. The so-called *drift pole*, a wooden pole a few metres long and weighted to float with only $\frac{1}{2}$ to 1 metre emergent, is often used to determine surface currents close to landmarks. Such a pole is simply allowed to drift with the water, its position being determined at intervals either from the shore or by approaching it in a small boat and fixing its position relative to the shore. Sheets of paper or patches of dye, such as sodium fluorescein, which can be photographed at intervals from a high point of land or from an aircraft are also used. Other near-shore drifters have been built as wooden crosses (similar to the "Chesapeake Bay Institute current drag" element shown in Fig. 6.2(a)) or even "hula hoops" covered with plastic garbage bags. In fact, the latter were found to most closely simulate oil being carried along by currents on the surface of the ocean.

The drift-pole idea has been extended to the open ocean by using a *freely drifting buoy* with a radio transmitter so that its position can be determined by radio direction finding from the shore. A more sophisticated and precise drifting-buoy technique, used a great deal in recent years, has been to fit the buoy with a VHF radio transponder (Fig. 6.1) which replies when interrogated from a satellite in orbit. By this means the buoy's position can be determined more accurately than with the lower frequency radio direction-finding technique above. The position of such a *satellite-tracked buoy* is calculated on board the satellite, using the Doppler shift of the buoy's VHF signal, and then transmitted to a ground station for recording. To extend their operating life, such buoys are often fitted with solar panels to keep their batteries charged.

Such buoys are relatively small and, by themselves, may not be perfectly coupled to the water motions. To ensure that the buoys do move with the water, and to minimize the effect of wind, they are frequently fitted with a subsurface drogue to provide additional water drag and more effective coupling with the water motions. The drifting buoys may also be instrumented to measure and transmit surface water properties, atmospheric pressure, etc. These buoys provide both track and speed of flow, i.e. a Lagrangian description, and have revealed many new details of eddies associated with ocean currents.

The most important feature of the drogued buoy system is the drag element that provides the coupling between the drifting "Lagrangian" particle (the buoy) and the water. Recent studies (Niiler *et al.*, personal communication) have examined a variety of drogue elements and have measured their slip through the water by installing current meters around the drogue element.

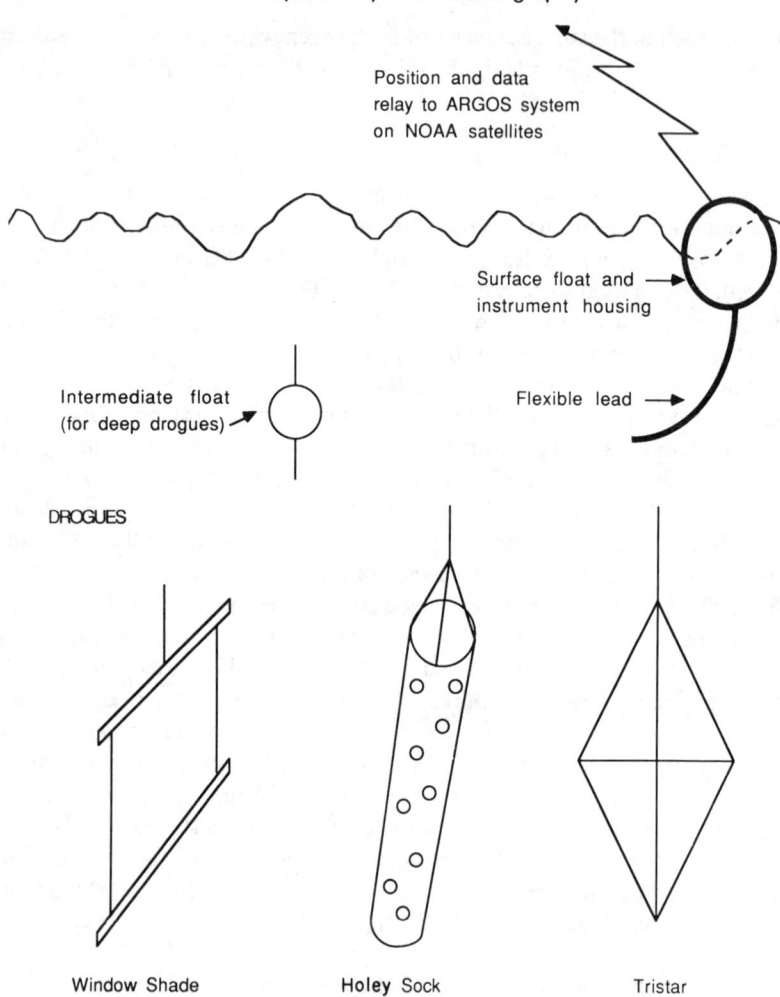

Fig. 6.1. Lagrangian free-drifting buoy elements: surface float with radio transmitter (top right), and three types of drogues.

These studies have shown that the standard "windowshade" drogue (Fig. 6.1) is not very effective since it has a tendency to "sail" up in the water column allowing the water to flow past it. The parachute drogue also proved to be ineffective in that it frequently collapsed and would not redeploy itself. The drogue shape that performed best was the "holey-sock", a wire cylinder covered with fabric with some holes in it. The study showed that a larger diameter cylinder was better than a longer cylinder of smaller diameter. Another effective drogue configuration, called the "tristar", consisted of a series of integrated diamond-shaped elements all rigidly connected in a shape similar to a radar reflector.

The most limiting factor in the ship-based buoy tracking is the ship time required. As mentioned above, satellite-tracked buoys use the Doppler shift of the radio-beacon signal transmitted by the buoy to quite accurately (± 1–2 km) locate the buoy's position. In addition, the satellite Data Collection System (DCS) can interrogate the buoys and transfer data sampled by the buoy instruments. The French ARGOS system, currently on the U.S. NOAA polar orbiting weather satellites, can handle over 200 data words per platform and can "talk" to over 1500 platforms at the same time. This "random-access" DCS system was developed by the U.S. as part of the NASA NIMBUS satellite programme. Today, drifting buoys are used to measure in situ variables such as the sea surface temperature (actually slightly below the surface as it is measured by a thermistor in or on the buoy hull), subsurface temperature profiles, air temperature, and wind speed/direction. All of these data are stored in the buoy and are transferred each time the buoy communicates with the NOAA satellite. Thus the oceanographer can sit at home and collect this data without going to sea. The life of the buoy as a data collector is limited by its battery endurance.

These satellite-tracked drifting buoys have been widely used in oceanography since the late 1970s and a lot of experience has been gained in working with such systems. During this time, and continuing today, the drogue elements and buoy configurations continue to develop, making it difficult to interpret all buoy tracks in the same way. Still it has become very clear that buoy tracks are related to both wind and ocean surface currents (Emery et al., 1985; Kirwan et al., 1978) and that over time and space buoy trajectories can be relied upon to provide a Lagrangian description of a large portion of the ocean surface current field. The data collected by such buoys have been considered sufficiently important that a separate system for providing integrated buoy data sets has been developed. The Canadian Marine Environmental Data Service (MEDS) has agreed to act as the Oceanographic Data Centre responsible for archiving surface drifting buoy data. The work of this data centre is simplified somewhat by the fact that many of the buoy data collected by the central ARGOS system are automatically loaded on to the Global Telecommunications System (GTS) which distributes a variety of weather information world-wide. Thus the data that are archived at MEDS are read daily from the GTS network. This data set covers about 60% of all buoys operating in the ocean. There are strong efforts underway to require all buoy operators to send out at least the position and surface temperature data on the GTS. The concern of many scientists is that by this dissemination technique, all other interested scientists can receive the same data without having gone to the effort and expense of buying and deploying the buoys.

A similar Lagrangian system has been developed to observe subsurface currents. For this purpose, John Swallow of the National Institute of Oceanography in England invented a *neutrally buoyant float* (Swallow, 1955). This "Swallow float" makes use of the fact that the density of the sea increases with depth. The float is adjusted before launching so that it will sink to a selected depth (in terms of density). It then remains at this depth and drifts with the water around it. The float contains equipment to send out sound pulses at intervals and it can be followed by listening to it through hydrophones from the ship. The ship chases the float and at the same time continuously determines its own position. It thereby determines the direction and speed of drift of the float and the water mass in which it is located. A limitation with this instrument is that one ship can follow only a very small number of floats at one time. If several floats are released, e.g. at different levels, it is quite likely that they will be perverse and drift off in different directions and the ship may not be able to keep track of them for long. Recently developed coastal and autonomous moored listening stations make it possible, using the SOFAR technique, to track these neutrally buoyant float without a ship. As with the satellite-tracked drifting buoys, interpretation is not simple since the Swallow floats are not perfectly coupled with the ocean currents. The Swallow float was really the first oceanographic instrument to give us reliable information on the speed and direction of deep currents, and as is usual when one ventures into a new area some of the results obtained have been unexpected (e.g. deep currents in the Atlantic, Section 7.342).

In the SOFAR float application (Section 3.7) (Rossby and Webb, 1970) the float must generate an acoustic signal that can be located and triangulated upon by the SOFAR listening stations. For regional studies far removed from the SOFAR sites, independent listening stations were developed that could be moored and operated autonomously to collect data. These data could then be retrieved by a research vessel as with any other moored buoy data. Recently, a reversed system (Rossby et al., 1986) was developed, called RAFOS (for SOFAR spelled backward) where the buoy was a simple listening device and the moored stations were strong acoustic sources. The buoys would listen and record the phase encoded information. At intervals, the buoy would come to the surface, report its data over the same ARGOS system used to track surface drifters, and then return to its preselected depth to collect more information. the RAFOS system is much cheaper since the listening systems are inexpensive while the sound sources are more costly and are a high repeating expense for the SOFAR system which does not recover the floats. An even lower cost alternative has recently been developed (Davis et al., personal communication) which is called the "pop-up" float. In this device, the float is again neutrally ballasted to float at a desired depth. After sinking to this depth at the deployed position the float travels around on the selected density surface. After some period of time the buoy inflates a bladder and goes to the surface to report its position (along with any in situ data collected) to the ARGOS system before sinking down again to continue its operation. The exact trajectories between reported locations are not known but the long-term patterns can be inferred from the series of positions. By removing all of the acoustic signal requirements, the cost per buoy can be dramatically reduced.

In the same class as the Lagrangian drifters is the procedure whereby, in fact, most of the information on open ocean surface currents has been obtained. It is a by-product of ship navigation. A ship which is sailing on a given course at a given speed will be deflected from its intended course by surface currents. Therefore a comparison of the actual track (checked by astronomical navigation, landfall, etc.) with that intended will give a measure of the surface current. This comparison may be made every time that a positional fix is obtained from the ship. Maury, about 1853, first suggested examining ship's navigation logs to extract such information on currents, in his case first for the Gulf Stream region off the eastern United States, and the method was subsequently extended world-wide. Most of the maps of surface currents presented in Marine Atlases, and shown in simple form in Chapter 7, are based on the accumulation of such ship-drift data.

To determine the path followed by a pollutant, such as sewage or industrial waste, it is often possible to use the substance itself as a tracer. Samples of water are collected from a grid of positions near the source and in likely directions of flow, and the pollutant concentration determined by chemical analysis. Radioactive materials seem attractive as artificial tracers of water movement, and they were successfully used as a by-product after some of the early Pacific atom-bomb tests. However, in the quantities needed in the sea the cost is often prohibitive, and there is always reluctance expressed by non-oceanographers to the release of radioactive materials in the neighbourhood of communities or commercial fisheries. A very convenient artificial tracer is the red dye rhodamine-B. This can be detected at extremely small concentrations (less than 1 part in 10^{10} of water) by its fluorescence, using relatively simple instruments, and it is also non-toxic at such dilutions. It is only practical to use it in coastal waters, as the quantities required to "tag" open ocean water masses would be impractically large. In all such studies it must be remembered that both advection and diffusion are acting three-dimensionally to spread the tracer and thus results cannot be interpreted solely in terms of advection by currents.

6.252 Eulerian methods; drag and propeller-type meters

The simplest Eulerian current meter is the *Chesapeake Bay Institute drag*. This consists simply of two crossed rectangles of wood (Fig. 6.2(a)) weighted and suspended by a thin wire. When the drag is immersed, the frictional force of the current pulls the wire to an angle with the vertical. The current speed is related by a simple formula (Fig. 6.2(a)) to the size of the drag, its weight in water, and the angle of the wire from the vertical. This angle is measured to determine the current. This device is simple, cheap to make and quick to use from an anchored ship. It is limited to depths of a few tens of metres because the current drag on the wire increases with length and complicates the interpretation of the wire angle at greater lengths.

Descriptive Physical Oceanography

(a) CHESAPEAKE BAY INSTITUTE DRAG

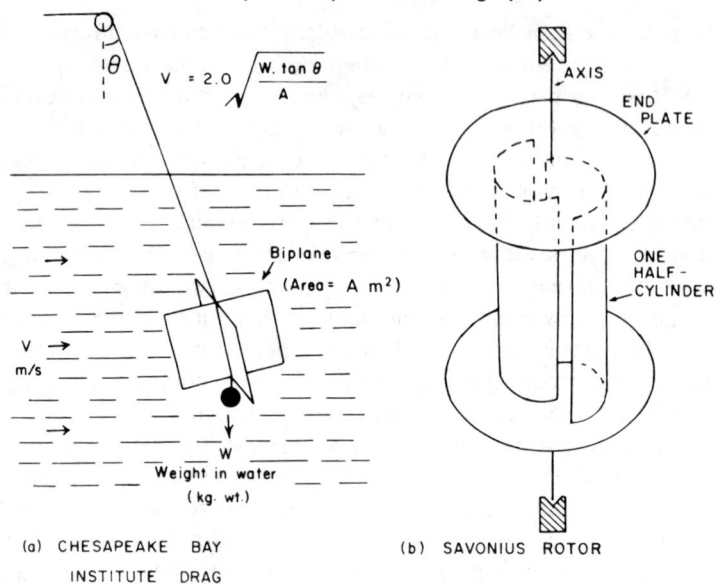

$V = 2.0 \sqrt{\dfrac{W \cdot \tan\theta}{A}}$

Biplane (Area = A m²)

V m/s

W Weight in water (kg. wt.)

(b) SAVONIUS ROTOR

AXIS, END PLATE, ONE HALF-CYLINDER

FIG. 6.2. (a) Principle of use of Chesapeake Bay Institute drag for current measurement, (b) simple form of Savonius rotor for current meters.

Before describing subsurface Eulerian current meters it should be mentioned that one of the fundamental difficulties is determining the *direction component* of the velocity. When the current meter is out of sight below the surface the only frame of reference available to it is the earth's magnetic field. This yields only a small torque to turn a direction indicator and herein lies one of the instrument designer's difficulties. This is particularly so at high latitudes near the magnetic pole where the horizontal component of the earth's magnetic field is small.

Before 1960 the most widely used Eulerian instrument was the *Ekman current meter*. This consisted of a 10-cm-diameter propeller mounted in a frame attached to the end of a wire and lower to the desired depth. A metal weight (*messenger*) was dropped down the wire to free the propeller to rotate and a second one was dropped after a measured time to stop it, the number of revolutions being recorded by a mechanical counter. The water speed was then proportional to the number of revolutions per minute. The current direction was recorded by the counting mechanism dropping metal balls at intervals along a magnetic compass into a tray with 10° sectors. This instrument had to be lowered and raised for each measurement—a tedious business. An improvement on the Ekman meter was the *Robert's current meter*, the forerunner of most present meters. In this meter, speed (from a propeller) and direction (from a compass) were transmitted electrically to the surface and recorded on shipboard or transmitted by radio from the supporting buoy to a mother ship.

Instruments and Methods 109

One disadvantage of the propeller-type current meter is that up-and-down motion, as when the ship rolls or the mooring moves, may cause the propeller to turn and cause inaccuracies in the speed measurement. A hollow cylinder with its axis horizontal mounted round the propeller minimizes this effect. An alternative to the propeller is the *Savonius rotor* which is less sensitive to vertical motion. It consists of two half-hollow cylinders mounted on a vertical axis with flat end-plates (Fig. 6.2(b)) and has the advantage of producing a large torque even in small currents. The rotor is made of plastic to be neutrally buoyant to reduce bearing friction so that it is sensitive to currents of as little as 2 cm/s. Even this low threshold value can be a problem in parts of the ocean where currents of this order prevail. The rotor carries several small magnets and as each passes a coil on the frame it induces a momentary electrical current pulse. The number of pulses per second is proportional to the current speed. The current direction is determined electrically with reference to a magnetic compass. The speed and direction information is either transmitted electrically up the supporting cable to the ship and there recorded or, more often, recorded within the instrument on photographic film or magnetic tape. The internally recording arrangement is used for current meters suspended from buoys for long-term measurements; the magnetic-tape system is most convenient for computer analysis of the records.

An example of a modern current meter is the Aanderaa (Pl. 4) which is about 54 cm high and 87 cm long. On magnetic tape it records current speed from 2.5 to 250 cm/s to $\pm 2\%$ or ± 1 cm/s, whichever is the greater, direction to $\pm 5°$ with a magnetic compass, depth to $\pm 1\%$ of range and time to ± 2 s/day for up to a year. It can also record temperature to ± 0.2 C° and conductivity to $\pm 0.1\%$ of the range from 0 to 7 siemens/m which corresponds to salinity from 0 to over 40, depending on temperature. Data may also be telemetered up an electrical cable or by acoustic link to the surface so that they may be observed in real time (i.e. as the events occur) to ensure that the meter is working correctly and so that all the data will not be lost if the current meter cannot be recovered. It was found that the Savonius rotor on the Aanderaa current meter was also subject to the effects of vertical mooring motion which caused an alternating artificial current by what is called "rotor pumping". This effect is most severe on shallow meters and in coastal regions where wave action becomes significant. To reduce this effect, Aanderaa now mount a semicylindrical shroud round one-half of the rotor and use flat rotor blades.

Many current meters record speed at short intervals but record direction only at longer intervals, which can lead to uncertainties when it is desired to know the (rectangular) components, e.g. to north and east, of the velocity. The *Vector-averaging Current Meter* (VACM) is designed to measure the velocity frequently, resolve it into components, record these separately, and average over discrete time intervals to give a more complete record of the velocity. *Vector-measuring Current Meters* (VMCM) use two propellers at right angles to measure the rectangular components directly (relative to a magnetic compass). The VCVM meter has been found to be particularly effective in the shallow near-surface layer where wave and wind effects become significant. The relatively low inertia of the multiple propellers also helps to make the VCVM the current meter of choice for resolving the details of the near-surface currents. The primary disadvantage at this time is its high cost compared with the Aanderaa meter.

6.253 Eulerian methods; non-propeller-type meters

A different method for current measurement is to use the rate of cooling of an electrically heated wire as a measure of the fluid speed past it. This is the *hotwire anemometer*. A thin wire or metal film about a millimetre long is exposed to the flow and maintained at a constant temperature by automatically adjusting the electric current through it so that the Joule heating is exactly equal to the rate of loss to the fluid. The magnitude of the electric current is then a measure of the fluid speed. This device has the advantage of small size and very rapid response to flow variations which makes it particularly suitable for the measurement of turbulent fluctuations of flow speeds. No reliable moored version has been developed.

Another technique for this purpose is to use a *sonic anemometer* in which the speeds of travel of pulses of high-frequency sound in opposite directions are measured, the difference being a measure of the component of fluid speed along the sound path. There are also *Doppler anemometers* in which the (Doppler) frequency shift of sound or light (laser) reflected from particles in the water is measured to give the water velocity, including short-period fluctuations characteristic of turbulence.

In their ocean application, these sonic anemometers are called "acoustic current meters" and have been used both to profile currents and as moored current meters. In a mooring configuration the acoustic system has the advantage of no moving parts to foul or provide inertial resistance to changes in ocean currents.

A recent development in the acoustic measurement of ocean currents has been the "Doppler log" which is a system to measure currents relative to a moving ship. (Originally, the speed of a sailing ship was measured by measuring the travel time of a log thrown into the water as it went from the ship's bow to its stern. From this practice, any speed-measuring device from a ship became known as a "log".) The Doppler log measures the speed of the ship by sending out an acoustic pulse which is then reflected back to the ship by particles in the water (such as plankton). The Doppler shift of the returned signal makes it possible to compute the ship's speed relative to the water. Reversed, this same system allows one to measure the water motion relative to the ship whose motion can be accurately computed from satellite and radio navigation information. By controlling the acoustic beam, the Doppler system reflects the currents at different depths below the ship. Alternatively, the acoustic system can be suspended from a moored buoy or mounted on the bottom (in shallow water). The systems have become known as Acoustic Doppler Current Profilers (ADCP's). In general, these systems are capable of profiling the currents to a distance of 300 m from the instrument. Using a 4-element sensor head, an ADCP is capable of resolving both speed and direction of the water movements relative to the sensor. For shipboard use, all recorded data must be corrected for the ship's motion determined from satellite or other radio navigation positioning. Most oceanographic research vessels today carry an ADCP system on board and operate it throughout a cruise.

The *electromagnetic method* uses a fundamentally different principle, first suggested by Faraday (1832), that an EMF will be induced in a conductor which moves across a magnetic field. In this case, sea-water is the conductor and when it flows across the lines of force of the earth's magnetic field an EMF, $E = B \cdot L \cdot v$, will be generated where v is the water speed, L the width of the current and B the strength of the earth's magnetic field component in a direction mutually perpendicular to the direction of both v and L. For a horizontal current, B would be the vertical component of the earth's field. Faraday was unsuccessful in applying his idea to measure the flow of the Thames because of problems with copper electrodes; some of the earliest reported measurements by this technique were of tidal currents in the English Channel (Young et al., 1920), and a long series of measurements was made of the Florida Current between Key West and Havana (Wertheim, 1954). The basic equipment required is a recording millivoltmeter and two electrodes to dip in the sea. The electrodes are best placed one on each side of the current and so a further requirement is an insulated connecting wire to the farther electrode. Unused commercial cable circuits have often been used for this purpose. One source of error is the finite, but usually unknown, electrical conductivity of the sea bottom which allows an electrical current to flow due to the induced EMF and so reduces the observed EMF below that to be expected from the formula. This introduces a constant scaling factor which must be determined by making some water-current measurements with another type of meter while the electromagnetic system is in operation.

A requirement for the electromagnetic method is stable electrodes so that electrochemical effects will not complicate the interpretation of the records. Silver wire with silver chloride deposited on it is the most satisfactory electrode for use in the sea.

An adaptation of this technique, by von Arx (1962), to permit underway shipboard measurement was called the *Geomagnetic Electrokinetograph* (*GEK*). Two electrodes were towed behind the ship and the EMF induced in the length of cable between the electrodes was recorded as a measure of the component, perpendicular to the ship's track, of the water velocity. To obtain the total water velocity, the ship was then turned at right-angles to the original track and a second component measured. Combining the two components gave the water velocity relative to the solid earth. The difficulty of reducing and interpreting GEK data led to a rapid decline in its use. A theoretical discussion of both the fixed electrode and the GEK principles was given by Longuet-Higgins, Stern and Stommel (1954).

The small magnitude of the earth's magnetic field together with electrical noise always present in nature makes the geo-electromagnetic method practical only with electrode separations of tens of metres or more. In 1947 Guelke described an electromagnetic current meter in which a local magnetic field was generated by an alternating current in a coil, the field being strong enough to give rise to a measurable EMF with electrodes only 20 cm apart when used to measure currents in Capetown Harbour. The use of an alternating magnetic field permitted amplification of the alternating EMF generated and largely eliminated long-period noise and electrode effects. Recently current meters employing this principle with even smaller electrode spacings have become available commercially. They have the advantage of having no moving parts but do need a significant electrical power supply.

6.254 Mooring current meters and other instruments

Even with the development of improved instruments, the measurement of currents from a ship has several disadvantages. A major one is that any ship movement introduces into the measured currents spurious components which are difficult to determine and to get rid of in analysis. In shallow water it is easy to anchor but wind and tide still cause ship movement. In deep water it is usually difficult to anchor at all, requiring special cables and large winches. Also, the fact that the anchor cable cannot be vertical leaves scope for ship movement which is usually difficult to measure with sufficient accuracy to apply corrections to the current data. The anchor cable may not be vertical because this would mean that the anchor would be pulled out of the bottom by any ship movement, allowing the ship to drift. Anchoring to a weight alone is impractical because an impossibly large weight would be needed. However, a small buoy may be anchored firmly to a weight which is small enough to handle from a ship. In this case, the anchor wire may be taut so that the buoy does not have scope to move significantly. Knauss (1960) developed a technique for continuously manoeuvring the ship so that it was in a known position (by radar) relative to the buoy at all times so that current measurements could be made. Even this technique has the disadvantage that measurements are made at one location only and that the ship and crew are tied up there for the duration of the current measurements at a cost of many thousands of dollars per day. The duration will be for a minimum of 15 days if one wishes to have a long enough time-series to be able to resolve even the major tidal constituents, and often longer series, e.g. a year, are desired.

For these reasons, techniques for the successful mooring and recovery of strings of current meters in the deep ocean, supported on a cable from a buoy beneath the surface to an anchor weight on the bottom, have been developed since the mid-1960s, and most current measurements are now made in this way. There are still problems associated with the movement of the current meter strings in strong currents, and the meters may not always be recovered, but the number of depths at which currents may be measured improves the description of the flow and the length of time-series observations permits resolution of the currents into mean and fluctuating components of various periods.

There are many different schemes for mooring current meters. Two are shown in Fig. 6.3, most employing a meter similar to the Aanderaa described earlier. An early development and recovery technique was to have a double anchor with a float at the surface marking the second anchor. This could be recovered and the system brought in, ending with the current meters and flotation. With this system, subsurface flotation could be used which greatly reduced mooring motion from surface effects. A modification of this system eliminated the surface marker for the second anchor and the ship grappled to find the cable between the two anchors. The absence of a surface marker is advantageous when there are ships or icebergs in the area. All of these methods are time-consuming and risky. Release systems were developed that could cut the cable, being actuated either by a timer system or fired by an acoustic signal sent from the ship. Due to weather and scheduling problems, it has become more common to use the command system and the acoustic release has become a usual component of the modern current meter mooring.

The acoustic release acts both as a release for the mooring and as a communications system for the scientists to "talk" with the mooring. Upon returning to the general location of the deployed mooring the scientist will reactivate the acoustic system on the release and use it to better locate the mooring and assure its condition as being ready for release. When ready, the release or wire-cutting mechanism is activated and the mooring is free to rise to the surface. There are many tense moments while waiting for the mooring to come to the surface; it may be difficult to spot as it floats low in the water so it usually carries a radio transmitter and a light to assist in locating it. The acoustic system has proved invaluable in working in ice-covered regions. Using the acoustic signal, the mooring can be located by triangulation from different locations on the ice and then divers can be despatched to retrieve the mooring from a hole cut in the ice.

As shown in Fig. 6.3, both surface and subsurface flotation systems are used in modern moorings. Frequently surface buoys are needed either for surface meteorological observations, such as wind, pressure, etc., or for data telemetry. Mooring lines themselves have consisted of steel cable, nylon rope and a recent synthetic called Kevlar. These different materials have their advantages and disadvantages. Steel cable is very strong and inexpensive but is subject to kinking while being deployed and this can dramatically weaken the cable. Nylon rope and Kevlar are both pliable and thus not affected by

114   Descriptive Physical Oceanography

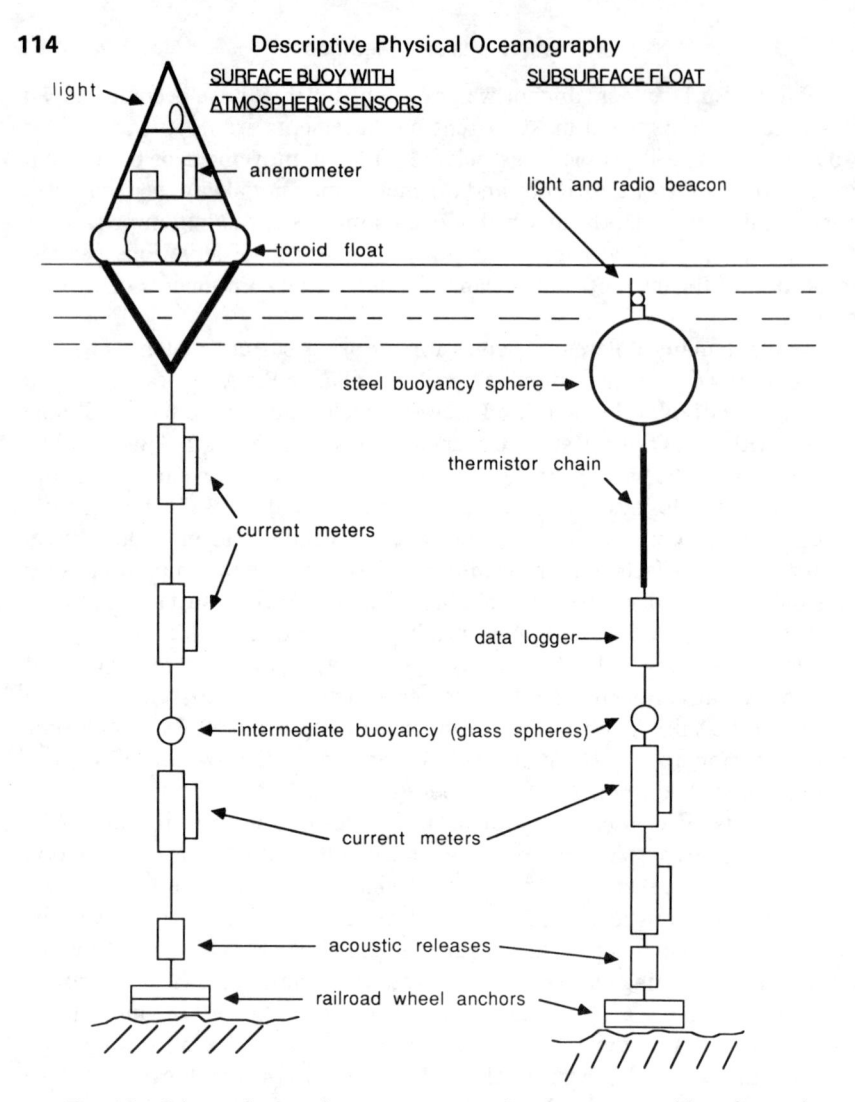

Fig. 6.3. Schemes for mooring current meters, data loggers, etc. with surface and subsurface upper floats.

kinking but are more expensive, especially for deep ocean moorings. Flotation devices also vary widely but one frequently used element is the glass sphere, because it is cheap and will withstand pressure. Glass spheres are particularly useful as intermediate flotation attached in deeper portions of the mooring line. The spheres are usually contained in a plastic container which is then attached to the line. Often a mooring is deployed in reverse by trailing the flotation and current meter string behind the ship and finally dropping the anchor weight when the ship is over the desired location. Many different anchor systems have been used, ranging from concrete blocks to used railroad

Instruments and Methods   115

wheels. The latter have become very popular since they are heavy enough, are abundantly available at low cost and conveniently have a hole in the middle for attaching the mooring line so that the wheel will sit flat on the bottom.

Recent technological efforts have produced a number of vertical current profiling devices. One, called the *Cyclosonde*, moves up and down a mooring line, measuring currents with a ducted propeller and other water properties. A similar current profiler developed by Düing (Düing and Johnson, 1972) consists of an Aanderaa meter in a streamlined body which is lowered from a ship anchored or drifting. Ship movements are carefully tracked and the current meter record is corrected for ship displacement.

6.255 Geostrophic method and dynamic topography

As mentioned earlier in this chapter, the *geostrophic method* is a procedure by which one can compute an important component of the ocean current from measured temperature and salinity profiles. Using this method we can learn something about the currents both at the surface and at depth. Strictly speaking this subject is in the domain of dynamical oceanography and a more complete description is given in Pond and Pickard (1983). Since the application of the geostrophic method to the description of ocean currents is an important part of descriptive physical oceanography we will present the essential elements of the method here. We will also endeavour to provide the reader with an understanding of how the method is commonly used in descriptive studies of ocean circulation.

Before we can discuss the meaning of *geostrophic currents*, it is necessary first to discuss the balance of forces that lead to the existence of these currents. In the ocean the distribution of mass is represented by the distribution of density over both the horizontal and vertical dimensions. Density is seldom directly observed in the ocean but rather is computed using an empirical equation of state (Section 3.55) from measurements of temperature and salinity as functions of depth. In the geostrophic method this computed density represents the distribution of mass in the ocean which can be thought of as creating gradients of pressure. We are familiar with the pressure exerted by sea-water in the vertical and we only need to realize that horizontal differences in density create horizontal pressure gradients.

To best relate these pressure differences, oceanographers have created a parameter called *dynamic height*. Clearly height differences in the ocean (which would normally be expressed by different heights at the sea surface) reflect pressure differences which can then be expressed as horizontal pressure gradients. The dynamic height is calculated from the density distribution and actually represents the ability of a column of water to do work due to differences in *geopotential* (in other words, the potential for gravity to do work because of the height of the water relative to some reference level). Thus geopotential height differences, expressed by changes in dynamic topography, provide us with a reflection of the horizontal pressure gradient force (*PGF*).

Due to the rotation of the earth, there is an apparent force called the *Coriolis Force* (*CF*) which acts on all moving bodies, including the moving ocean and atmosphere. Geostrophic currents exist due to the balance between the *PGF* and the *CF*. As shown in Fig. 6.4(b) the *PGF* and *CF* act oppositely to each

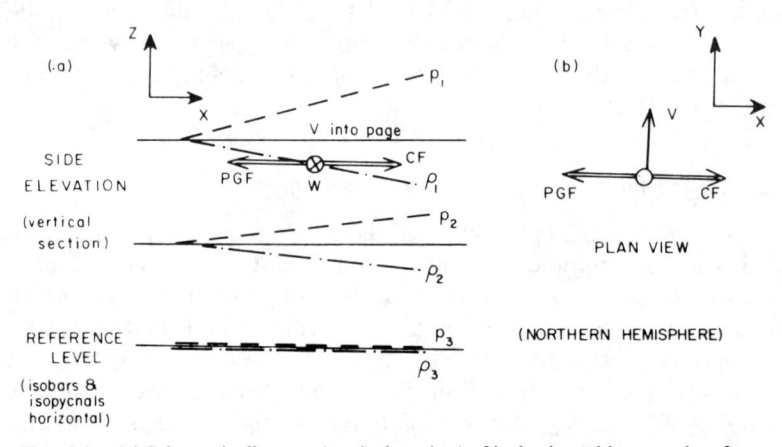

FIG. 6.4. (a) Schematic diagram (vertical section) of isobaric and isopycnal surfaces and corresponding steady-state, no friction, geostrophic flow direction (northern hemisphere) with pressure gradient force (*PGF*) and Coriolis force (*CF*) in balance, (b) plan view of the balanced forces and corresponding flow direction (*V*).

other and the resulting *geostrophic current V* is oriented perpendicular to these forces. In the northern hemisphere this balance is such that the higher levels (Fig. 6.4(a)) are on the right (while in the southern hemisphere they are on the left). In the situation described in Fig. 6.4, imagine a particle of water *W* in a region where the pressure surfaces p_1, p_2, p_3, etc (dashed lines) are inclined to the horizontal or level surfaces (solid lines). In this case the pressure on the right of *W* will be greater than on the left and the *PGF* will be to the left. In a stationary system this would cause the particle to move down the gradient to the left but as soon as it starts to move on the rotating earth the *CF* acts upon it and the steady-state balance will be a geostrophic velocity *V* perpendicular to the line of action of the forces.

This balance is expressed mathematically in the *geostrophic equation*:

$$V \cdot 2\Omega \sin \varphi = -\frac{1}{\rho}\frac{\partial p}{\partial x}$$

where V = speed of flow, Ω = angular speed of rotation of the earth ($= 7.29 \times 10^{-5}$ seconds^{-1}), φ = geographic latitude, ρ = water density and $\partial p/\partial x$ = horizontal pressure gradient. Using this equation the oceanographer can compute geostrophic currents knowing only the pressure gradient derived from the horizontal density distribution. It is important to point out that the ρ, in this equation, is a single-valued average density and does not reflect the point-to-point changes in density which give rise to the pressure gradient.

As stated above we do not measure pressure, or even density, directly but rather compute them from measured profiles of temperature and salinity. There are two ways of describing the pressure field in the ocean: (a) to give the pressure on a series of level (with respect to the earth's geopotential surfaces) or (b) to give the geopotential height of equal pressure surfaces. As suggested by the previous discussion of dynamic height, common oceanographic practice is to use the latter method and to give the dynamic topography (the geopotential height) of equal pressure (equal to equal depth) surfaces. In Fig. 6.4 the *isopycnal* (constant density) *surfaces* are represented by the dash-dot lines. Note that these lines appear to be mirror reflections of the upward-sloping *isobars* (equal pressure surfaces) that define the horizontal pressure gradient. If we plot a vertical section of density, the slope of the isopycnals will tell us whether the current is flowing into the section (as in Fig. 6.4(a)) or out of it (if the isopycnals sloped the other way). In the upper waters of tropical and mid-latitudes, where density is determined primarily by the temperature structure, a temperature section can also be used to infer the geostrophic current direction (from the isotherm slope) and speed (the steeper the isotherms the faster the current). A useful rule-of-thumb is that in the upper layers of the northern hemisphere the "*light*" water (the higher sea surface and hence greater pressure gradient) is on the *right* of the direction of geostrophic flow. (In the southern hemisphere then, light water is on the left of the flow direction.)

If the slopes of the isopycnal surfaces vary with depth, this indicates that the currents vary with depth. We say that, in this case, there is 'current shear" or "geostrophic shear". This is characteristic of the *baroclinic* situation in which the slopes of the isopycnals vary because the water density depends on water properties (temperature and salinity) as well as on pressure (depth). The *barotropic* situation is one in which the density depends on depth only, as in an isothermal fresh-water lake. In this case the isopycnal surfaces must be parallel. If the water is stationary they will be horizontal, if the water is moving they will all be at the same angle to the horizontal (determined by the geostrophic relation) and there will be no current shear, i.e. the current will be the same from top to bottom. This is called barotropic flow and the geostrophic equation gives us no information about it. The total current in the sea then may be a combination of baroclinic and barotropic components.

A limitation, in principle at least, to the use of the geostrophic relation is that it only gives us relative currents, i.e. the current at one level relative to that at another. To convert these relative currents into absolute currents, which are what we need, we must determine the absolute current at some level (*reference level*). The most usual way has been to assume that the absolute current is zero at some depth (*depth of no motion*) and use this as a reference level. Then the geostrophic equation can give the absolute current at all other levels. The selection of a reference level is one of the key problems in using the geostrophic method to compute currents. Early users of the method hoped to find a single depth of no motion as a reference level where all ocean currents were zero. Experience has shown that there is no global reference level and that this level must be chosen to fit the region being studied. Since in general the strength of ocean currents decreases from the surface downward, a common practice has been to arbitrarily choose some lower level as a reference level of no motion. The selection of the depth of the reference level is often determined by the maximum common depth to which temperature and salinity measurements have been made in the particular region and values of 500, 1000, 1500 or 2000 m have frequently been used. While it is clear that there may be currents at and below these depths they will generally be small and will only contribute a small error term to the geostrophic currents computed above the selected reference level. (Note, however, that such a small error term in the individual currents may add up to a large error in the total *volume transport* integrated from the surface to the selected reference level.) An alternative to arbitrarily assuming a depth of *no* motion as a reference level is to use a depth of *known* motion. For example data from current meter measurements, or the tracks of sub-surface acoustic floats, may be used to define the current at some level for which one has temperature and salinity profiles from which to calculate density. By assuming that the measured motion is all in geostrophic balance it can be used to define the slope of the isopycnals at that depth, from which the slopes at other depths can be adjusted.

To compute the pressure gradient we need data from at least two oceanographic stations. If we have a spatial grid of stations we can compute the gradients over the entire domain from pairs of stations.

A simple example illustrates the application of a reference level to the geostrophic current computation. In Fig. 6.5 the dynamic depths of isobaric surfaces between a pair (A, B) of three oceanographic stations (A, B, C) are shown as solid lines. The slopes of these lines are relative to a flat (relative to the geopotential) sea surface. In reality the surface has a topography associated with it that represents the dynamic topography related to the geostrophic currents. Thus to find the slope at the surface, and in the layers below the surface, we convert the relative isobar slopes to absolute slopes by selecting a reference surface which we then set to zero height, i.e. rotate it to become level. All other isobars are then rotated through the same angle to give the isobars indicated by the dashed lines which now show a slope at the sea surface between stations A and B. This correction of relative to absolute geostrophic currents can also be done by first computing the dynamic values relative to a "flat" sea surface, computing the horizontal differences, calculating the velocities using the geostrophic equation, and then inverting these velocities relative to the reference level. Such a computation is shown in Table 6.1 which addresses the dynamic heights at stations A and B and hence computes the geostrophic current between them for a reference level set at 1500 m.

Instruments and Methods   119

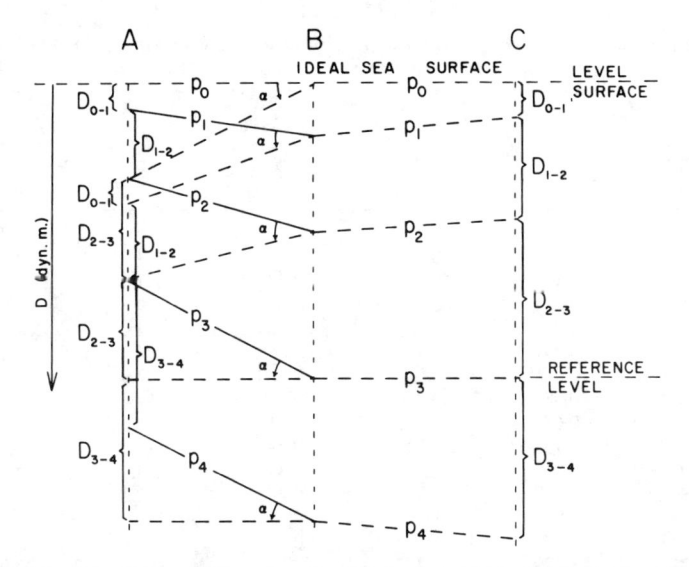

Fig. 6.5. Schematic vertical section showing the dynamic depth of isobaric surfaces relative to the ideal (level) sea surface and the absolute dynamic depth. Relative dynamic isobaths are shown by full lines and absolute dynamic isobaths by dashed lines.

In regions where the water density in the column is low the isobaric surface will be high, and where the water density is high in the column the surface will be low. The result will be a contour map of the isobaric surface, like a topographic map of the land, i.e. a *dynamic topography map*. Flows on this surface will be *along* the contours with the "hills" on the right of the flow in the northern hemisphere (on the left in the southern) and the speed of flow at any point will be proportional to the steepness of the slope at that point, i.e. inversely proportional to the separation of the contours.

The height of the upper isobaric surface will depend on the total density distribution in the column between it and the depth of the reference level below. The quantity used for contouring is called the *geopotential distance* (*dynamic height* in oceanographic jargon) between the surfaces. For two isobaric surfaces p_2 (upper) and p_1 (lower), the geopotential distance is:

$$\Phi_2 - \Phi_1 = \int_1^2 \alpha \cdot dp = -\int_1^2 \alpha_{35,0,p} - \int_1^2 \delta \cdot dp.$$

The first integral term is called the *standard geopotential distance* ($\Delta\Phi_s$ or $D_{35,0,p}$) as it depends only on the pressure difference; the second ($\Delta\Phi$ or ΔD) is the *geopotential anomaly* (or *dynamic height anomaly* in older literature) and it contains the effects of the actual temperature and salinity distributions relative to the arbitrary standard values of 0°C and 35 for the standard distance. In practice we only calculate the anomaly and refer to it as the "dynamic height".

TABLE 6.1
Computation of geostrophic currents from the field of mass between Atlantis stations 5298 and 5299 across the Gulf Stream

Depth (m)	D_A $\sum \Delta D_{5295}$	D_B $\sum \Delta D_{5299}$	$D_B - D_A$	SPEED (cm/s) Rel. current	Abs. curr. with reference to 1500 m
0	0	0	0	0	207.6
25	0.093	0.102	0.009	3.7	203.9
50	0.173	0.190	0.017	7.0	200.6
75	0.241	0.264	0.023	9.5	198.1
100	0.298	0.332	0.034	14.0	193.6
150	0.391	0.450	0.059	24.3	183.3
200	0.466	0.546	0.080	33.0	174.6
300	0.581	0.720	0.139	57.4	150.2
400	0.675	0.891	0.215	89.1	118.5
500	0.753	1.042	0.289	119.3	88.3
600	0.815	1.164	0.349	144.0	63.6
800	0.919	1.362	0.443	182.8	14.8
1000	1.011	1.496	0.485	200.1	7.5
1200	1.097	1.594	0.497	205.1	2.5
1500	1.229	1.732	0.503	207.6	0
2000	1.449	1.947	0.498	205.5	2.1

Distance between the two stations = 28.06 km, $\phi = 36.3°N$, and assuming that the absolute current is zero at 1500 m.

(Note that although the quantities are called "distance" and "height", and the latter is quoted in units of "dynamic metres" or "dynamic centimetres", they have physical dimensions of energy per unit mass as they represent work done against gravity. Their correct SI units are $J/kg = m^2/s^2$.) For numerical convenience, oceanographers have used the dynamic metre such that 1 dyn m = 10.0 J/kg. To indicate that the dynamic metre (dyn m) is being used, the symbol D is used for geopotential, and the geopotential distance $(D_2 - D_1)$ is then numerically almost equal to the depth difference $(z_2 - z_1)$ in metres, e.g. relative to the sea surface and taking the acceleration due to gravity $g = 9.8$ m/s²:

	SI units	Mixed units
at a geometrical depth in the sea	= +100 m	= +100 m
then	$z_2 = -100$ m	= −100 m
and the pressure will be about	$p_2 = +1005$ kPa	= +100.5 dbar
and the geopotential distance relative to the sea surface	$(\Phi_2 - \Phi_1) = -980$ J/kg,	
	$(D_2 - D_1)$	= −98 dyn m.

(See the Appendix for definitions of the pressure units kPa and dbar.)

Two examples of such dynamic topography maps are shown. For the Pacific (Fig. 6.6), Wyrtki (1974) computed the annual average field of dynamic

Instruments and Methods 121

ΔD 0/1000 dbar
in dyn. cm.

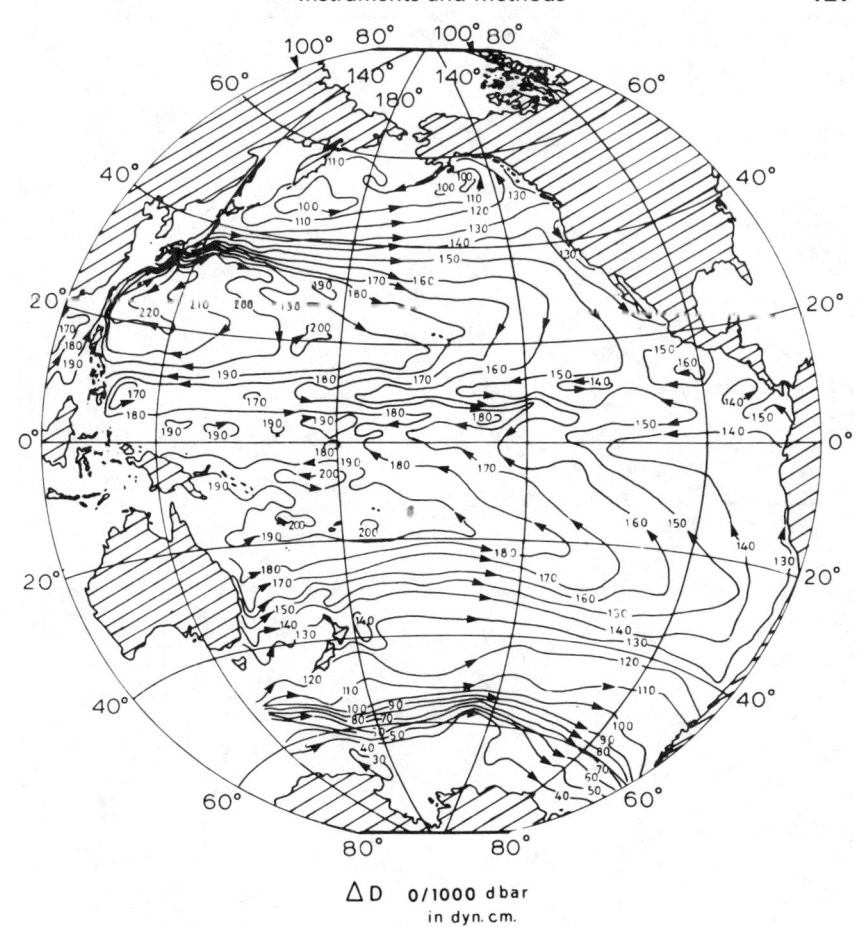

FIG. 6.6. Mean annual dynamic topography of the Pacific Ocean sea surface relative to 1000 dbar in dyn cm ($\Delta D = 0/1000$ dbar); 36;356 observations.

height of the zero pressure surface (sea surface) relative to the 1000 dbar surface (10,000 kPa) ($\Delta D = 0/1000$ dbar) from all available oceanographic stations. For the North Atlantic, Stommel, Niiler and Anati (1978) computed $\Delta D = 100/700$ dbar as shown in Fig. 6.7. (Arrows have been added to the dynamic height contours to assist the reader in interpreting the direction of flow.) Note the similarities, especially the clockwise circulations north of the equator and the strong western boundary currents (contours very closely spaced), and the anticlockwise circulation of the South Pacific bordered on the south by the Antarctic Circumpolar Current. It should be mentioned that while the speeds in the Pacific at the 1000-dbar level are probably quite small in most regions, this is probably not the case at the 700-dbar level chosen for the Atlantic map. Such maps provide considerable detail to a description of

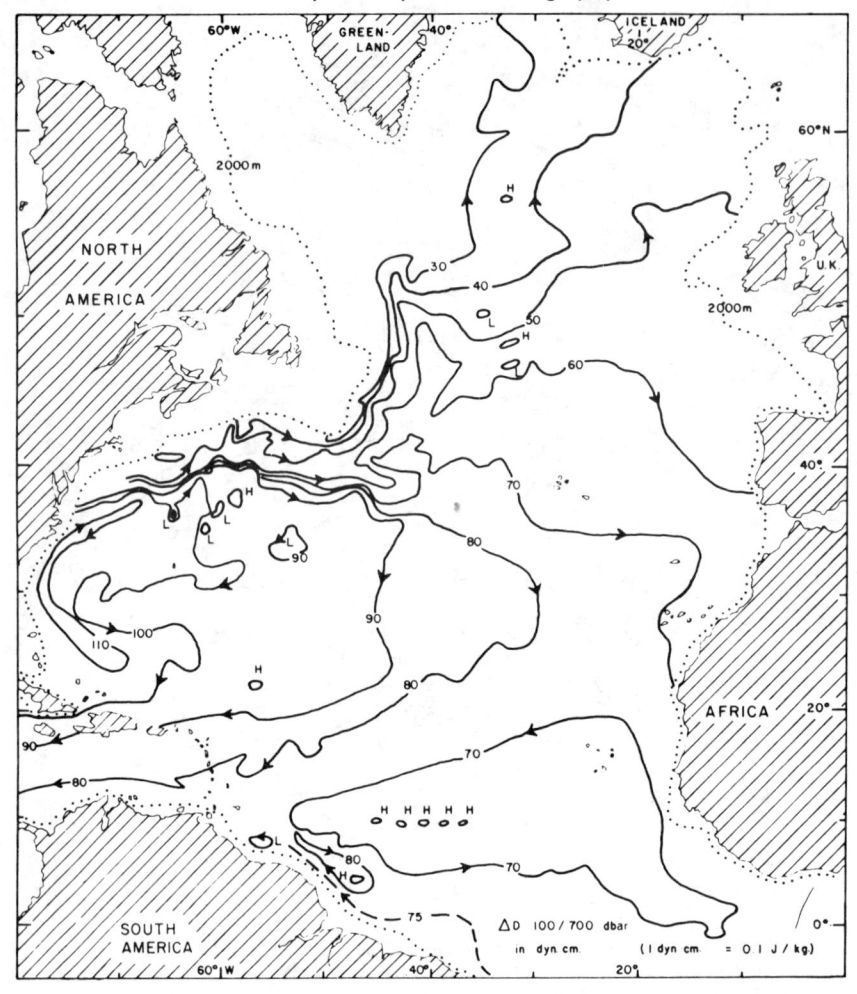

FIG. 6.7. Dynamic topography of 100-dbar surface relative to 700-dbar surface ($\Delta D = 100/700$ dbar) in dyn cm, Atlantic Ocean.

part of the current field in the ocean. However, it should be remembered that the observations used were not "synoptic", i.e. not made simultaneously, and any variation from the steady state which occurred during the period covered by the observations would introduce irregularities or "noise" into the maps. In most cases, dynamic topography maps (and plots of water properties) for ocean size areas have to be prepared from data collected by various expeditions, in different parts of the area, and over many years. The fact that well-defined features appear in such maps, or are repeated when there are sufficient data to prepare maps for the same area for several seasons or several individual years, indicates that these must be permanent features of the ocean.

6.256 Acoustic Tomography

As discussed in Section 3.7, a feature of the ocean is that electromagnetic radiation penetrates only short distances, ~ 100–200 m for light, but the ocean is essentially transparent to sound waves and the speed of sound in sea-water is mainly dependent on temperature. This fact led to the suggestion that measurement of the travel times of sound pulses between a source and a receiver in a particular region might be used to obtain information on the temperature distribution. The first large-scale experiment (Cornuelle et al., 1985) used a moored array of sound sources (S_1, S_2, etc.) and receivers (R_1, R_2, etc.) in an array as in Fig. 6.8 in a 300 km square. Each source and receiver

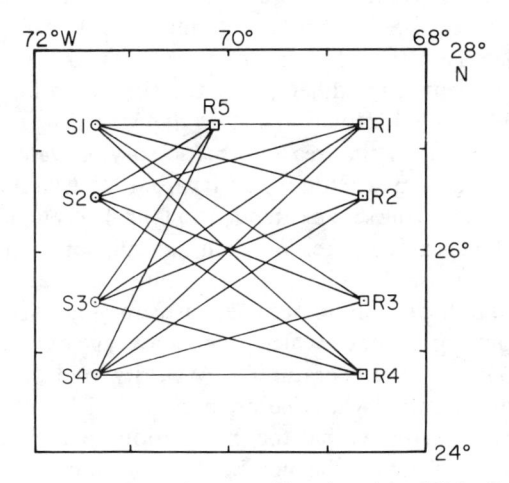

FIG. 6.8. Moored array of sound sources (S) and receivers (R) for first large-scale acoustic tomography trial.

had accurate clocks so that travel times for sound pulses could be determined along each of the 20 possible paths joining the sources and receivers which were situated at about 2000 m depth in the sound channel. Along each of the source/receiver directions there are 5 to 10 possible sound paths in the vertical plane so that information would be available about sound speeds over a range of depths between about 500 m and 3500 m. Analysis of travel-time data by inverse methods then led to data on sound speed and from this could be inferred the temperature structure in the volume of ocean included within the area of the array and the depth range stated. This experiment was essentially a first test (over a period of four months) of the viability of the method from a technological point of view but comparisons with the temperature structure determined directly by CTD surveys showed good correspondence. Sound speed variations of about 2 m/s corresponded to temperature variations of 0.4K at 700 m depth. This technique is referred to as *ocean acoustic temperature tomography*.

An extension of this idea, called *ocean acoustic velocity tomography* uses measurements of the differences in travel time between pairs of transceivers (co-located transmitter/receiver pairs) to provide information about the velocity of the water itself (e.g. see Howe and Worcester, 1987). For instance, if the water between transceivers A and B is moving, e.g. from A toward B, then the travel time for a sound pulse is less from A→B than from B→A. The difference Δt in travel time is proportional to the component Δu of the water speed along the path AB. Then if a number of transceivers (A, B, C, etc.) are moored in an ocean area, components of water velocity along a variety of paths AB, BC, AC, etc. may be determined and then combined to provide the field of total velocity within the volume enclosed by the transceivers. It is estimated that water speeds may be determined to a few centimetres per second. A special case is that of three moorings A, B and C at the vertices of a triangle. Measurement of the difference of travel times in the two directions ABCA and ACBA yields information on the *relative vorticity* (rotation about a vertical axis) of the water in the area, information of great value in studying the dynamics of the ocean (e.g. see Pond and Pickard, 1983, Chapter 9). The three-mooring study can, of course, be extended to four, five, etc. moorings at the vertices of a square, pentagon, etc., and then transmissions across the area can also add information on the water motion within the area.

One feature of both versions of ocean acoustic tomography is that they are essentially integrating methods which tend to average out small scale (time and space) variations of temperature and velocity, and thus remove these features which are just noise when one is concerned with larger scale features of the ocean. Another feature is that the information obtained in tomography increases geometrically with the number (N) of moorings, i.e. as N^2, in contrast to that for moored current meters for which information only increases directly as N.

6.26 Radiation measurement

Direct measurements of Q_s are made with a pyranometer. The sensing element of the Eppley pyranometer consists of two flat plates of copper, one painted with a flat black paint and the other whitened with magnesium oxide. The two plates are placed horizontally with a clear view of the sun and sky and are shielded from draughts by a clear hemispherical cover. The black paint absorbs all radiant energy, short- and long-wave, falling upon it and is thereby heated above the surrounding temperature. The white plate reflects practically all of the energy between 0.3 and 5 μm (short-wave radiation) but absorbs all long-wave energy. The white plate is consequently heated less than the black one and the difference in temperature between them is a measure of the short-wave radiation (Q_s) falling on a horizontal surface in the locality of the instrument. The difference in temperature is measured by connecting the "hot" junctions of a group of thermocouples to the black plate and the "cold" junctions to the white plate. The difference in temperature gives rise to a thermoelectric EMF which is measured by recording galvanometer. The instrument is calibrated by exposing it to a standard source of energy, such as a standard electric filament lamp.

The downward directed component of the long-wave radiation term Q_b is measured with a radiometer. The Gier and Dunkle instrument consists of two horizontal plates of black material separated by a layer of material of known heat conductivity. The upper sheet of black material absorbs all the radiation falling upon it from above and is thereby heated above the temperature of the lower sheet which is screened from radiation from below by a sheet of polished metal. The difference in temperature between the upper and lower sheets is measured by thermocouples and is a measure of the rate at which the sum total of long- and short-wave energy is coming down from above. To determine the value of the long-wave component itself it is necessary to subtract the short-wave radiation rate as measured with a pyranometer. An alternative procedure is to omit the polished metal screen from below the black horizontal plate and arrange the instrument so that the upper plate "looks at" the atmosphere and the lower plate "looks at" the sea below. In this "net radiometer" arrangement the difference in temperature between the upper and lower plates is a measure of the net amount of radiant energy reaching a horizontal surface, i.e. it is a direct measure of $(Q_s - Q_b)$.

To determine the transmission of visible light through the water the simplest device is the *Secchi-disc*, a white plate about 30 cm in diameter fastened to hang horizontally on the end of a rope marked in metres. The disc is lowered into the sea and the depth at which it is lost to sight is noted. This depth decreases as the vertical attenuation coefficient of the sea-water increases. In very clear water the depth may be over 50 m, in coastal waters 10 to 2 m, and in some river estuaries less than 1 m. The Secchi-disc is only a semi-quantitative device but being simple it is often used. When one first starts using the device it is rather surprising to find that after a very little practice it is possible to obtain consistent readings to better than 10% with little variation from individual to individual.

Preisendorfer (1986) reviewed in detail the applications and the hydrological optics of the Secchi disc and concluded that its principle purpose is to provide a simple visual measure of water clarity and it may also provide an estimate of $(\alpha + K)$, where α and K are respectively the photoptic beam attenuation coefficient and diffuse attenuation coefficient of the water due to dissolved and particulate material therein, by means of the relation: Secchi disc depth $Z_{SD} = \Gamma/(\alpha + K)$ where the coupling constant Γ has a value between 8 and 9. (In Section 3.8 we combined $(\alpha + K)$ as k.) It was pointed out that a number of factors enter into the observed value for Z_{SD} which Preisendorfer states as his "Ten Laws of the Secchi disk" which include that Z_{SD} varies inversely with (1) the amount of attenuating material in the water, (2) the optical state of the sea surface, (3) the amount of light reflected from the sea surface, (4) the reflectance of the sea surface and (5) the height of the observer above the sea surface, while Z_{SD} varies directly with (6) the reflectance of the disc, (7) its diameter, (8) the solar altitude. The value of Z_{SD} is larger if the line of sight in the water is in shadow. (The effects of items (6) and (7) are small.) Preisendorfer pointed out that to separate the values of α and K requires the use of additional, generally electronic, instruments. He concluded that the primary function of a Secchi disc is to provide a *simple* visual index of water clarity and that to extend the use of the disc with auxiliary objective measurements (of α or K) is to risk obviating or abusing this primary function.

A more quantitative instrument for determining the (beam) attenuation coefficient is called a *transparency meter* or a *turbidity meter*. In this, the sensing head which is lowered into the water has a lamp and a photoelectric cell in separate watertight housings mounted 0.5 to 2 m apart and opposite to each other. The principle of the instrument is that the current from the photocell is a measure of the amount of light falling upon it. The lamp has a lens and collimator to ensure a parallel beam of light which passes through the water and falls upon the photocell. The light output is kept constant by regulating the current through the lamp. The light beam is usually arranged horizontally because the absorption characteristics in the sea are often horizontally stratified. Also the photocell must not face upward or it would pick up direct sunlight which would give too high an apparent transmission by the water. The photocell is connected by an electrical cable to a microammeter on deck. The current from the photocell is recorded when the sensing head is on deck with the light passing through air whose attenuation is taken to be zero. In the water the photocurrent decreases due to the attenuation of light by the water. The decrease from the air path value is a measure of the attenuation by the water. If the initial light intensity is I_o and that in water after a path length x is I_x, the beam attenuation coefficient C may be obtained from the formula $I_x = I_o \cdot \exp(-C \cdot x)$ or in logarithms to base 10 as $C = 2.3$ (log I_o − log I_x)/x. The length x used in these formulae is the length of the water path between the lamp and photocell in metres (usually from 0.1 to 2 m). The instrument is lowered in stages and the attenuation coefficient measured at a series of depths to give a vertical profile for this water property. The attenuation is often associated with particulate material present in the water independently of its temperature and salinity and therefore may be useful as an independent tracer of water masses. The attenuation coefficient C, or the 'turbidity" as it is sometimes called, has been used to distinguish water masses both in the open ocean and in coastal fjords.

It must be noted that the parallel beam attenuation coefficient C is not identical with the vertical attenuation coefficient k (Section 3.8) where light at a point in the sea may be arriving from a variety of directions. Both C and k represent loss of light both by absorption (i.e. conversion to heat or chemical energy) and of scattering (i.e. change of direction while remaining radiant energy). For both beam and vertical attenuation, absorption is effective in reducing the light intensity with distance. However, scattering is more important for beam attenuation because *any* scattering represents loss of energy from the beam whereas for vertical attenuation we are concerned with energy arriving at a point in the sea where light is travelling in a variety of directions and a significant part of the energy arriving there will have been scattered from light travelling in other directions. If we express the beam attenuation coefficient C as the sum of absorption (a) and scattering (b), i.e. $C = a + b$, then theory and observation suggest that roughly $k = 1.4a + 0.03b$, i.e. scattering contributes relatively little to k compared with absorption (e.g. see Jerlov, 1976).

For some purposes it is desired to know the amount of daylight which reaches a particular depth. This is done by mounting a photocell behind a filter in a watertight housing to face upward. A microammeter, connected to the cell, is read first with the cell on deck facing upward to give a measure of the sun and sky light falling on the sea surface in the spectral range passed by the filter. The cell is then lowered to the depth of interest and the photocurrent again measured. The ratio of the photocurrent with the cell in the water to that on deck is then a measure of the ratio of light intensity at depth to that at the surface. If the photocell indications have been calibrated against a standard light intensity meter then the actual light intensity at depth will be known. It is usual to measure the light intensity at a series of depths below the surface in order to obtain a profile of light intensity as a function of depth. It is usual to mount an opal glass or hemisphere in front of the photocell so that it will collect light from all directions above its plane.

In the above instruments, the Weston type of photovoltaic cell is often used because its response can be made substantially linear with light intensity at a particular wavelength, because it responds over the full range of visible light and because it does not require a separate source of power. However, the Weston cell has a limited sensitivity and it may be inadequate if measurements are required of the *in situ* light intensity at considerable depths. In this case it is necessary to use the more sensitive photomultiplier type of cell. The disadvantage of this type is that it requires a high-voltage supply and precautions have to be taken to avoid electrical leakage when working in seawater.

6.3 Platforms

6.31 Sea- and airborne

The majority of oceanographic measurements have been made from research ships, with auxiliary measurements from coastal stations (tide gauges, wave staffs, lighthouse temperature and salinity observations, etc.). Today the research vessel continues to play an essential role in oceanographic research but the rising costs of operating these vessels, along with rapid improvements in technology, have brought a variety of new options to the ocean measurement community. In addition, the advantages of observing the earth from aircraft and satellites have further motivated the continued development of these measurement technologies.

The basic research vessel has changed dramatically in recent years. As discussed previously, early oceanographic research vessels were built for long expeditions at sea. Even as recently as the 1970s research vessels were being built to carry a large complement of scientists and to perform a large variety of research tasks during long periods at sea. More recently there has been a shift toward smaller ships that can't carry quite as much equipment or spend quite as much time away from port. These smaller ships are ideal for more coastal related projects where the ship can return to port after a few days or weeks. The computerization of most data collection systems has made it possible to have a lot of data collection activities running at the same time on these smaller ships. Also automation has made it possible to operate the ship with far less manpower and research vessel crews have decreased considerably in size making the operation of the vessel cheaper.

Some common characteristics of a good research vessel are its manoeuvrability, its stability at sea, its comfort in living and working spaces and its reliability in operation. Whether large or small, research vessels require excellent ship handling capabilities and good navigation equipment. A very significant part of taking measurements at sea is knowing exactly where you are and what the behaviour of the ship has been while you have been collecting your samples. These navigational chores are commonly the responsibility of the ship's captain and officers who generally strive to provide the calibre of information needed for geophysical mapping. Another common characteristic of research vessels is the abundance of support systems to get equipment on or off the ship as well as to get them over the side into the water and back again on to the deck. Generally a couple of cranes are located to cover the working deck area for this purpose. Winches are placed in appropriate locations for handling samplers and sensors over the side.

Physical oceanographic sampling activities on a research vessel can range from a long series of sampling bottle or CTD casts to deploying moored current meters. Making up moorings and deploying current meters are very tiresome and laborious tasks requiring lots of people and deck space. CTD stations can be long (for deep casts) and monotonous when taken regularly over a long watch period. While these make up the majority of most physical oceanographic observations there are other tasks such as dropping XBT's, deploying satellite tracked buoys, etc.

A relatively rapid sampling alternative to working from a research vessel is to use sensors mounted on aircraft. Most of these airborne systems sense remotely using radiation either emitted or reflected from the ocean's surface. Infrared sensors are used to map sea-surface temperature (SST) patterns while visible light sensor channels are used to measure patterns of ocean colour

Instruments and Methods   129

related to biological productivity and the amount of chlorophyll in the surface waters. Multi-spectral scanners are capable of simultaneously measuring radiation in both the visible and thermal infra-red channels. Recently, passive and active microwave sensors have also been flown on aircraft to sense ocean parameters. One of the most useful of these is the Synthetic Aperture Radar (SAR) which uses the motion of the aircraft to synthesize a larger antenna than an aircraft could carry, making it possible to greatly improve the ground resolution. Aircraft SAR imagery has been particularly useful in the detailed mapping of sea ice and its motion.

Another important use of aircraft is in the collection of AXBT data. As described earlier in Section 6.24 the AXBT contains both the XBT probe and a small radio transmitter to send to the aircraft the temperature versus depth data being collected as the probe falls through the water. These AXBT probes are usually dropped from altitudes between 300 and 1000 m but testing has shown that they will survive the impact of being dropped from 6000 m. The designs from the different manufacturers of the AXBT's vary but all of them have some type of atmospheric drag element (parachute, wings, etc.) to slow their descent and soften the impact with the sea. The combination of probe and radio transmitter makes the cost per AXBT probe much higher than that of the normal ship XBT but the relatively lower cost per area surveyed of using an aircraft often offsets the higher cost per probe. Also if near-simultaneity is a requirement of the sampling programme, AXBT observations are one of only a very few methods of carrying out this type of sampling.

Ground effect vehicles ("Hovercraft") are found to be very effective as platforms in shallow waters such as estuaries and mud-flat regions and in ice-infested waters.

6.32 *Satellites*

6.321 *Satellite remote sensing*

One of the biggest problems in physical oceanography is the spread of time and space sampling that results from the normal ship-based measurement system. Because of ship speed limitations (10–12 knots) it is impossible to observe a large area in a "synoptic" (i.e. near simultaneous) fashion and oceanographers have had to be satisfied with long-term averages of data collected over a period of time. For many stationary and longer period phenomena such averages were adequate but as the study of time-variable oceanographic processes has intensified the need for more nearly simultaneous sampling has increased.

6.322 *Weather satellites*

Earth observing artificial satellites offer a unique perspective in this regard and while measurements are limited to the surface of the ocean the almost synoptic sampling from satellites makes it important for oceanographers to effectively utilize information sensed from space. At present much of this information comes from operational weather satellites both in polar and in geostationary orbits. At altitudes of about 500–800 km polar orbiting weather satellites are generally deployed in "sun-synchronous" orbits to maintain the same local equatorial crossing time each orbit. This is a requirement for many studies which require the same arrangement between solar and satellite scan angles. Thus each polar orbiter passes over each location at least twice per day (once at night and once during the day). Geostationary satellites fly at 36,000 km altitude and can scan the earth almost continuously.

Polar orbiting weather satellites have progressed dramatically from the late 1960s when the television camera on the Television Infra-Red Observing Satellite (TIROS) could only point at North American latitudes during its orbit due to the spin stabilization of the spacecraft. After the first TIROS a series designated the ESSA satellites mounted a camera looking out along the radius of the spinning satellite and took a successive set of pictures of the earth to provide the first global view of the earth's surface. Today, 3-axis stabilization on the TIROS-N satellites makes it possible to keep a single scanner, the Advanced Very High Resolution Radiometer (AVHRR), pointed continuously at the earth's surface. The AVHRR is a 5-channel filter-wheel spectrometer with channels in the visible (0.58–0.68 μm), the near-infra-red (0.725–1.10 μm) and three in the thermal infra-red (3.55–3.93 μm; 10.3–11.3 μm; 11.5–12.5 μm). This combination of channels has proven to be useful in meteorological studies of cloud patterns and temperatures, land-surface vegetation and sea-surface temperature (SST). The thermal infra-red channels also provide meteorological images at night when there is no visible radiation to reflect from the earth. Once operational the TIROS-N satellites are referred to as NOAA 1–11 satellites to indicate their operation and control by the National Oceanic and Atmospheric Administration (NOAA).

The multiple AVHRR thermal infra-red channels make it possible to estimate the atmospheric attenuation, by atmospheric water vapour, of the infra-red signal emitted from the ocean's surface. Using the difference between the radiation received in channel 4 versus that on channel 5 it is possible to estimate the amount of infra-red energy lost to atmospheric moisture before reaching the AVHRR sensor optics. The relatively weak surface temperature gradients in the ocean make it necessary to carry out the best atmospheric moisture correction possible when working with SST's computed from AVHRR data in an effort to get the precision of the satellite measurement to the 0.3K accuracy recommended for modern climate studies.

One of the most important aspects of working with the AVHRR data, and any other satellite data, is the correction of imagery for earth distortion and the remapping of the corrected image to a selected map projection for intercomparisons between images and with other *in situ* data. This step is generally referred to as "image navigation" and is essential to being able to

Instruments and Methods

employ AVHRR imagery quantitatively. Without this image navigation step, satellite infra-red imagery can only suggest the complexity of scales on which the ocean varies. Quantitative studies can only be carried out with properly navigated imagery.

The AVHRR provides three different types of image data. The most commonly available form is the direct readout mode which is called High Resolution Picture Transmission (HRPT) and is usually directly read-out by a tracking antenna at a ground station. These data have the approximately 1 km spatial resolution of the AVHRR system. Total images depend on the altitude of the satellite over the receiving station and each station receives between 4 and 6 passes per satellite per day. During each orbit the satellite system records a lower spatial resolution product (approximately 4 km square) called the Global Average Coverage (GAC). These data are only read out at receiving stations at Wallops Island, Virginia and Gilmore Creek, Alaska operated by NOAA, the agency responsible for the U.S. operational weather satellites. The value of the GAC data is that each day a satellite will provide a day and night image of the entire globe. Finally, each satellite has a number of tape recorders that can record the full 1 km image data during a short part of the orbit. These Local Area Coverage (LAC) data are recorded at areas selected and commanded by NOAA and are then downloaded (received at) one of the two NOAA stations. In this way it is possible to "see" parts of the earth that are out of range of the NOAA operated receiving stations.

Examples of the 1 km imagery available from the AVHRR are shown in Plates 11 and 12. The former is an infra-red (channel 4) AVHRR image from October 2, 1980 of the North American west coast region. This unusually cloud-free image stretches from the northern-tip of Vancouver Island (top, left of centre) to northern California (bottom right). The grey shades are enhanced so that white represents the colder temperatures while the warmer temperatures appear darker. The infra-red SST reveals the presence of cold water along the coast extending in narrow tongues westward into the North Pacific Ocean. There is also a trend from warmer water at lower latitudes to (lighter coloured) cooler water at the higher latitudes. Unlike the smooth temperature maps of Figs. 4.1 and 4.2, the SST gradients in this image are quite complicated with eddies and meanders dominating the scene. These mesoscale features are generally larger at lower latitudes, and smaller-scale features populate the boundaries between warm and cold water. This truly synoptic sample clearly indicates the complex spatial variability of many features being studied by oceanographers.

The image in Plate 12 from the east coast of the U.S. is also dominated by the mesoscale SST structures associated with the Gulf Stream. The warm central ribbon of the Gulf Stream core is clearly visible and the pattern depicts the marked southward meander of the Stream. Just to the south of this meander there are three cold, cyclonic eddies (see also Section 7.344) that have broken off the Stream. North of this meander there is a weaker expression of a warm Gulf Stream eddy. Temperature gradients are much stronger in this area of the Gulf Stream than they are off the west coast as seen in Plate 11. In regions where SST gradients are stronger, AVHRR imagery provides very useful pictures of the SST patterns. Still one must be cautious since rapid heating and cooling events can quickly eradicate the surface thermal signatures of subsurface warm or cold features.

6.323 The Coastal Zone Colour Scanner (CZCS)

One of the most important developments in optical remote sensing was the Coastal Zone Colour Scanner (CZCS) which flew on the NASA satellite NIMBUS-7. Launched in 1978 for a 2-year lifetime the CZCS continued to provide useful image data up through mid-1986 providing an 8-year time-series of ocean surface colour imagery. The CZCS was constructed to supply surface imagery in a number of narrow visible bands which could then be separated into the individual colour components. Since biological productivity, in the ocean's surface layer, is directly related to the colour at the sea surface, the CZCS sensor was intended to provide information on chlorophyll and primary productivity.

In spite of some problems with drifts of the sensor system the CZCS produced some very valuable images of the ocean's productivity and made it possible for the first time to make a global map of primary productivity in the ocean. With a 1 km spatial resolution like the AVHRR the CZCS provided visible channel data with features very much like those seen in the thermal image of Plate 11.

6.324 SEASAT

The first and only satellite dedicated to ocean remote sensing was the short-lived SEASAT satellite. Launched in mid-1978, SEASAT failed after 90 days of operation (due to a power supply failure) and has never been replaced. Even during this very short life, SEASAT proved the concept of a number of unique microwave instruments designed to measure various parameters of the ocean. The first of these was the Scanning Multi-channel Microwave Radiometer (SMMR) which was intended to map all-weather images of sea-ice and SST. Since microwaves are only slightly attenuated by atmospheric moisture they are an excellent candidate for observing an earth's surface that is often obscured by clouds. This is particularly important for the ice-covered polar regions which are frequently cloud-covered. In addition the passive SMMR responds uniquely to the changes in surface emissivity related to ice age and thickness. As for the SST, SMMR demonstrated that the infra-red systems are still better for measuring SST. After the demise of SEASAT a second SMMR was quickly prepared and launched on NIMBUS-7 which also carried the CZCS.

Instruments and Methods

The systems described above are all "passive" systems in which the satellite sensors detect thermal radiation from the earth's surface or reflected/reradiated solar energy; "active" systems are those in which the satellite radiates (microwave) energy downward and then senses such energy reflected/reradiated from the earth.

An example of an active system was the unique microwave instrument on SEASAT called the RADAR Scatterometer which was designed to measure wind speed and direction from the RADAR backscatter from the small wavelets that form at the ocean's surface when the wind blows. Using a pair of antennae to resolve the direction, verification studies demonstrated the accuracy of the system for measuring wind stress over the ocean both in terms of magnitude and direction. This appears to be the only possible way to resolve the wind patterns, and their changes, over the open ocean and it is hoped that another scatterometer will soon fly.

The (active) Synthetic Aperture RADAR (SAR) on SEASAT was the first SAR flown on a satellite. In its short period of operation the SEASAT SAR produced some very interesting images of the ocean's surface which are still being analysed today, although it is clear that we do not yet completely understand the SAR signatures of ocean features. SAR has also proven very useful for the detailed mapping of sea-ice and its motion. Again the all-weather capability of an active microwave sensor makes it possible to see through persistent cloud cover. Also the antenna synthesis available with SAR makes it possible to attain high spatial resolution (25 m on SEASAT) with a small satellite antenna.

Another active microwave system on SEASAT was the RADAR Altimeter which monitored the height of the sea surface, and its changes, during the satellite orbits. Again a lot of experience was gained in working with satellite altimetry obtained from the short life of SEASAT. The altimeter also provided the first truly global map (Cheney et al., 1983) of eddy energy from fluctuations of the ocean's surface height. More recently, a new SEASAT altimeter has been operating on the GEOSAT satellite since 1986. The first year of the GEOSAT mission was classified in order to provide the U.S. military with a new estimate of the earth's geoid. In November, 1986, GEOSAT as manoeuvred into an Exact Repeat Mission (ERM) which followed the earlier SEASAT orbit and thus made the altimeter data available for general study and analysis by oceanographers. These altimeter data have provided new insight into the recent El Niño event.

Finally, SEASAT carried a (passive) visible-infra-red radiometer to provide imagery simultaneously with the microwave data. All of the instruments functioned well during the short lifetime of the satellite and it was only the failure of the power supply which terminated the mission. The concepts behind all of the microwave sensors were established by the SEASAT mission and the way that remote sensing must go was made clear.

6.325 Other sensors and derived products

In addition to the operational weather satellites, the CZCS on NIMBUS-7 and SEASAT, there are other satellite systems that can be used in the study of oceanography. For example the high spatial/spectral resolution imagery from the multi-channel visible sensors on the LANDSAT satellites can be used to map sea-ice cover and some plankton blooms. Other sensor systems flying on the meteorological satellites are also useful, particularly in the study of air-sea interaction. A list of some of the more important directly observed or derived quantities follows:

(a) Sea surface temperature is routinely computed from infra-red imagery. The primary problem at present is the correction of these infra-red radiances for atmospheric attenuation due to the presence of moisture in the troposphere. Accuracies presently range from 0.8–1.0K. Coincident visible images are often used for cloud filtering before the computation of the infra-red SST. In the future it is hoped that cloud-independent measurements will be possible from advanced microwave sensors. At present the relatively low emissivity of sea-water makes it impossible to retrieve accurate SST's from microwave data.

(b) Radiant energy components are estimated to about ± 3 W/m^2 from both the visible and infra-red radiances. Visible radiances are used to estimate the instantaneous effects of clouds on solar insolation to correct for the total amount of incoming radiation from sun and sky. Infra-red imagery can be used to compute out-going long-wave radiation (Q_b).

(c) Rainfall over the ocean can be inferred from the presence of highly reflective cloud seen in both geostationary and polar-orbiting satellite imagery. This provides a fairly crude estimate because there is no definite relationship known between the amount of highly reflective cloud present and the level of rainfall experienced. Rainfall can also be estimated directly from microwave radiances from cloud liquid water.

(d) Atmospheric water vapour can be directly measured as a vertical integral by microwave channels or can be computed from a moisture profile derived from primarily infra-red channels. The TIROS Operational Vertical Sounder (TOVS) on the NOAA polar-orbiting weather satellites uses a combination of infra-red and microwave channels to measure atmospheric moisture and temperature profiles. Both of the microwave and infra-red methods produce atmospheric water vapour values accurate to around 2 g/cm^2; the microwave data are cloud independent while the infra-red sensors are limited by the amount of cloud cover.

(e) Upwelling events in the sea can be located and monitored by both their surface thermal signatures and their expression by increased primary productivity in ocean colour imagery. One must be careful to separate

Instruments and Methods 135

the *in situ* effects such as plankton blooms and heating and cooling from changes due to upwelling alone.

(f) Currents can be estimated from the displacement of features in sequential imagery. Early studies used the visual method of feature tracking (Vastano and Borders, 1984) while more recent efforts have computed the maximum cross-correlation location between images to estimate advective displacements between the images (Emery *et al.*, 1986). Surface current vectors computed with this maximum cross-correlation (MCC) technique are superimposed on the Gulf Stream image in Plate 12. This same procedure can be applied to sequences of visible satellite images to compute the motion of sea-ice (Ninnis *et al.*, 1986). Applied to highly resolved sea-ice SAR imagery, this method produces very detailed maps of the ice displacements (Collins and Emery, 1988). Finally the same technique can be applied to sequences of ocean colour imagery to estimate surface advection.

(g) In addition to estimating the motion of sea-ice, satellite imagery can be used to monitor the presence of sea-ice cover and to estimate its concentration. Optical systems are limited in this activity by the frequent presence of clouds at polar latitudes but microwave sensors have proven extremely useful in mapping sea-ice cover and concentration (Gloersen *et al.*, 1978). In addition the microwave imagery can also be used to estimate sea-ice parameters such as thickness and ice age.

(h) Mean wave-height estimates accurate to ± 1 m or 25% of the actual wave height are possible from RADAR altimeter backscatter.

(i) Surface wind speed can also be inferred from microwave brightness values in terms of the change in emissivity due to the surface roughness. Accuracies are around ± 2.6 m/s. The scattering cross-section of a nadir RADAR altimeter return also provides an estimate of wind speed at the ocean's surface accurate to around ± 2.0 m/s.

It is important to remember that all of these measurements are from remotely sensing platforms and thus are not direct observations of a variable. Thus considerable validation with "ground-truth" data is needed to determine the levels of accuracy and reliability possible with the remote measurements. Still the advantages of synoptic sampling of large regions makes it worthwhile to understand and use these data.

There are many other more specific satellite-based observational systems that are not detailed here. A good review of the field can be gained by reading *Methods of Satellite Oceanography* (Stewart, 1985) which is listed in the Suggestions for Further Reading. Students interested in a more detailed exposure to the field of ocean remote sensing are encouraged to seek out the many texts presently available that address themselves to specific aspects of satellite remote sensing such as passive and active microwave sensing. There are also a few other texts that address themselves specifically to the topic of satellite oceanography.

6.326 Other satellite applications in oceanography

Satellites play important roles in ocean research other than in the remote sensing of surface parameters. One important function again of polar-orbiting satellites is the collection of data from instrumented platforms in the ocean. The most useful of these platforms are freely drifting buoys which are usually instrumented to measure parameters such as surface and sub-surface temperatures, atmospheric pressure, etc. The system which interrogates these instrumented platforms is called ARGOS and is a French built instrument operated by a French company called "Service ARGOS". This system is capable of accurately (± 1 km) locating the buoy in the ocean using the Doppler shift of the transmitted signal; the ARGOS system can also receive data from the buoys up to 32 data words per transmission.

Another important function fulfilled by satellites is the accurate location of ships at sea. Satellite navigation has been one of the most accurate methods of positioning ships at sea since the early 1970s. The first system used a single shell of polar orbiting satellites to position the ship when the satellite passed overhead, again based on the Doppler shift of the radio signal. Called the NAVSTAR programme, this system was completely supported and operated by the U.S. Navy but was an open system available to ships from all nations everywhere in the world. Commercial receiving units quickly developed into low-cost systems for accurate positioning. Coupled with good estimates of ship's speed and heading the satellite navigation system could provide an excellent means for mapping the course of a vessel at sea. Used in conjunction with a shorter-range accurate system, such as LORAN, the satellite navigation system provided a very precise check. At low latitudes, satellite position fixes were possible every couple of hours depending on the number of satellites in operation. At higher latitudes, where the orbits of the polar orbiting satellites nearly overlap, fixes were much more frequent.

Recently a new Global Positioning System (GPS) has been introduced that uses three shells of navigational satellites to provide continuous access to three satellites simultaneously. Having three independent Doppler shifts makes it possible to fix the ship (or other platform) position very accurately (± 10 m) in both the horizontal and vertical directions. Thus this system can also be used for aircraft navigation to report altitude as well as geographic position. At present the GPS is only partially operational and the full complement of 18 satellites has not yet been deployed. It is interesting that the Navy had planned two levels of signals from the GPS satellites. One would be open to the public and would be less accurate than the encrypted code. Only the less accurate signal has been in operation for the past two years and researchers have demonstrated that by using differential GPS positioning it is possible to achieve very high accuracies with the less-precise public signal. Differential GPS uses some fixed position transmitters to compute the target position relative to an accurately known reference location, thus increasing the accuracy of the position solution. Future deployment of the GPS system will make it possible to routinely position both research vessels and drifting buoys to a high level of accuracy.

6.327 The future

Satellite remote sensing is about to embark on a new era with the proposed Earth Observing System (EOS) of the U.S. National Aeronautics and Space Administration (NASA). As part of the U.S. national Global Change research programme the EOS project is an effort to bring a comprehensive suite of new remote sensing instruments to bear on the study of the earth as a system and of its changes. Part of that earth system is the ocean and any EOS programme must include the study of the ocean as a part of the global whole. Central to the EOS project are two large polar-orbiting spacecraft called "platforms". These will carry a large complement of multiple instruments for measuring earth phenomena on the surface and in the atmosphere. The total EOS plan also calls for three additional geostationary platforms to provide shorter time-scale sensing of the earth's surface and atmosphere. In addition to these U.S. sponsored spacecraft there are plans for two European supported polar platforms and two polar platforms provided by the Japanese. To augment the large platforms there are plans for a series of smaller free-flying satellites to carry specific sensors such as an ocean colour sensor.

6.4 Age of ocean water

The term *age* as applied to ocean water means the time since the water mass was last at the surface and in contact with the atmosphere. The importance of this characteristic is that it gives some indication of the rate of overturn of ocean water. This is of interest in connection with the rate of replenishment of nutrients in the upper layers and of the use of the ocean basins as places to dump noxious material, particularly radioactive waste. If the average time of overturn is much less than the half-life of such materials it would be dangerous to dump them in the ocean because they would be brought to the surface while still active and might be picked up by fish and so conveyed back to man. We could easily calculate the age of ocean water if we knew the speed of travel of deep currents but this is exactly what we know so little about. For this reason, attempts have been made to develop other methods for estimating the age, and some of these will be described.

One method, devised by Worthington (1954), was to use the rate of consumption of dissolved oxygen. He observed that at depths of about 2500 m (in the North Atlantic Deep Water) the average oxygen content decreased by

0.3 mL/L between 1930 and 1950. He attributed the decrease to steady consumption of oxygen by chemical combination with detritus, the rate being 0.3 mL/L in 20 years = 0.015 mL/L per year. Assuming that the water had been saturated with oxygen when it was last at the surface it would have decreased in oxygen content from its saturation value of 7.6 mL/L at the surface to the 5.8 mL/L observed in 1930 in (7.6–5.8) mL/L ÷ 0.015 mL/L per year = 120 years (assuming the consumption rate to have been constant). This implied that the water was last at the surface in 1810. This was in a period of very cold climate and Worthington suggested that much of the present North Atlantic Deep Water may have been formed cataclysmically at that time, and that relatively little had been added since. This use of the oxygen-consumption rate to measure the age has been criticized on the grounds that the accuracy of measurement of dissolved oxygen may not be sufficient for the figure of 0.015 ml/L per year to be very reliable, and on the grounds that the assumption of constant oxygen consumption may be unreal. However, the method is a legitimate attempt to wrest information from the distribution of properties.

Another method, first reported for ocean water by Kulp et al. (1952), was to use the decay rate of ^{14}C in deep waters. It was assumed that the atmosphere at the sea surface was the only source of ^{14}C to ocean waters. Away from the surface the ^{14}C content would not be replenished and it would decay with its half-life of about 5600 years. The early measurements suggested an age of the order of 2000 years for water at 2000 to 5000 m in the North Atlantic, but this was subsequently shown to be too high on account of contamination in the chemical processing. The most recent estimates for replacement or residence times in the ocean waters deeper than 1500 m, based on GEOSECS studies (to be described later in this section) of ^{14}C distributions (Stuiver et al., 1983) are: Atlantic, 275 years; Pacific, 510 years; Indian, 250 years; and for the world ocean as a whole, 500 years. These times correspond to an average upward motion at the thermocline level of: Atlantic, 4 m/year; Pacific, 5 m/y; and Indian, 10 m/y; and annual contributions from the Antarctic Circumpolar waters of 25 Sv to the Pacific and 20 Sv to the Indian Oceans, while the Atlantic contributes 4 Sv to the Circumpolar Water. (1 Sv = 1 sverdrup = 10^6 m^3/second is a commonly used unit for expressing volume transports of ocean currents. The "sverdrup" is not an SI unit but will be used for brevity.) These estimates are related to rates of production of North Atlantic Deep Water of 30 Sv and of Antarctic Bottom Water of 20 Sv. (These are names for distinct water masses to be described in Sections 7.35 and 7.231, respectively.)

Apart from the experimental uncertainty in the method of measuring the ^{14}C activity in a sample of sea-water (a 200-litre sample is needed) there are complications associated with exchange between living and dead matter in the sea and even greater ones resulting from changes in ^{14}C concentration in the atmosphere in the present century. The rapid increase in the use of fossil fuels since about 1900 has reduced the ^{14}C concentration in the atmosphere

Instruments and Methods

because the CO_2 from the fuels contains mostly the stable isotope ^{12}C, the fuels having been away from the atmosphere for so long. In the opposite direction, since 1954 the effect of nuclear explosions in the atmosphere has been to cause a huge increase in ^{14}C.

Measurement of the ^{90}Sr content of ocean water has revealed significant amounts at depths to 1000 m. As the only source of this isotope is presumed to be the residue from atom bombs, starting in 1954, this indicates that the rate of vertical mixing in the upper waters may be quite rapid.

Another radioactive tracer which is being used is tritium (^{3}H) with a half-life of about 12 years. It occurs in the upper layers of the ocean at concentrations of the order of only one tritium atom for 10^{17} or 10^{18} ordinary hydrogen atoms (^{1}H) but with new techniques it can be measured quantitatively for age determinations of sea-water which have been routinely measured in the systematic redetermination of the distribution of the main elements in the world oceans in the GEOSECS (Geochemical Ocean Sections Study) which was a multi-national, multi-institutional study of the main oceans whose objective was the study of the geochemical properties of the oceans with respect to large-scale circulation problems. It was a part of the International Decade of Ocean Exploration.

6.5 Graphical Presentation of Data

6.51 Variation in space; profiles and sections

6.511 Vertical direction

The synoptic oceanographer cannot rely on the distribution of one property alone to determine the full story of the ocean circulation. The discussion in Section 7.35 of the four sections of Fig. 7.15 will make this apparent. He must use as much independent information as he can obtain, but he must be careful that his interpretation of the circulation is consistent with all the property distributions.

The distribution of properties with depth is best shown on the temperature/depth or salinity/depth *profiles* (e.g. Figs. 4.4 and 4.11) which are usually drawn as the first stage in examining oceanographic data. When taking oceanographic stations at sea the oceanographer tries to make observations at standard depths to facilitate comparison between stations. However, the drift of the ship and effects of currents generally cause the oceanographic wire to be at an angle to the vertical and the observations often come out at other than the desired standard depths. If this is the case it is possible to interpolate on the profiles to get values at the required standard depths. A more sophisticated way to do this is to feed the "raw" (observed) data into a computer which has been programmed to interpolate between the observed points according to some standard procedure and to give out the values at the required depths.

The computer is usually programmed to do more than this. Not only does it interpolate for the desired depths but prepares, as output, a complete master sheet with observed and interpolated data, station position, date, meteorological observations, etc., in a standard format ready to be duplicated for the *data report* which is one of the first products from an oceanographic cruise. This data report is simply a list of the numerical values for the observations made at sea, corrected for instrument errors, etc., but without any attempt being made to interpret them.

Vertical profiles are one way of displaying data from a single oceanographic station. For a number of stations, and when other parameters enter, such as time or a range of geographical positions, it is necessary to use further displays. For example, in Fig. 4.6(a) are presented vertical profiles of temperature at a fixed station for a number of months. The variation of temperature as a function of time is then shown in two ways in Figs. 4.6(b) and 4.6(c). These plots were discussed in Section 4.24.

Vertical sections, such as those of Fig. 7.12, 7.15, etc., are used to display an aspect of the geographic distribution of properties in the vertical direction, e.g. along a line of stations, usually a straight or substantially straight line. Such sections, if well chosen, can be useful in showing the path followed by a water body and the changes which take place along such a path. The "core method" developed by Wüst (Section 6.543) uses this display.

6.512 Horizontal direction

The horizontal *sections* (*maps*) of properties such as those of Figs. 4.1, 4.2 and 4.9 are most useful in displaying the geographical distributions of individual properties. They are not generally used to show the interdependence of the properties because it is difficult to show more than two properties simultaneously on one diagram, even using multiple colours. Even with only two properties, the relations between them cannot be appreciated as well, for instance, as on the characteristic diagrams described in Section 6.54.

6.52 Variation in time

6.521 Time-series plots

To display the variation of properties as a function of the fourth parameter, time, we may initially use plots which are basically similar to those already described for spatial variations. The commonest is the *property time-series* in which the property value is plotted along the y-axis and time along the x-axis, e.g. Fig. 6.9(a). To show the relations between the time variations of several properties, e.g. temperature, salinity, such time series plots may be arranged one below another down a page with the time scales aligned vertically. A variation is a *time-series of profiles* in which a temporal series of profiles is

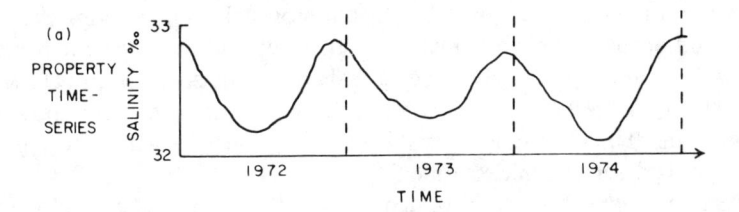

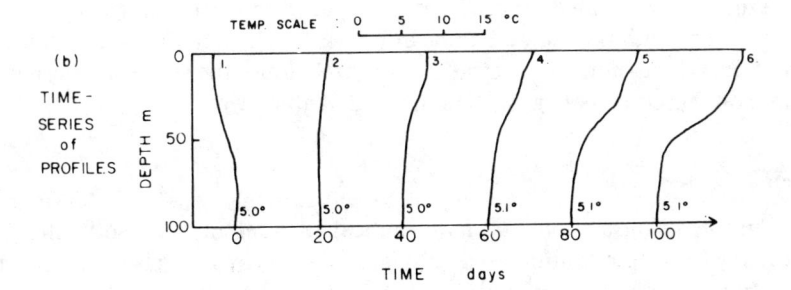

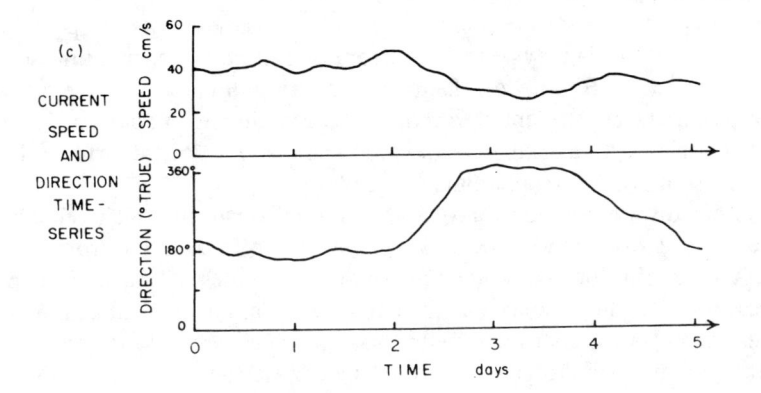

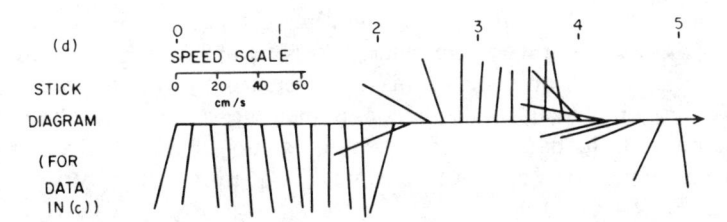

FIG. 6.9. Examples of time-series plots: (a) property/time, (b) time-series of profiles, (c) current speed and direction, (d) stick diagram for data of (c) above.

plotted across a page at positions along a time axis corresponding to their time of observation, e.g. Fig. 6.9(b). (The location of such profiles along the time axis would probably be made relative to the deep end of the profile because temporal variations are likely to be least there.) If the successive profiles are markedly different, they may be plotted on a single vertical profile diagram as in Fig. 4.6(a). The simultaneous display of two properties with time, e.g. the (TS_t) plot, is described in Section 6.544.

A *time section* plot would have depth along the y-axis and time along the x-axis with numerical values of the property initially entered on the plot and then contoured as in the isotherm plot of Fig. 4.6(b).

A more sophisticated procedure is to prepare *spectral analysis* of property fluctuations with time to determine any periodic characteristics, e.g. diurnal, tidal. For discussions of spectral analysis techniques the reader is referred to more advanced texts (e.g. see Jenkins and Watts, 1969).

6.552 Current data plots

Current being a vector, is more difficult to present because it must be described by two quantities, magnitude (i.e. speed) and direction, in contrast with the scalar quantities considered so far (e.g. temperature, salinity) which have magnitude only.

For observations at one point, a pair of time-series plots of speed and direction respectively may be presented, one above the other, with time scales aligned vertically (Fig. 6.9(c)). Data from a string of current meters at one mooring can then be presented with the speed and direction of plots one above another. Related data such as wind speed and direction or tidal rise and fall may be displayed on parallel time-series plots.

Another display method for vectors is the *stick plot* in which velocity is represented in speed and direction by a line drawn to scale out from a time axis. A series of velocities is then represented by a series of "sticks" drawn in succession from the time axis at their time of observation as in Fig. 6.9(d). Again, currents from a string of meters on a mooring can be plotted one above another down a page to give an idea of the correlation between depths.

An alternative procedure is to represent the velocities as time-series of two *rectangular components* either (a) east-west and north-south or (b) parallel and perpendicular to a shore-line or the centre-line of a passage.

Verticle sections of current components perpendicular and parallel to the geographic direction of a section may be prepared by entering numerical values of the components on the sections at their measurement positions and then contouring the data. The north–south equatorial section of east–west current components in Fig. 7.33.U showing the Equatorial Undercurrent is an example.

Horizontal maps of current observations are often used, as in Fig. 7.9 and 7.31, to show compilations of current observations or to show the current pattern derived from the dynamic topography of a surface as in Figs. 6.6 and 6.7.

Time-series current data are usually subjected to spectral analysis to determine the periodic components, e.g. tidal or due to internal waves, leaving the residual non-periodic components.

6.53 Isentropic analysis

Although it will not be used in this introduction, a student who goes further in descriptive oceanography will soon come across another technique, analogous to horizontal sections, which was introduced for oceanographic use by Montgomery (1938) and referred to as *isentropic analysis*. He argued that the flow of ocean waters may be expected to occur most easily along surfaces of constant entropy. Because of the complex thermodynamic nature of sea-water it is difficult to determine surfaces of constant entropy as such and Montgomery presented arguments to show that surfaces of constant potential density (σ_θ) would be a close approximation, while for the upper layer of the ocean the more easily determined surfaces of constant σ_t (which we will call isopycnal surfaces) would be adequate. He argued, therefore, that flow should occur preferentially along isopycnal surfaces and that mixing between water masses would occur more freely parallel to these surfaces (called "lateral" mixing) than across them (which would be hindered by stability). A comparison of Figs. 7.15 and 7.44 shows that flows indicated by cores (see Section 6.543) of salinity and of oxygen are approximately along σ_θ surfaces in the former or σ_t surfaces in the latter. (In the deep water of the Atlantic it is better to use σ_4 surfaces as discussed in Section 7.352.) The procedure is first to plot T–S, T–O_2, etc., diagrams (see the next section) for each station in the region under study. Then for each selected σ_θ or σ_t the salinity and temperature at each station are read off the T–S diagram, and the oxygen value for the temperature is read off the T–O_2 diagram. The salinity and oxygen values are plotted on charts of the region and the data contoured. The tongues of extreme values of properties then show flow directions, and mixing is indicated by changes in values of the properties along the flows. As an example, Montgomery applied this method to study the flow patterns in the southern North Atlantic, and showed that in the upper few hundred metres, the flow patterns on σ_t surfaces as delineated by salinity and dissolved oxygen distributions were generally simpler and more continuous than on surfaces of constant geometrical depth. (The σ_t surfaces are generally not horizontal over substantial distances and so surfaces of constant geometrical depth cut through them and therefore through the flow patterns.) This method of presenting property distributions on constant σ_t (or constant $\Delta_{S,T}$) surfaces has been widely used, although it is now recognized that there are situations in which mixing across σ_t surfaces does occur. Notice that as σ_t is determined principally by temperature in the upper layers of the ocean, a plot of temperature on σ_t surfaces is redundant.

6.54 Characteristic diagrams

6.541 Two characteristics, e.g. T–S, T–O_2

Although one can plot separate vertical profiles and sections of temperature, salinity, dissolved oxygen, etc., one must be careful not to consider the distributions of these properties to be independent. Physically the properties are independent of each other in the sense that in a sample of sea-water it is possible to alter the value of any one without altering the others. (Note that density is not included in the above list because it is dependent on temperature and salinity.) However, observations indicate that in the oceans the properties do not occur widely in all possible combinations. The reason for this is that most ocean water masses acquire their characteristics at the surface of the sea in particular localities. The water properties are determined there by the local climate, and when the water sinks along density surfaces it carries these properties with it. The result is that instead of all possible combinations of temperature, salinity, oxygen, etc., occurring in the ocean, we find that only a limited number of particular combinations occur, and these occur in different regions. In consequence, we can often recognize a water mass by its characteristic combination of water properties.

To show these combinations, it is usual to plot *characteristic diagrams* of the water properties. Helland-Hansen (1916) first suggested this technique by plotting temperature against salinity (T–S plots) for individual oceanographic stations, in addition to the separate plots of temperature, salinity, etc., against depth. To prepare a T–S plot for a station, a point is plotted for the temperature/salinity combination for each bottle depth, or for a selection of depths from CTD data, and the points then joined in order of depth by a smooth (but generally not straight) line which is called the *T–S curve* for that station. (Figures 7.19 and 7.20 contain smoothed, average examples for the main oceans; they will be discussed in Chapter 7.) Other characteristic diagrams such as T–O_2, S–O_2, O_2–phosphate, etc., may also be plotted.

It is not necessary to plot density as a property on a characteristic diagram because it depends on temperature and salinity and therefore appears automatically on the T–S diagram. Each point on the diagram corresponds to a particular combination of temperature and salinity and therefore to a particular density. However, the same density may be attained by different combinations of temperature and salinity, and these combinations lie on a smooth curve on the T–S plotting sheet and can be drawn in as shown by the full lines on Fig. 3.1 or 6.10.

When discussing the T–S diagram, a water body whose properties are represented by a point is called a *water type*, while one represented by a line is called a *water mass*. These are ideal definitions. In practice the points on a T–S diagram representing a realistic water type would have some scatter about the ideal point while the points representing a water mass would have some scatter

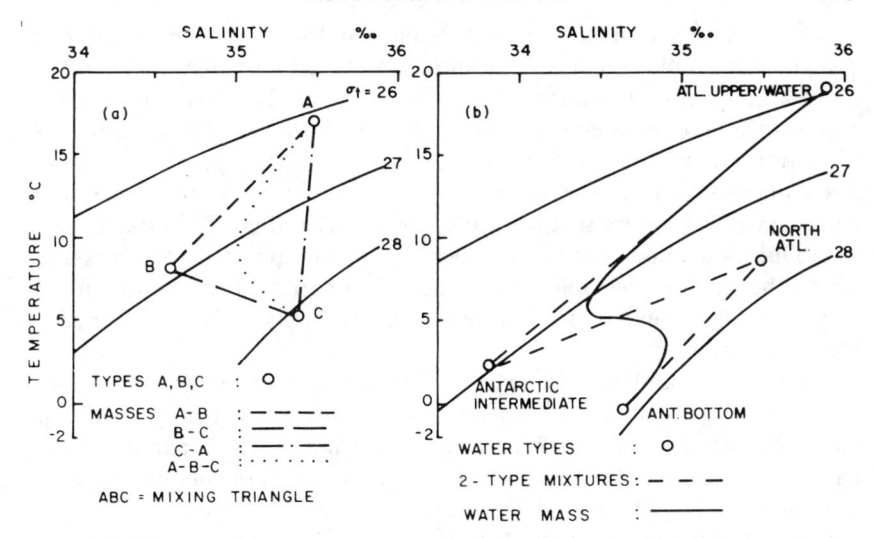

Fig. 6.10. (a) T–S diagram showing three water types and four water masses, (b) example of realistic T–S water mass diagram (Atlantic) and basic water types.

about an ideal line. Some examples of these scatters will be given when discussing Pacific and Atlantic Ocean water masses in Chapter 7 (e.g. Fig. 7.21). Climatic processes tend to form water types at the surface; a water mass results from the mixing of two or more water types. For instance, in Fig. 6.10(a) the small circles A, B and C represent arbitrary water types while the straight dashed lines AB, BC and CA represent water masses made up of mixtures of A and B, B and C, and C and A respectively. The curved dotted line represents one example of a water mass made up of the three types A, B and C. Lines within the *mixing triangle* ABC represent all possible masses from types A, B and C. This interpretation implies that temperature and salinity are conservative quantities in the ocean, i.e. that no processes exist for generating or removing heat or salt. This is true within the body of the ocean, but near the surface it is not so. Here, the sun may heat the water or evaporation cool it, rain may decrease the salinity or evaporation increase it. Therefore in plotting the T–S diagram it is usual to regard the points corresponding to shallow depths within the influence of the surface effects as less conservative, or even to omit them altogether.

The T–S diagram turns out to be a powerful tool for the study of ocean waters (e.g. see Mameyev, 1975). The shape of the T–S curve is often characteristic of water from a particular locality in the ocean, and individual features of the curve may indicate mixtures of different types of water. For example, the Atlantic water masses could be represented to a close approximation as a mixture of four types as shown schematically in Fig. 6.10(b) which may be compared with mean T–S curves for that ocean in Fig. 7.20.

Salinity minima or maxima such as B and C in Fig. 6.11(a) are common on T–S diagrams but, except for the surface and bottom values, temperature minima or maxima are uncommon. The reason is that except in polar or coastal regions, density depends chiefly on temperature and less on salinity. A temperature minimum, such as in Fig. 6.11(b), would imply that the water below the minimum, i.e. toward H, was less dense than that at the temperature minimum and therefore that the water between G and H was unstable. Such instabilities are found occasionally, such as in the complicated water masses at the northern side of the Gulf Stream where the Labrador Current joins it, but only over very limited depth ranges and probably only as a transient phenomenon.

On the other hand, in polar regions where the salinity and temperature are both low the density depends chiefly on the salinity. Under these conditions the isopycnals (σ_t) on the T–S diagram are more nearly parallel to the temperature axis as in Fig. 6.10(c) and a temperature minimum (K) or maximum (L) is quite possible from the surface to the deep water.

The T–S curve is also useful for checking data to determine if individual points may be in error. If a point lies well off the expected curve for the region one tends to regard it with suspicion.

The T–S diagram, however, suffers from two disadvantages. One is that it gives a poor indication of the distribution in depth of the different water properties because the depth scale along the T–S curve not linear (see Figs. 7.19, 7.20 or 7.28(b)). For this reason it is not possible to interpolate temperature or salinity values for depth on a T–S curve, except very roughly. The other disadvantage is that the T–S diagram gives no indication of the spatial distribution of water properties; this is better done on vertical or horizontal sections.

6.542 Global T–S characteristics of ocean waters

In Chapter 7, Figs. 7.19 and 7.20, which show mean T–S curves for some of the main ocean water masses, are based on the curves presented by Sverdrup et al. (1942). Since that publication, many more oceanographic observations have been collected and analysed, and Emery and Meincke (1986) have reviewed and summarized the available material published between 1936 and 1982 to present a revised table of water masses and sets of T–S curves for the three oceans. In their tabulation, they recognize nineteen masses in the upper waters (0–500 m), twelve intermediate masses (500–1500 m) and four deep and abyssal waters (1500 m to bottom). They present a water-mass location map for each of these layers and T–S curves for most of the water masses for each of the oceans, Atlantic, Indian and Pacific.

As our present text is intended as an introduction, we have limited ourselves to the simpler Sverdrup T–S plots (which are not substantially different from the corresponding updated ones). A student desiring more detail is then

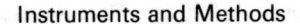

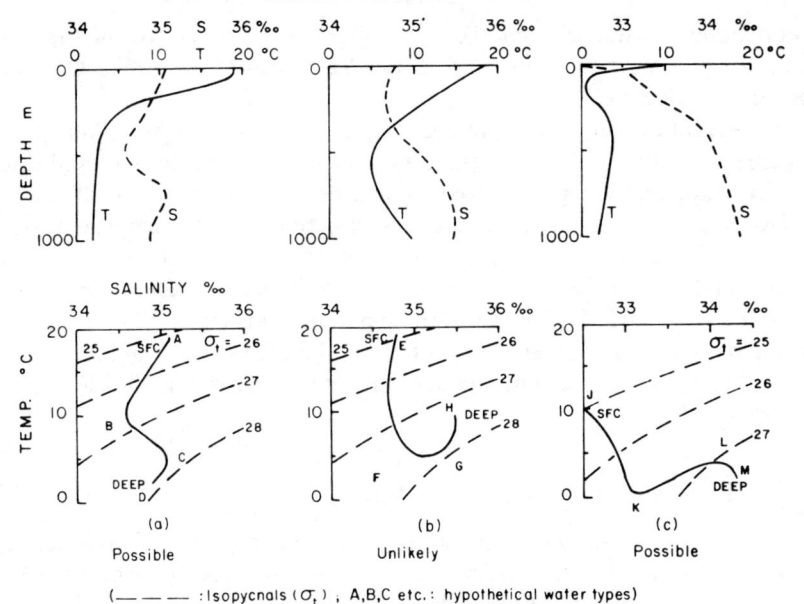

FIG. 6.11. Temperature and salinity profiles and corresponding T–S diagrams.

referred to the figures in Emery and Meincke (1986) and to their tabulation which names the various water masses and gives numerical values for their temperature and salinity ranges based on the more extensive data which they reviewed.

6.543 Core method

Tongue-like distributions of water properties in the vertical plane are very helpful in indicating the direction of movement of the water. Wüst (1935, 1957) developed the *core method* as a technique for determining water-flow paths in the ocean. A "core" is defined as a region where a water property reaches a maximum or minimum value within a tongue-like distribution, e.g. the salinity minimum of the Antarctic Intermediate Water or the salinity and oxygen maxima of the North Atlantic Deep Water (Fig. 7.15). A vertical section of the water property is used to locate such a core which is taken as the centre of the flow. A core gradually weakens along its length as a result of mixing with the surrounding water. A T–S curve may be plotted from the temperature/salinity values for a series of samples of water along the core itself from its start at the region of formation of the water type, usually at the surface, to its end where it can no longer be detected. The point on the diagram representing the original water type is taken to represent 100% concentration and the end point represents 0% concentration of the original type. The position of any point on the T–S line then represents the proportion of original water remaining by its proportionate distance from the starting point. Another use of the core method for tracing the flow of Atlantic Water round the Arctic Ocean is described in Section 7.522.

Care must be taken in interpreting tongues or cores in the horizontal plane as being indicative of flow direction. If the water property does not influence density significantly, e.g. dissolved oxygen or small amounts of silt, then a horizontal tongue may well indicate flow direction but a temperature tongue may not. This is because temperature has a strong influence in determining density and, as can be appreciated from the discussion of the geostrophic method (Section 6.255), flow tends to be parallel to isopycnals, i.e. along the isotherms outlining a tongue and not along the core which would be across the isotherms/isopycnals. The flow should therefore be around the tongue, not along it.

6.544 Three characteristics, e.g. T–S–V, T–S–t

An interesting development of the T–S diagram was developed by Montgomery (1958) and used by him and his colleagues (Cochrane, 1958; Pollak, 1958) to display the distribution of temperature and salinity of the ocean waters in proportion to their volume. Essentially a T–S plotting sheet was divided into a grid of squares, say 2K by 1 unit in salinity. On each square was entered the volume of water whose property values lay within those of the square. For this purpose the potential temperature was used rather than the *in situ* temperature. The volume information was arrived at from observed oceanographic data in the following manner. Oceanographic stations were selected for which measurements of water properties extended from the surface to the bottom. The stations were as evenly spaced as possible and each was taken to represent a horizontal area around it so that the sum total of all the station areas was equal to the total area of the ocean being considered. At each station, the temperature and salinity observed at each depth were taken to represent the layer of water extending from half-way to the observation above to half-way to the observation below. The product of the interval of depth with the horizontal area represented by the station gave the volume of water for which those values of temperature and salinity were representative. This volume was entered in the appropriate grid square on the T–S diagram. This procedure was carried out for all the temperature/salinity values for all the stations and then the volumes in each square were totalled. The final result was the gridded volumetric T–S diagram in which each square showed the volume of water in the ocean having the temperature and salinity values of that square. It will be called a T–S–V diagram. On a typical such diagram, some squares have large volumes and many have none at all.

As an example of the result, a T–S–V diagram for the Atlantic and adjacent seas is shown as Fig. 6.12(a). This figure is slightly simplified from Montgomery's original diagram; the unit of volume is 10^5 km^3. The number in

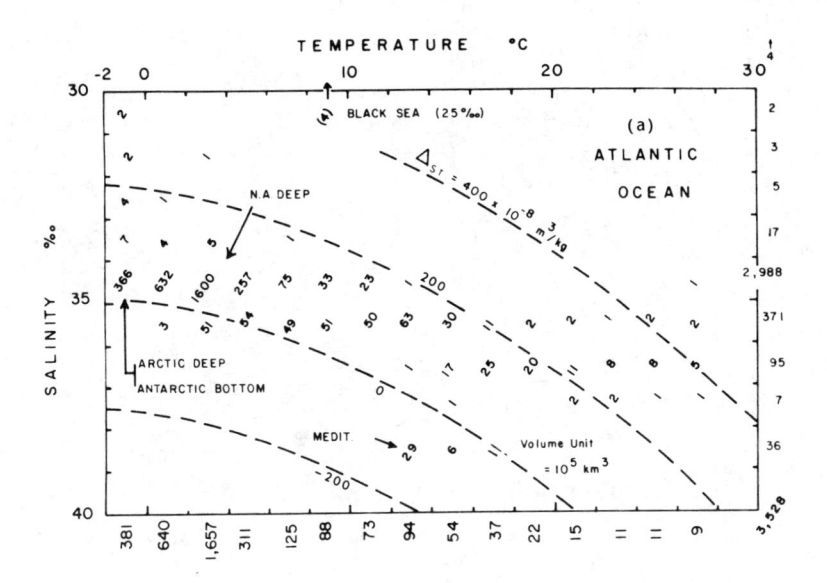

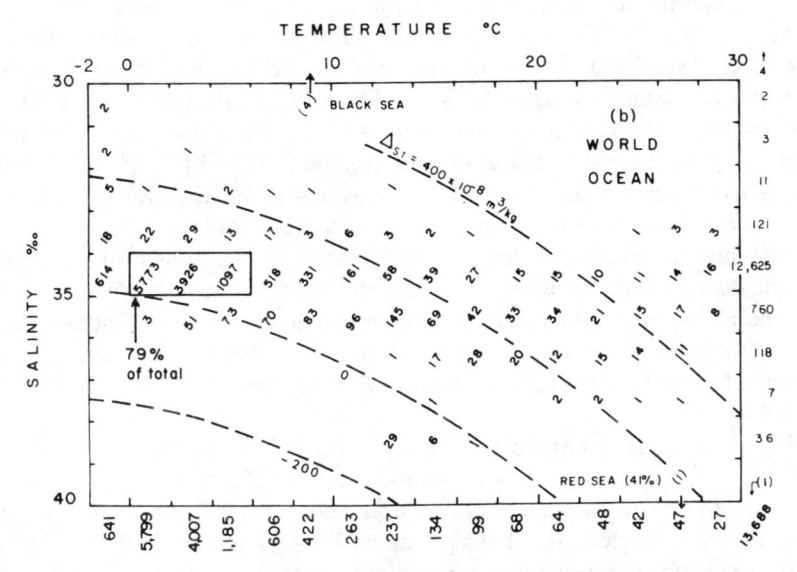

FIG. 6.12. Temperature–salinity–volume (T–S–V) diagrams for the Atlantic and the World Ocean.

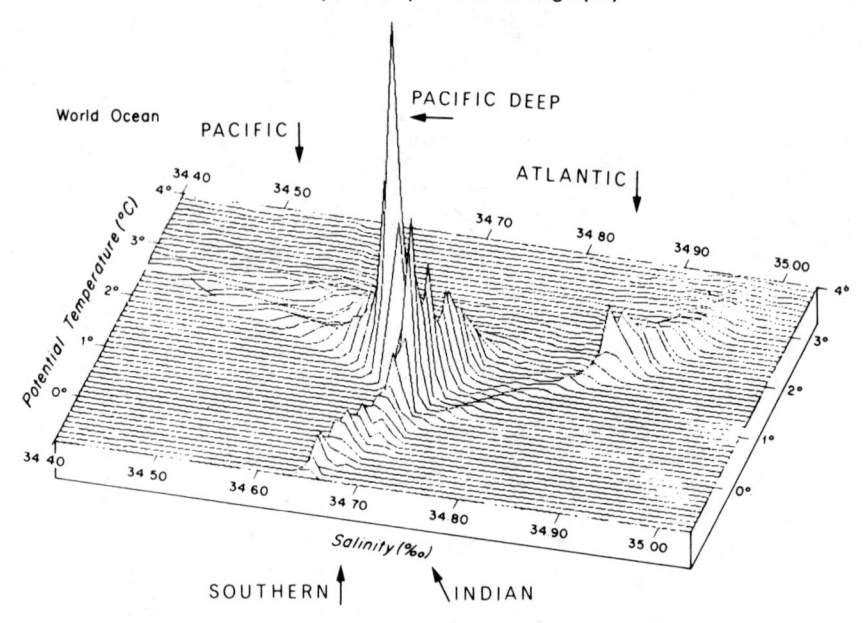

FIG. 6.13. Simulated three-dimensional T–S–V diagram for the cold water masses of the World Ocean.

each square then shows the number of these units of volume which have the temperature and salinity within the range of that square. It is seen that the water properties are by no means uniformly distributed but that there is a concentration of volume near 2°C and 35. This North Atlantic Deep Water, with a volume of 1600×10^5 km^3, is seen to comprise some 45% of the total volume of the Atlantic (3528×10^5 km^3). The Antarctic Bottom Water and the Arctic Deep have similar characteristics and fall in the same square on the diagram (near 0°C, 35) with a combined volume of 366×10^5 km^3. Two other distinctive water masses are the Mediterranean Water with characteristic high salinity and temperature, and the Black Sea water with low salinity.

The numbers outside the axes of the diagram are the sums of the volume numbers in the horizontal rows or in the vertical columns respectively on the diagram. Down the right-hand side, this series of numbers shows that the great bulk (85%) of the Atlantic water has a salinity between 34 and 35. In temperature, 47% of the water is between 2° and 4°C and 76% is between $-2°$ and 4°C.

Figure 6.12 has examples of coarse-scale T–S–V diagrams. Fine-scale diagrams were also prepared with grid squares of 0.5°C and 0.1 side to show more detail (see the Montgomery, Cochran and Pollak references) and, more recently, Worthington (1981) has presented T–S–V diagrams with 0.1°C and 0.01 squares for the deep waters. Some of his results are shown below.

Similar diagrams were prepared to show the T–S–V characteristics of the

Instruments and Methods

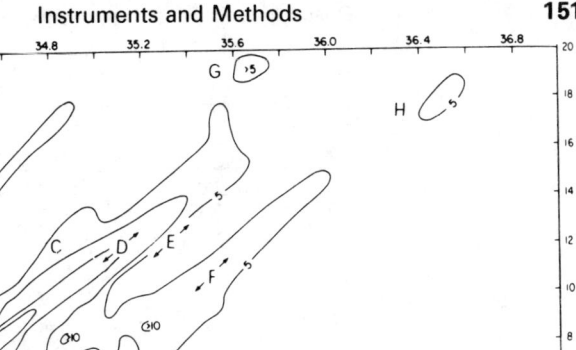

FIG. 6.14. Contoured T–S–V diagram for the warm water masses of the World Ocean: (A) Subantarctic, (B) North Pacific and Eastern South Pacific, (C) South Pacific and Subtropical North Pacific, (D) South Atlantic, South Pacific and Indian, (E) South Atlantic and Indian, (F) Western North Atlantic, Red Sea Water in Indian Ocean, (G) South Pacific, (H) 18° Water in Western North Atlantic. (The black shaded area at the bottom represents the cold water of Fig. 6.11.)

world ocean (Fig. 6.12(b)). Montgomery pointed out that the mean values calculated from the 1958 statistics did not differ significantly from the values determined by Krümmel in 1907. One of the chief reasons for the small differences between Krümmel's averages based on few data and Montgomery's based on many more is the small range of temperature and salinity found in the bulk of the oceans. The oceanographic "climate" is so uniform below the surface layer that comparatively few observations sufficed to enable Krümmel to make a good estimate of the mean values. It must not be concluded, however, that little had been gained in oceanographic knowledge between 1907 and 1957. Mean values conceal much detail, and the very fact that deep-water characteristics have a small range of values makes it difficult to distinguish one water mass from another. In fact, the development in recent years of the electrical conductivity salinometer, with its increased sensitivity and precision, has helped a great deal to distinguish between different water masses, even though the increased sensitivity may not result in much change in mean values. Worthington's 1981 census of the world ocean water masses was based on such high-quality salinity data, which is presumed to be the reason why some of his volume figures differ somewhat from those of Montgomery. Worthington pointed out that while the North Atlantic had plenty of high-quality stations, they were sadly lacking in a large part of the Pacific and in the South Atlantic and South Indian Oceans.

152   Descriptive Physical Oceanography

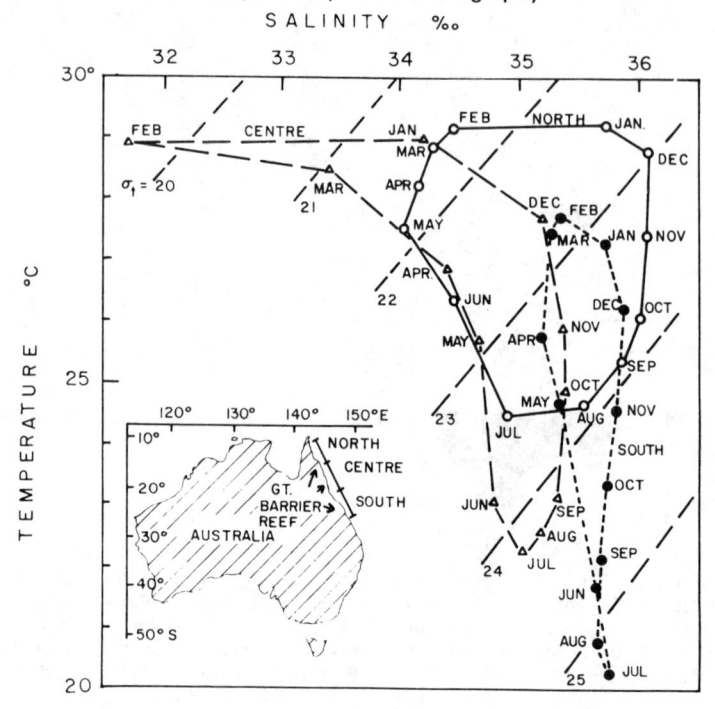

Fig. 6.15. Temperature–salinity–time (T–S–t) diagrams for lagoon waters inside the Great Barrier Reef.

The usual way to present the T–S–V diagram on paper is to display temperature and salinity along two axes and volume as numbers, as in the examples above. As the diagram presents relations between three quantities we would like to present it in three dimensions. It would not be difficult to build up a three-dimensional T–S–V model by placing on each square a column whose height represented volume, although such a model would obviously only be suitable for demonstration purposes. An alternative nowadays is to show a computer-generated simulation view of such a three-dimension model. Figure 6.13 is an example of such a simulation for the cold waters of the oceans from the new census of world ocean waters by Worthington (1981) for classes of $0.1K \times 0.01$. The ridge extending to the lowest temperatures at about 34.6 represents water in the Southern Ocean, the ridge extending to high temperatures and salinities is for the Atlantic, while the Pacific Ocean water is represented by the ridge extending toward 3°C and 34.4. The central spike, representing about 2.6% of the cold water is for the deep water in the North Pacific. Indian Ocean waters are represented in the central ridge from about $-0.5°$ to $2°C$.

While Fig. 6.13 shows the volume distribution of the cold (deep) waters, Fig. 6.14 shows the distribution of water properties for the upper, warm ($>4°C$)

water which constitutes about 24% of the total ocean volume. The contours show the volumes (in 10^3 km^3) in classes of 0.1K × 0.01. The shaded area at the bottom represents the remaining 74%, the cold water shown in Fig. 6.13. The ocean areas where much of the warm water occurs are indicated on the three prongs of the contours in the figure.

Before leaving this description of the T–S–V diagrams, the reader should note that the temperature used, although indicated by T in the abbreviation, is usually the potential temperature θ as the water volumes usually come from a wide range of depths.

The last form of three-dimensional characteristic diagram considered is the temperature–salinity–time (T–S–t) diagram which is a compact way of showing the sequence of combinations of water properties with time. As an example, Fig. 6.15 shows monthly mean values for three zones of the Australian Great Barrier Reef lagoon (Pickard, 1977). In the south, the annual variation is mainly in temperature, in the north there are large variations of both temperature and salinity, while in the centre zone there is an extreme salinity variation. The reason for the differences is that the north and centre zones are subject to heavy monsoonal rains in the austral summer (January to April) while the south zone escapes these. The very low salinity in the centre is due to the rivers there which drain much larger inland areas than do the smaller rivers in the north. The temperatures used for these surface T–S–t diagrams are the *in situ* values t.

6.55 Conclusion

From the above descriptions, it will be seen that it is not practicable to recommend any one form of graphical presentation as the standard for oceanographic data. An oceanographer who wishes to present his data and his analysis of it in a report has to decide for himself which form best displays the features which he considers to be important. He may well have to use all of the methods mentioned to make all his points, or he may have to invent new ones.

CHAPTER 7

Circulation and Water Masses of the Oceans

WE NOW come to the circulation of the oceans, first discussing some of the general features of the main ocean circulations and then proceeding to describe the circulation and character of the water masses in the individual oceans and in some seas.

7.1 Introduction; mechanisms

It must be stressed that in one way or another the energy from the sun's radiation is responsible for these circulations. In fact, both the atmospheric and the oceanic circulations are driven by this energy and we should perhaps really study both together (e.g. Perry and Walker, 1977). As our book is primarily about the ocean we will concentrate on the water circulation but it will be apparent that the atmosphere plays a considerable part in driving ocean currents.

The ocean circulation can be divided into two parts, the thermohaline and the wind-driven components. Another way to describe these is to say that the ocean circulation is in part due to changes in density caused by weather or climatic changes and in part due to the wind stress. The manner in which these factors determine the circulation is discussed in texts on dynamical oceanography and all that will be done here is to describe the physical processes in a qualitative fashion.

7.11 Thermohaline circulation

The term *thermohaline circulation* is used to refer to the movement of water that takes place when its density is changed by a change of temperature or of salinity in a suitable part of its bulk. A standard laboratory demonstration of thermal circulation is to take a large beaker of water and heat it at the bottom by means of a bunsen burner. The water which is heated expands and rises (convection) in accordance with Archimedes' Principle. A few crystals of dye dropped into the water will dissolve and colour the bottom water and show it rising over the source of heat and sinking elsewhere. The sinking in this case is simply a consequence of continuity of volume. In the case of the atmosphere the sun's energy is absorbed by the land, this heats the air near to it and a similar circulation to that in the beaker takes place. Cumulus clouds developing on a summer day often reveal where the upward flow of this kind of circulation is taking place.

The laboratory experiment with the beaker of water and the heating of the atmosphere have this in common—the heat is supplied at the bottom of the fluid. The situation is different in the ocean in that the sun's heat is supplied at the top; this makes a great deal of difference. A liquid warmed at its top surface cannot rise and therefore a circulation does not develop. If we were to take a long tank of water and heat the surface at one end, then some motion would take place. The heated water would expand and raise the water level slightly at this end. The warmer surface water would then flow toward the cold end over the top of the cold water but no circulation in the full sense would develop. Some of the early theories of the ocean circulation ascribed the currents to a thermal flow of this sort. The stronger heating of the sea at low than at high latitudes was recognized and it was suggested that the water then flowed north and south toward the poles. While it is possible that the differential heating between low and high latitudes may have some effect of this nature it is now believed to be a minor one.

Since there is no significant source of heat at the bottom of the oceans a thermal circulation like that in the beaker and in the atmosphere plays little part in the main ocean circulation, although very local convective circulations probably take place in the vicinity of hydrothermal vents in the sea floor.

However, heating at low latitudes is only one part of the heat budget. It has already been shown that there is a net heat loss from the oceans at high latitudes. The result of the cooling of the water is an increase in density which may be sufficient to cause it to sink and so displace the deeper water. The laboratory experiment to illustrate this is to float an ice cube in a beaker of water; the water cooled by the ice will sink. (Allow the water to sit in the beaker for a few hours to allow any motion to die out before adding the ice cube, and put a drop of dye on the latter to reveal the cooled water sinking to the bottom.) If a little salt has been dropped to the bottom of the water some hours beforehand, it will have dissolved and increased the density of the bottom layer. Then the dyed water will stop sinking at the top of this "bottom water". The thermohaline circulation of the oceans is due to an increase of density at the upper surface, either directly by cooling or indirectly when ice freezes out, ejecting salt and thus increasing the density of the remaining water. In the North Atlantic the cooling effect in the winter is considered to be responsible for the sinking of water to considerable depths. In the Antarctic, the freezing effect is important. The sea-ice will not be pure ice, as some salt is usually trapped among the ice crystals, but it is usually less saline than the sea-water from which it was formed. The remaining sea-water is thus more saline and more dense than before, and this will cause it to sink. One may summarize then by stating that a characteristic of the thermohaline circulation of the oceans is that it originates as a vertical flow sinking to mid-depth or even to the ocean bottom, followed by horizontal flow.

An increase of salinity may also occur due to evaporation in the tropics but generally this is accompanied by solar heating and the decrease in density due to heating prevails over the increase due to evaporation. The more saline water stays near the surface. Evaporation is therefore not a primary factor in causing water to sink but it does act indirectly as will be described for the Mediterranean and the Red Sea.

Although the details of the *thermohaline circulation* or *ventilation* of the deep ocean are not yet well understood, it is just as important as the upper-layer circulation which is easier to study and to which more effort has been devoted. A generally accepted model suggested by Stommel (1958) as the basic mechanism for the large-scale thermohaline circulation and the maintenance of the thermocline is that sinking takes place chiefly in the northern and southern limits of the Atlantic, and deep water then flows into the three oceans and eventually upwells toward the surface. Two mechanisms causing sinking are now recognized—*near boundary sinking* and *open-ocean sinking*.

The major example of near-boundary sinking occurs in the Weddell Sea portion of the South Atlantic sector of the Antarctic where ice freezes out and rejects salt to increase the salinity and hence the density of the already cold water there. This water sinks along the shelf and down the slope into the South Atlantic. The mixture of cold and relatively saline shelf water mixes further with upwelling Deep Water to form the densest water in the Atlantic, sinking to the bottom and then spreading. Most of this dense water formation takes place over the Weddell Sea shelf but there is also some formed by the same process in the Ross Sea and on the shelves off the Adélie Coast, Enderby Land and Wilkes Land of Antarctica.

In addition, the open-ocean process is effective in the deep water of the Weddell Gyre in the North-eastern Weddell Sea. This rather complex process basically requires a cyclonic circulation and surface cooling and takes place in stages as described for the formation of Mediterranean Deep Water in Section 7.41. There is also the possibility of deep water formation associated with the Weddell polynas described in Section 7.544.

In the high-latitude North Atlantic, the open-ocean mechanism appears to be the main one, chiefly in the Greenland Sea where deep water is formed from the surface by cooling (without ice freezing out) and then sinks and overflows over the sills between Greenland and Scotland into the North Atlantic. It is also possible that surface cooling in the Labrador Sea generates water dense enough to sink toward the bottom by the same process.

The dense waters from the South and North Atlantic sources then flow north and south respectively. For dynamical reasons (conservation of vorticity) these flows are constrained to the western side of the ocean (see Pond and Pickard, 1983, Sec. 10.1). (Actually, the presence of the cold water on the western side of the South Atlantic, suggesting northward flow there, had been observed by Wüst (1957) in his Atlas of the water properties in the Atlantic prepared from the data from the 1925–27 *Meteor* Expedition.) From these deep western flows, slower geostrophic currents spread outward into the main body of the deep Atlantic and supply the cool water required for the slow upward flow over most of the low- and mid-latitudes to maintain the thermocline at a relatively constant depth (Section 4.22). For dynamical reasons again, the directions of these deep geostrophic flows in the ocean interior must be cyclonic (anti-clockwise in the northern hemisphere and clockwise in the southern), i.e. the meridional components of these flows must be directed away from the equator in both hemispheres.

It is more difficult to obtain information on the volume flows of these deep currents than for upper-layer ones but some estimates have been made. In the north it is estimated that the average flow over the Greenland–Scotland sills into the Atlantic is 1–2 Sv (1 Sv = 1 sverdrup = 10^6 m^3/s), which is increased to 10–20 Sv by mixing with North Atlantic Deep Water before flowing south in the eastern North Atlantic and southwest in the western North Atlantic. This amount decreases to one-half by the time it reaches the South Atlantic. The rate of formation of bottom water in the Weddell Sea is estimated at 2–4 Sv which increases to 20–40 Sv by mixing with North Atlantic Deep Water as it flows down the slope into the South Atlantic.

There is no production of bottom water in the Indian Ocean and it is suggested that a northward flow of about 5 Sv from the Antarctic Circumpolar Current occurs along the African boundary and also a flow of about 4 Sv west of Australia along the east side of the Ninetyeast Ridge in the middle of the eastern basin of the Indian Ocean. In the interior, away from the boundary flow, the geostrophic flow south of the equator is clockwise and north of the equator there may be small flows into the Arabian Sea and the Bay of Bengal.

In the South Pacific it is now believed that a small amount of bottom water flows out from the Ross Sea but the main northward flow, again from the Antarctic Circumpolar Current, of some 20 Sv makes it the largest of the deep western boundary currents. In the South Pacific, a basically clockwise circulation flows around that ocean and rejoins the Antarctic Circumpolar Current. There is some possibility, but no conclusive evidence, that the western boundary flow penetrates across the equator into the North Pacific. There is evidence of a southwestward flow in the North Pacific off Japan under the Kuroshio. This is not a result of deep water formation but is required for continuity, the geostrophic flows in that ocean being from tropic latitudes in the west, anticlockwise around the main body of the North Pacific and back to the west, north of Japan.

The upward flow necessary to preserve the thermocline comes from the cyclonic interior flows. Stommel and Arons (1960b), from the Robinson and Stommel theory (1959), estimated a global average upward flow through the thermocline of 50 Sv giving a mean vertical speed of 3 cm/d but felt that this might be too high on account of the assumptions made in that theory and concluded that about 0.5 cm/d would be more likely. The western boundary deep flows all appear to be wider than the corresponding western boundary upper-layer flows.

In addition to the above deep flows of cold water originating in the North and South Atlantic, there are also flows to middle depths in the main oceans of warm waters rendered dense by their high salinity resulting from evaporation in marginal seas. Water from the Mediterranean, made dense by evaporation in the eastern basin, flows westward below the surface along that sea and out through the Strait of Gibraltar into the Atlantic. There it spreads westward across the ocean, being evident as a high salinity tongue between about 500 and 3000 m. The outflow through the Strait is at least 1 Sv which increases to some 10 Sv by mixing with Atlantic water as it flows westward along the shelf and down the slope. In the Red Sea, evaporation increases the salinity in the summer and then winter cooling of the saline upper water causes it to sink to mid-depth as will be described in Section 7.8. Then some of this high salinity water from the Red Sea (and similarly formed water from the Persian Gulf) flows out and southward in the Indian Ocean as a high salinity tongue between 500 and 2500 m. Outflow from the Red Sea is of the order of 0.5 Sv and from the Persian Gulf only about 0.1 Sv.

An increase in salinity occurs by evaporation in the tropics in the open ocean due to solar heating but the increase of density is largely offset by the temperature rise and the more saline water sinks only to moderate depths to give rise to high salinity tropical water masses at depths of only one to three hundred metres.

7.12 Wind-driven circulation

The *wind-driven circulation* is principally in the upper few hundreds of metres and therefore is primarily a horizontal circulation in contrast to the thermohaline one. In texts on dynamical oceanography the mechanism is described by which the wind blowing over the sea surface causes it to move. Here we will just accept this as a consequence of fluid friction, i.e. the air/water frictional stress from the wind acts on the sea surface to cause it to move horizontally, and the communication of this motion downward through the sea by friction causes deeper layers to move in turn. The consequences are on two scales. In the vertical, the effect of the wind stress extends through the upper hundred metres or so. In the horizontal, the main gyral circulations of the oceans are a consequence and extend to an ocean-wide scale (e.g. Figs. 7.9, 7.31, 7.45). We will discuss these effects in turn.

Considering first the vertical distribution of the effect of the wind stress, an important feature is that the upper-layer motion resulting from a steady wind stress is not in the direction of the wind but at a substantial angle to it, to the right in the northern hemisphere, to the left in the southern. Historically, Nansen had observed qualitatively that icebergs drifted to the right of the wind in the north and reasoned that this was because the current caused by the wind affected their large bulk below the surface more than the wind directly affected their smaller exposed area. He communicated this idea to Ekman who developed a quantitative theory of the upper-layer wind-driven circulation. Ekman (1905) showed that the behaviour observed by Nansen resulted from the Coriolis force, i.e. the effect of the rotation of the earth on bodies moving relative to the earth. Imagine a parcel of water P (Fig. 7.1(a)) in the surface

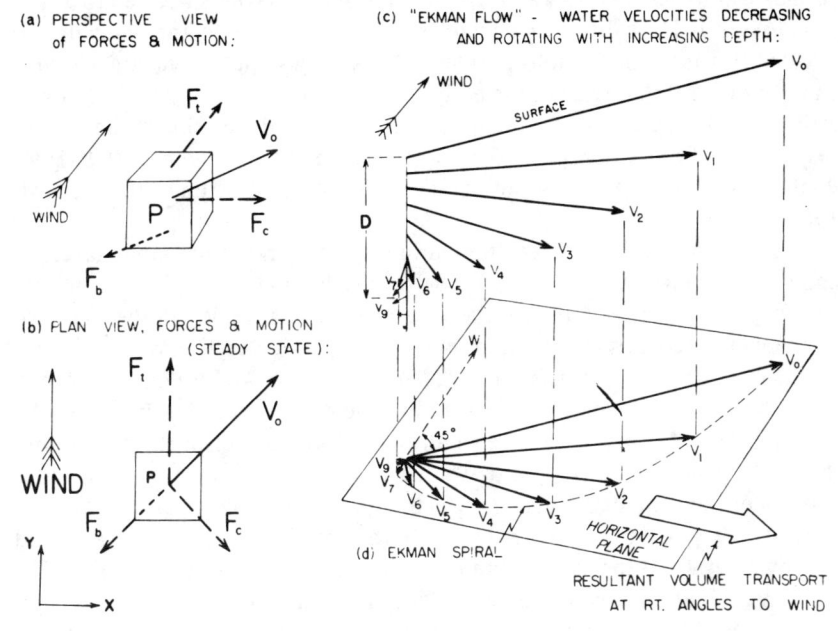

FIG. 7.1. Wind-driven currents—northern hemisphere: (a) forces and motion of surface parcel P, (b) plan view of forces and motion, (c) water velocity as a function of depth, (d) Ekman spiral.

layer. When the wind W blows over the surface, it causes a frictional stress F_t in the direction of the wind on the top of the parcel which starts to move in the wind direction. Immediately the Coriolis force F_c acts at right angles to the direction of motion, tending to cause the flow V_0 to be redirected to the right of the wind direction (in the northern hemisphere, to the left in the southern). At the same time, as the surface parcel commences to move a retarding friction stress develops between it and the next layer below. This retarding force F_b on the bottom of the parcel is opposite to that of the water motion V_o. In the simple application of the theory, a steady state develops in which the surface water velocity V_0 is at 45° to the right of the wind direction W in the northern hemisphere (45° to the left in the southern hemisphere) (Fig. 7.1(b)). A similar argument can be applied to successive layers beneath the surface layers and each of these will move at a slower speed and in a direction further to the right of V_o. In a simulated three-dimensional view (Fig. 7.1(c)) the arrows represent the velocities in successively deeper layers, showing how the speed (length of arrow) decreases with increased depth and how the direction changes clockwise (in the northern hemisphere). If the arrows are projected on to a horizontal plane their tips form the *Ekman Spiral* (Fig. 7.1(d)). The depth D_E at which the water velocity is opposite to the surface velocity V_o is called the *depth of frictional influence* and is taken as a measure of the depth to which the surface wind stress affects water motion. This upper layer is referred to as the *Ekman Layer*. The value of D_E depends on the eddy friction in the water and on latitude (through the Coriolis parameter). Making some simplifying assumptions about the character of the eddy friction, for a wind speed of 10 m/s at latitudes 10°, 45° and 80°, the values of D_E would be 100, 50 and 45 m respectively. For a wind speed of 20 m/s the value of D_G would be 200, 100 and 90 m respectively. (For a more detailed discussion see Pond and Pickard, 1983.)

It can be shown that the resultant volume transport in the Ekman layer (as shown by the large arrow in Fig. 7.1(d)) is at right angles to and to the right of the wind direction W in the northern hemisphere (to the left in the southern hemisphere). It is this resultant flow which gives rise to vertical motion in and below the Ekman layer, referred to as *upwelling* and *downwelling* which can occur both near ocean boundaries and in the open ocean depending on the divergence or convergence of the wind-driven Ekman transports in the upper layer. We will discuss coastal upwelling in Section 8.2 together with other aspects of the Ekman circulation which apply in shallow water.

At the *top* of the Ekman layer, both upwelling to the surface and downwelling from the surface can occur in the open ocean as a result of the Ekman circulation as will be described in some detail for the equatorial region in Section 7.614. In addition, upward or downward motion can be forced at the *bottom* of the Ekman layer as a result of the horizontal motion in the layer. For example, this can occur as follows. If the wind stress over an area of ocean were to the north but were strongest in the western part of that area, then the horizontal Ekman flow would be to the east everywhere but would be stronger in the western part of the area than in the eastern. A result would be convergence in the Ekman layer which would give rise to downward motion at the bottom of that layer. Conversely, an increase of wind stress to the east would cause upward motion at the bottom of the Ekman layer. This vertical

Circulation and Water Masses of the Oceans 161

motion in response to the Ekman divergence/convergence is referred to a *Ekman pumping*.

Using the monthly climatological wind stress estimates of Hellerman and Rosenstein (1983), Levitus (1988) has calculated the Ekman volume fluxes (horizontal and vertical) for the world ocean. In Figure 7.2 we summarize the directions of the calculated meridional (horizontal) flows in the Ekman layer (across the full width of each ocean) and their values. The values for the Pacific and Atlantic are qualitatively similar because of the similarity of the wind fields, with larger values in the Pacific. In both oceans there is also a subsidiary minimum flow near 8.5°N centered in August/September. The northern Indian Ocean shows a marked annual cycle at all latitudes with maximum flows in July/August, associated with the monsoonal cycle of the winds.

Associated with the convergences and divergences of the horizontal Ekman flows are vertical flows (Ekman pumping) which were calculated for the bottom of the Ekman layer and are summarized in Figure 7.3. Again the Pacific and Atlantic have similar distributions with the Indian Ocean showing a strong annual variation north of the equator (and even a semi-annual variation) at about 8°N.

In summary, the easterly trade winds (blowing westward) are responsible for the poleward horizontal Ekman flows in the tropical Atlantic and Pacific and the westerlies (blowing eastward) cause the equatorward flows at higher latitudes. The monsoonal wind variations in the Indian Ocean, north of 5°S, give rise to the pronounced seasonal variations in the currents and water property distributions in that ocean.

It must be appreciated that the above description is only of the elementary theory of the Ekman circulation in deep water. For instance, one assumption made, that the eddy viscosity in the water column is constant with depth, is probably not true. It is probably highest in the upper layers (because of the wind waves generated at the surface and the resulting turbulence) and probably decreases with depth (as the effect of the waves diminishes). Also, it was assumed that the wind was constant in time which is rarely the case except approximately in the trade wind regions. The speed with which the Ekman circulation develops depends on latitude, because the Coriolis force depends on latitude, and it has been calculated that it takes about one pendulum day for an essentially Ekman circulation to develop. (One pendulum day = (one sidereal day/sin ϕ) where ϕ = latitude, e.g. for $\phi = 10°$, 45°, 60° then 1 pendulum day = 5.7, 1.4, 1.2 solar days, so that an Ekman circulation should develop from rest in only a day or so in mid- and high latitudes but develops more slowly in low latitudes.)

Although a pure Ekman spiral circulation has never been observed in the open ocean, there are sufficient current observations to indicate that the theory is basically correct and it is essentially accepted as such. In the atmosphere, a bottom-layer Ekman distribution (Section 8.3) of the wind vectors is well known and supports the theory.

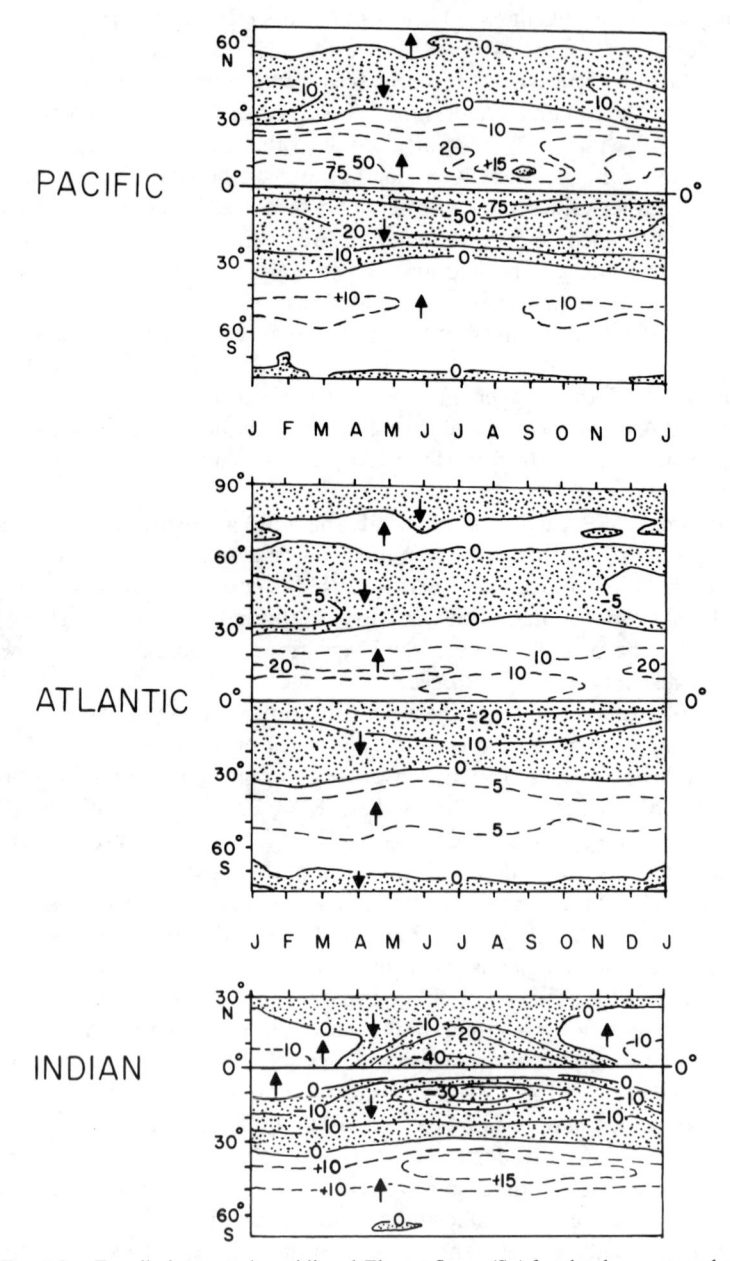

FIG. 7.2. Zonally integrated meridional Ekman fluxes (Sv) for the three oceans by latitude and month. (Flow in clear areas is northward, in stippled areas is southward.) (Adapted from Levitus, 1988.)

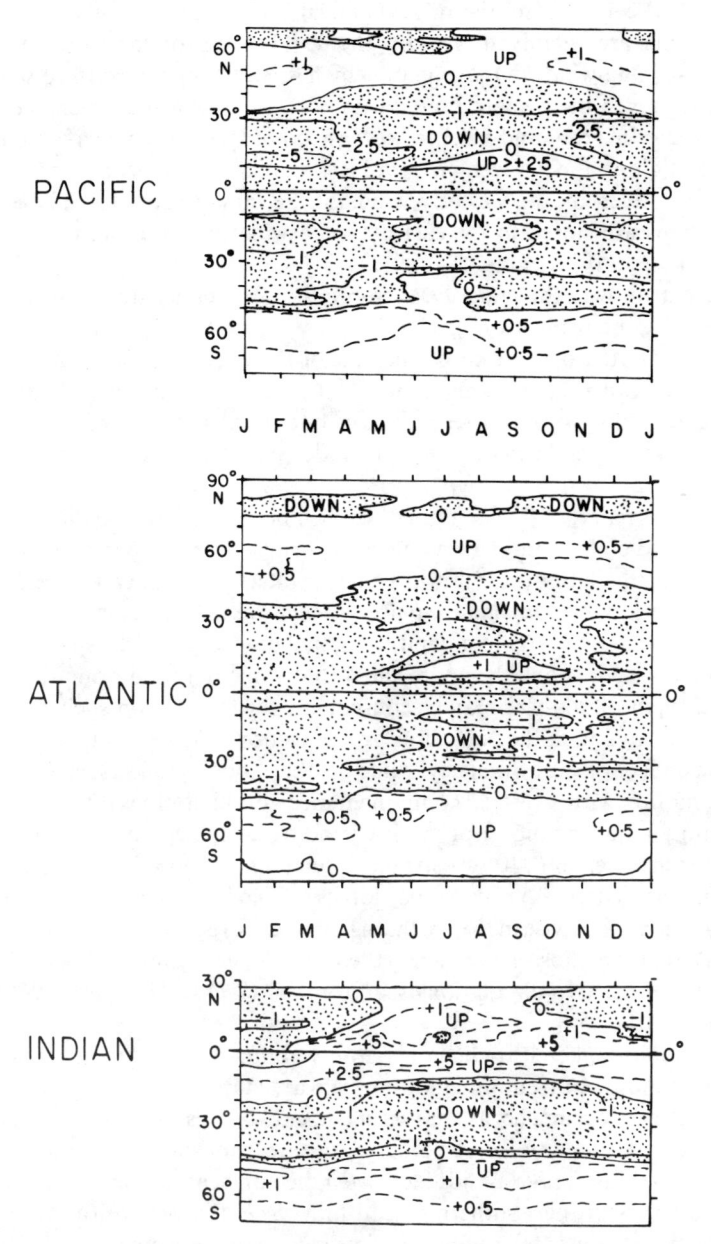

FIG. 7.3. Zonally integrated vertical Ekman volume flux (Sv) at the base of the Ekman layer per 1° latitude belt by latitude and month. (Flow in clear areas is up, in stippled areas is down.) (Adapted from Levitus, 1988.)

In the open ocean, the wind stress is the major cause of the gyral circulations which prevail in the upper layers, clockwise in the northern hemisphere and anticlockwise in the southern. Dynamical studies have shown that these circulations are mainly driven by the wind stress of the northeast and southeast trade-winds in the tropics, which drive the upper layers to the west until they reach the western boundaries of the oceans which divert the flows poleward in narrow and intense western boundary currents. After moving poleward along the western geographic boundaries the mean currents turn eastward driven by the westerly winds of the middle latitudes. These gyral circulations are closed to the east by much wider, weaker and more diffuse eastern boundary currents.

The reason why the main gyral circulations are concentrated in the west ("westward intensification"), e.g. Fig. 7.9, with the equatorward return flows distributed over much of the rest of the oceans, arises from the fact that the Coriolis force increases with latitude. Why this should cause the intensification is not intuitively obvious but was demonstrated by Stommel (1948) with a simple mathematical model (see Pond and Pickard, 1983, Section 9.8).

Ocean currents are then a result of the combined effects of the thermohaline motions and of the wind-driven ones. The former probably prevail in the deep water while the latter prevail in the upper waters. In both cases, the motion usually continues far beyond the place where it was initiated, just as water may be made to circulate round a hand basin by blowing along the surface at one side only. Also it should be appreciated that while the wind primarily causes horizontal motion, there may be consequent vertical motion such as the up- and down-welling in the equatorial zone (Section 7.614) and upwelling near the coast (Section 8.2). Similarly, while the thermohaline effects primarily give rise to motion with a vertical component, horizontal motions may result from continuity requiring inflow or outflow, and from changes in the field of mass (i.e. density) leading to geostrophic currents. Theoretical studies have demonstrated that thermohaline forces alone can give rise to major anticyclonic (clockwise in the northern hemisphere) gyres which are similar to, but weaker than, those generated by the wind alone. For further discussions of the roles of the two circulations see Stommel (*The Gulf Stream*, 1965) and Wyrtki (1961).

We should make clear the conventions about statements of current and wind directions. Current directions are always stated as the direction *toward* which the flow occurs, i.e. a "westward current" travels *to* the west. For winds there are two conventions. The older is to state from whence the wind comes, i.e. the Northeast Trade Winds blow *from* the northeast *to* the southwest, and the Westerlies in the Southern Ocean blow *from* the west *to* the east, i.e. a *westerly* wind comes *from* the west. This is a holdover from sailing ship days when the sailor was probably most conscious of where the wind came from. We will generally retain this convention for names of winds but when we wish

Circulation and Water Masses of the Oceans   165

to emphasize the direction *toward* which the wind blows we will refer to it as, for example, a *westward* wind or wind component.

7.13 Mixing processes

Mixing on scales above the molecular diffusion level is generally a result of turbulence in a fluid, and the generation of turbulence requires an energy input which may come from (a) kinetic energy, e.g. conversion of linear motion into the chaotic motion of turbulence as in a shear flow (i.e. where speed varies perpendicular to the direction of motion) or (b) from potential energy, i.e. convection when the density distribution becomes unstable (e.g. when the surface layer of a fluid cools and becomes more dense). The majority of mixing processes of significance in the ocean are initially on the centimetre and greater scales but there is one type in class (b) in which molecular diffusion is important initially. Turner (1981) has reviewed the extensive theories and experiments concerned with mixing and we will follow his treatment, with an emphasis on the oceanic applications (because there are many other fields in which the study of this process is important (e.g. Turner, 1974)). Turner (1981) also suggested a second classification of mixing processes in which energy for mixing comes from outside (e.g. wind stress on the sea surface) or internally away from boundaries (e.g. from velocity shear). We will first review mixing in the upper mixed-layer and then internal mixing processes.

There are two sources of energy supply to the upper layer—the surface stress of the wind and heat exchange through the surface. If the upper layer is initially stably stratified, e.g. density increasing linearly with depth (Fig. 7.4(a)), the effect of a wind stress (τ) on the upper surface is to generate motion which is maximal at the surface and decreases with increase of depth, i.e. vertical velocity shear develops. If this is sufficiently large, turbulence can develop and a result will be the mixing of the upper layer so that a substantially uniform density mixed-layer will develop (Fig. 7.4(b)). If the surface stress causes waves these will add more turbulent energy to increase mixing (particularly if they break) and, in addition, Langmuir circulations (Section 4.23) can be set up which will promote mixing, possibly through the full depth of the mixed layer.

A second possibility for mixing in the upper layer is when an appropriate "buoyancy flux" occurs through the surface. *Buoyancy* is defined as $-g\rho$ = weight/unit volume of fluid, and the *buoyancy flux* $B = -g\overline{\rho'w'}/\bar{\rho}$ where $\overline{\rho'w'}$ is the mean value of the product of the density change and vertical velocity fluctuation at any instant in the fluid and is referred to as the (vertical) "Reynolds flux" and $\bar{\rho}$ = mean density of the fluid. The change of density ρ' may be generated by heat transfer through the surface or by evaporation/condensation at the surface. If Fig. 7.4(c) represents the initial (stable) density distribution and then heat is lost from the top layer (Fig. 7.4(d)), the water will become unstable (negative buoyancy influx) and mixing in the upper layer will result in a deepened mixed-layer (Fig. 7.4(e)). Or, for a similar initial state (Fig.

166 Descriptive Physical Oceanography

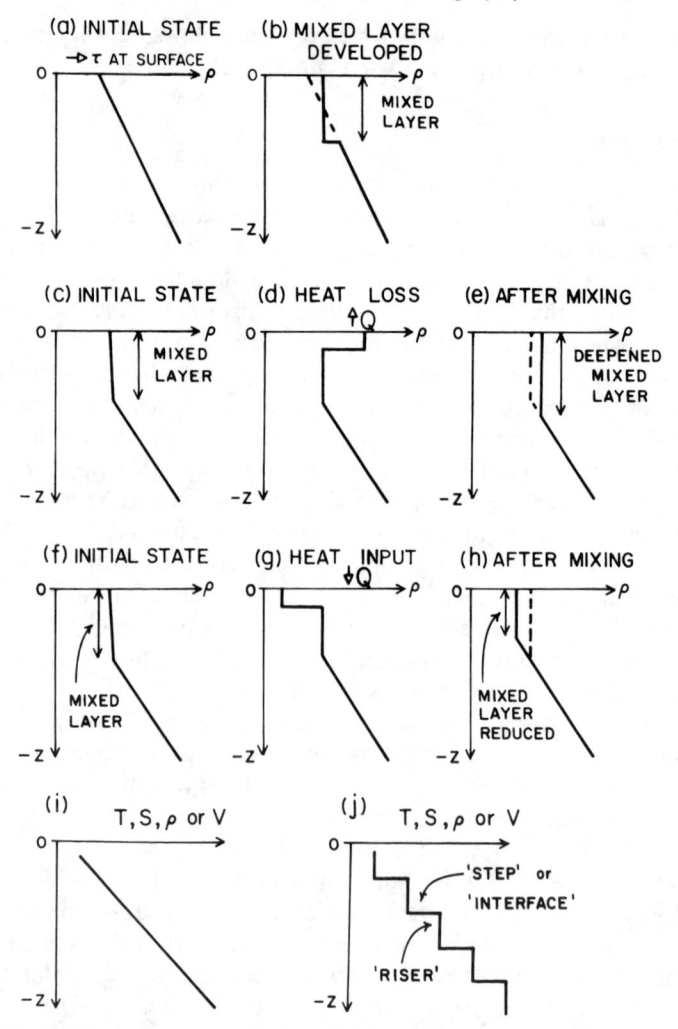

FIG. 7.4. Schematic diagrams of (a, c, f) initial stably stratified upper layer, (b) mixed layer developed, (d) heat loss through surface increases top layer density (i.e. negative buoyancy input) causing instability and vertical mixing to give (e) a deepened mixed layer; (g) heat input through the surface causing positive buoyancy input, then (h) if upper layer mixing occurs, mixed layer depth is reduced, (i) initial smooth profile of temperature, salinity, density or velocity, (j) stepped profile developed and revealed by continuous profiling instrument.

7.4(f)), if heat is added then the top layer will become less dense (positive buoyancy influx) and more stable (Fig. 7.4(g)) but subsequent mixing of the upper layer could result in a reduced mixed-layer depth (Fig. 7.4(h)).

A second boundary surface is the ocean bottom where turbulence, and hence mixing, can be generated by currents or current shear near the bottom.

Circulation and Water Masses of the Oceans   167

This type of mixing has been studied much less than other mechanisms. In shallow (e.g. coastal) waters, the generation of turbulence due to tidal currents over a rough bottom can cause mixing through the entire water column. An empirical criterion for complete mixing of the column is that $H/U^3 < 70$ m^{-2} s^{-3} where $H=$ water depth (m) and $U=$ tidal current (m/s). If there is a buoyancy flux (B) through the surface, e.g. heat or fresh water influx, the criterion becomes $BH/U^3 < 100$ (from field measurements around the British Isles), assuming that the buoyancy flux is limited to the upper 10–30 m.

Currents due to density differences can also cause mixing. An example is when a submarine earthquake stirs up bottom sediment into the water and thereby increases its effective density and so causes a turbidity current down an underlying bottom slope. Another example is that of the dense Mediterranean water flowing out from the Strait of Gibraltar into the less dense Atlantic water where it mixes with the lighter upper-layer and forms a stable tongue of intermediate density but relatively high salinity water which can be traced across the Atlantic.

In the interior of the ocean (i.e. away from boundaries), a feature that was discovered when continuous profiling instruments were developed and used extensively was that vertical profiles of water properties, e.g. temperature and salinity, and hence density, were often not smooth (Fig. 7.4(i) but 'stepped" (Fig. 7.4(j)). The vertical scale of the steps in such profiles could be decimetres to many metres, too small to be identified from the data from routine bottle casts where bottle spacings would be tens of metres and bottles themselves were half-a-metre long. The basic explanation is that the small-scale turbulence in water with an initially smooth profile will only act over small distances and will generate the vertical "risers" (representing well-mixed, nearly homogeneous layers) between the horizontal "steps" or "interfaces" (Fig. 7.4(j)).

One source of the energy required for mixing is considered to be internal waves, which are particularly effective in promoting mixing when they break like surface waves. Another mechanism which is important in causing mixing is that associated with interfacial shear. If velocity and density profiles are initially similar and the velocity shear over a thin layer (interface) is increased, a parallel flow becomes unstable when the *gradient Richardson Number* falls below 1/4. (The gradient Richardson Number $Ri = N^2/(du/dz)^2 = (-g\, d\rho/dz)/\rho_0 (du/dz)^2$ where $N=$ Brunt-Väisälä frequency (Sec. 4.43) and (du/dz) is the velocity gradient (shear). *Ri* is equivalent to the ratio of the stabilizing buoyancy to the destabilizing shear.) That is, instability occurs when the velocity shear becomes greater than 4 times the buoyancy as measured by N^2. The fastest developing instability is in the form of *Kelvin-Helmholtz billows*, which are peaked waves in the interface, which may break and cause intense mixing. This mechanism can also be effective where internal waves occur at sharp interfaces, e.g. the bottom of the upper mixed-layer because in such waves vorticity is concentrated, i.e. strong velocity shear occurs at density interfaces.

A quite different mechanism can occur even in the absence of mechanical turbulence. Although the effects of molecular diffusion are so small as to be negligible for the large-scale transfer of properties (scales of tens of metres or more), it was discovered that at the centimetre to metre scale, molecular diffusion may be important, and can result in mixing because of the different rates of diffusion for different solutes. Because two rates are concerned, this process is known as *double diffusion*. Turner (1974) described a variety of situations in which double-diffusion is important such as in the ocean, lakes, atmosphere, stars, liquid metals and in chemical processes. In order to illustrate double diffusion we will describe two oceanic situations in which double diffusion has been shown to be effective.

The first is where the Mediterranean water flows out from the Strait of Gibraltar into the Atlantic, particularly in conditions when the turbulence level is low. As described above, warm, saline Mediterranean water intrudes as a mid-depth layer into the cooler, less saline Atlantic water (Fig. 7.17(b)). At the bottom of this layer, the Mediterranean water is slightly less dense than the Atlantic water below it, so that the two layers are in stable (static) equilibrium. Because of the temperature and salinity differences between the layers, heat and salt will both tend to diffuse molecularly across the interface between the layers. If we assume that vertical turbulent mixing across the interface can be neglected in this case, the transfer of properties will be by molecular processes. Because the rate of molecular diffusion for heat is about 100 times that for salt, the upper layer loses heat and the lower gains it faster than the upper loses salt and the lower gains it. In consequence the upper water becomes denser and tends to sink into the lower layer while the lower becomes less dense and tends to rise into the upper layer. In laboratory experiments with such layers the water movements occur in small columns or "fingers", a few millimetres across and up to 25 centimetres long. Then lateral diffusion occurs between the fingers and gives rise to a uniform layer. The process is referred to as *salt fingering* or *double diffusion*. Then the process may start again at the two interfaces which are now present and eventually a number of individually homogeneous layers develop with sharply defined interfaces in terms of temperature and salinity (as in Fig. 7.4(j)). The layers may be metres to tens of metres thick separated by thinner interface zones of sharp gradients of temperature and salinity.

Although in principle these layers could have been detected with bottle casts by placing sampling bottles close together, the step structures generally occur at depths where bottles would be 100 m or more apart. When the temperature or salinity data points were plotted for vertical profiles they would be joined by smooth curves and the layering was unsuspected.

Step structures in temperature/depth and salinity/depth traces were first identified in CTD (i.e. *continuous* recording) casts below the Mediterranean water in the Atlantic (Tait and Howe, 1968, 1971) and were taken as the first evidence of the double-diffusive process in action. Numerous other observations have confirmed the occurrence of the process and Williams (1975) has observed them by optical means in the ocean.

Note that the double diffusive process described above which occurs when warm, saline water overlies cool, fresh water is one in which water from the upper layer penetrates across the interface into the lower layer and vice versa. In the second situation, *above* the Mediterranean water, the diffusive process is slightly different. Here there is cool, less saline water over warm, more saline water, the system being in static equilibrium. Again assuming that turbulence may be neglected, the upper (cool) layer will gain heat from the lower (warm) layer by molecular diffusion without gaining much salt and so will become less dense and parcels of water will tend to rise within the upper layer. The lower layer, losing heat through the interface but not much salt, will become denser and water will tend to sink, again within its own layer. The difference from the previous case then is that fluid does not cross the interface—water stays within its own layer, mixing and carrying heat up or down respectively.

For laboratory experiments on double diffusion, a layer of salt solution over one of sucrose, is convenient. The difference in diffusive rates is not as great as for salt/heat but the convenience of not needing to prevent lateral heat exchange, as must be done for the salt/heat case, is great. The condition for fingers to form is that $\beta \, \Delta S/\alpha \, \Delta T > \tau^{3/2}$ where β and α are the changes of density for unit changes of salinity and of temperature respectively and $\tau = (\kappa_S/\kappa_T)$, i.e. the ratio of the molecule diffusivities for salt and heat respectively, has a value of $1/100$ for salt and heat (but only about $1/4$ for sucrose and salt).

Other studies have suggested that in some cases the step structure may be a result of interleaving along density surfaces when water masses of different T–S characteristics come together laterally, the masses having different density/depth profiles. Such interleaving has been observed in a number of situations including the neighbourhood of the equator in the western Pacific (Jarrige, 1973) where the saline South Pacific Subtropical Water moving north meets less saline North Pacific Water.

An account of these various so-called *fine-structure* observations, including effects of double diffusion and density layering, has been given by Federov (1976).

Another mechanism, called *cabbeling*, which can give rise to buoyancy effects internally arises from the non-linear relation between temperature, salinity and density of sea-water (Fig. 3.1), particularly at low temperatures. In Fig. 7.5, we show isopycnals in the neighbourhood of $0°C$ and 34.6 salinity and identify two water types A and B of different temperature and salinity but the same potential density. If the two water types are neighbours in the sea and start to mix, the properties of the mixture will lie on the straight line AB which lies on the high-density side of the isopycnal ($\sigma_\theta = 27.8$ here), i.e. negative

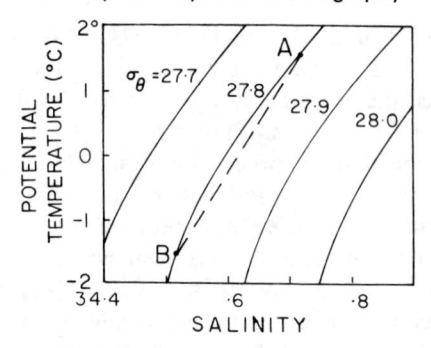

FIG. 7.5. Isopycnals in the vicinity of 0°C and 34.6 salinity to show their curvature and that mixing of water types A and B ($\sigma_\theta = 27.8$) can form water denser than that value.

buoyancy is generated, and consequently any such mixture will tend to sink. We have to assume some initial mixing process which generates the initial negative buoyancy and hence further mixing. The formation of Antarctic Bottom Water is probably related to this process as pointed out by Fofonoff (1956).

A rather specialized aspect of mixing is where buoyancy is added at the bottom of a layer of water as from a sewer outfall where the discharged fluid is less dense than the sea-water and mixing takes places as the discharge rises to the surface due to its buoyancy. This is a situation which has been much studied by marine engineers but is beyond the scope of our present text.

7.14 Circulation and water masses

When one attempts to describe the *circulation* and *water masses* of the oceans one is faced with a problem. If one wishes to present the details one can only do so for a region of limited extent and so one must divide the world ocean into regions for this purpose. In practice, this is not too difficult to do; the oceans almost divide themselves into regions. For instance, the Atlantic can be described regionally as the North, the Equatorial and the South Atlantic, together with the adjacent seas such as the Mediterranean, the Labrador Sea, etc. A professional oceanographer who is studying a particular region for a specific purpose, e.g. as part of a study of the fisheries, will be concerned with the details of that region. But almost invariably when he selects a particular area for study he finds that the waters within the region are influenced by water without the region, e.g. by currents entering the area of study from elsewhere. Then to understand the particular area he has to extend the study beyond that area, and sometimes it is difficult to stop this extension before reaching the opposite boundary of the ocean. In other words, to understand the part one must understand the whole.

Since the present book is intended only as an introduction to Descriptive

Oceanography, and since the total volume of detailed data available is far too great to include or even summarize within its pages, the main endeavour in the following description will be to acquaint the reader with the major features of the ocean circulations and water masses, with some description of details for a few areas as examples. It is hoped that this approach will provide him with enough feeling for the character of the ocean circulation and water masses as a whole that he will have sufficient background to study smaller areas in detail later on, e.g. using Tchernia's text *Descriptive Regional Oceanography*.

Again, when describing the circulations and the water masses one is faced with the question of which to present first. It is the old problem of "which came first, the chicken or the egg?" In this book the authors have decided to take an operational approach. For the upper waters, the circulation will be described first but for the deep waters the water masses will be described first. The reason for this procedure is that we have a good deal of direct information about the circulation of the upper layers, but the movements of the deep waters have largely been inferred from the distributions of properties.

From the analysis of ship's navigation logs, as described in Section 6.251, we have a knowledge of the surface circulation of much of the oceans. This knowledge is substantial and detailed in the regions much travelled by ships, such as the main traffic routes across the North Atlantic and North Pacific, but is scanty in other regions such as the eastern South Pacific and the southern Indian Ocean. As has already been stated, the upper-layer circulation is driven primarily by the winds and from the results of studies in dynamical oceanography we can obtain from the surface-layer circulation a very good idea of the whole upper-layer circulation down to the thermocline. Therefore, in describing the upper layers of the ocean, the circulation will be described first and the water properties second as these are chiefly determined by the history of the water, i.e. where it has been carried by the currents.

For the deep water, the water mass characteristics will be described first and then the movements, since most of our knowledge of the deep-water circulation has been obtained by interpretation of the distributions of properties. In addition to the deep water flows having been inferred from the property distributions the flows are mostly driven by the distribution of density which is itself determined by the distributions of temperature and salinity. Only in recent years has the invention of the neutrally buoyant float and the development of mooring techniques and current meters permitted satisfactory direct observations of the deep water movements. To date, the information available from these sources is limited in amount and is from only a few regions.

Some indication of what is meant by "upper" and "deep" waters must be given before going further. It is not easy to give an exact figure for the thickness of the *upper water* but it is usually taken to extend from the sea surface to the depth at which the decrease in temperature with depth becomes small. This may be between 300 and 1000 m (Fig. 4.4). The upper layer contains the surface or *mixed layer* of 50- to 200-m depth in which most of the seasonal variations of properties occur and which is usually fairly homogeneous due to mixing by the action of the waves caused by the wind. The remainder of the upper layer is usually well stratified and stable, and includes the thermocline. The *deep water* includes all that below the upper water, and is less stable than the latter. If the layer in contact with the sea bottom has properties distinct from those of the deep water above, it is referred to as the *bottom water*.

Before proceeding to the more detailed description of the ocean areas, the main features will be presented briefly so that the reader may appreciate how the details fit into the whole. The *circulations in the major ocean areas* show considerable similarities; the differences are largely in detail. In the upper layer there is a major clockwise circulation or "gyre" in both the North Atlantic and the North Pacific and a counter-clockwise gyre in the southern parts of the Atlantic, Pacific and Indian Ocean. These "anticyclonic" circulations dominate the low- and mid-latitude portions of these oceans. In the North Atlantic and Pacific a very conspicuous feature is that the currents are narrower and swifter on the west side of each ocean than elsewhere (*westward intensification* of the currents). There is evidence of the same phenomenon in the South Atlantic and Indian Oceans but the western South Pacific circulation is rather complex and the intensification is not clear. In the equatorial regions of all three oceans there are similar current systems consisting of a westward-flowing South Equatorial Current at or south of the equator and a westward-flowing North Equatorial Current further north. In the Pacific these two are separated by an eastward-flowing North Equatorial Counter Current along the full width of the ocean; in the Atlantic the Counter Current is only significant in the eastern part. These equatorial counter-currents owe their existence to variations in the wind stress from north to south across the equator. In the Indian Ocean, the three-current pattern is present for part of the year and a two-current system for the remainder. Another component of the equatorial current system is the subsurface Equatorial Undercurrent which flows eastward along the equator, at about 100 m depth, carrying water built up in the west by the westward-flowing Equatorial Currents. Defant called the surface equatorial current systems the "backbone of the circulation" to emphasize our belief that the northern and southern gyres are driven mainly by the trade winds in low latitudes. The facts that the regions have maintained their identity of circulation and of water mass characteristics for more than a hundred years indicates that the climatic processes which determine them must be continuing to act.

In the deep water the major flows are north and south, not necessarily evenly distributed across the widths of the oceans but probably stronger on the west sides.

The two polar regions show marked differences which are due in part to the differences in character of the driving forces but principally to the difference in topography of the two basins.

Circulation and Water Masses of the Oceans   173

Since the Southern Ocean is openly connected with all the other oceans, and water which acquires its characteristics here has a profound influence on the deep waters in the other oceans, we will start with this region.

7.2 Southern Ocean

7.21 Divisions of the Southern Ocean

The Southern Ocean has the land mass of the Antarctic continent to form its southern boundary but has no land boundary to its north and is continuous with the other oceans. However, the surface waters of the region have well-defined characteristics whose isopleths run roughly parallel to lines of latitude (i.e. are zonal) and features of these characteristics are used to define the northern boundary of the Southern Ocean. The classical picture of the upper layer zones is as follows. Going north from the Antarctic continent the average sea surface temperature increases slowly until a region is reached where a relatively rapid increase of 2 to 3K occurs. The surface water from south of this region is moving north and sinks when it reaches the region, continuing north below the surface. At the surface therefore the water is converging to this region which was known as the Antarctic Convergence but is now called the *Antarctic Polar Front* (APF) (Fig. 7.6). Continuing north from this APF the temperature rises slowly to a second region where it rises rapidly by about 4K and the salinity by about 0.5. This is referred to as the *Subtropical Convergence* (Fig. 7.6).

The APF is found at about 50°S in the Atlantic and Indian Oceans and at about 60°S in the Pacific (Fig. 7.6). The Subtropical Convergence is at about 40°S round most of the Antarctic but its position has not been well determined in the eastern South Pacific which is a poorly known region oceanographically. The two convergences divide the surface waters of the Southern Ocean into two zones, the *Antarctic zone* from the continent to the APF and the *Subantarctic zone* from there to the Subtropical Convergence (Fig. 7.6). In the Antarctic zone, the surface temperature is between $-1.9°$ and $1°C$ in winter and between $-1°$ and $4°C$ in summer, while in the Subantarctic zone it is between $4°$ and $10°C$ in winter and up to $14°C$ in summer. (Note that unless a minus sign is placed before a temperature it is understood to be positive.)

The above description of the upper layer of the Southern Ocean was based on data available to the mid-1940s. Subsequently more data have become available, both from oceanographic expeditions and from supply ships to the Antarctic continent, particularly in the sector from western Australia east to South America. It now appears that the characteristics, particularly in the 50–60°S zone, are more complicated than described above. In the first place a single, clearly defined convergence of surface water at the boundary between the Antarctic Surface Water and the Subantarctic Surface or Upper Water (both described later) (Figs. 7.6 and 7.7) is not confirmed; rather the transition

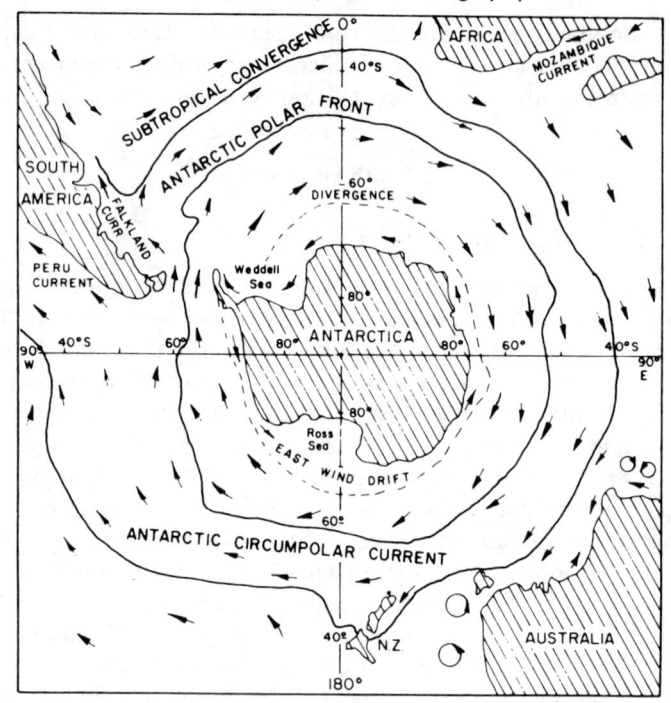

FIG. 7.6. Southern Ocean circulation and mean positions for the Antarctic Polar Front and Subtropical Convergence.

zone between these water masses extends over a range of latitudes with seasonal and longitudinal changes in width and position. The term *Antarctic Polar Frontal Zone* (APFZ) is preferred to "Antarctic Convergence" and from studies of both station data (temperature and salinity) and XBT sections (temperature only) Emery (1977) suggested the terms *Antarctic Polar Front* for the southern boundary and *Subantarctic Front* for the northern boundary of this zone. The most important result of this and other studies (Gordon et al., 1977) was that the Antarctic zonation that had typified the area south of Australia was also present in the other sectors of the Southern Ocean but in a more compressed form. In the Drake Passage, both the Antarctic Polar Front and the Subantarctic Front are present but the APFZ between them is much narrower than in the Australian sector due to the constriction of the Drake Passage. In this zonation pattern the northern front limits the southward extension of Subantarctic Surface Water and is marked by a strong temperature gradient as well as a shift in T–S values. A change in T–S characteristics also occurs at the Antarctic Polar Front which, as the northern limit of Antarctic Surface Water, coincides with a subsurface temperature minimum in summer. These fronts are generally associated with zonal current jets. The structure of the zone itself is quite variable and complex, with evidence of eddies and meanders. Much of this variability, and indeed the width of the Frontal Zone itself, is due to instabilities in the zonal current jet associated with the APF. This front has been observed to meander, forming northward extensions which entrain and capture Antarctic Surface Waters. These meanders then separate to form cold, cyclonic eddies which populate the APFZ. Cold eddies "export cold" from the Antarctic polar latitudes to higher subantarctic latitudes and thereby participate in the meridional heat balance of the Southern Ocean. The evidence for the location of formation of these cold eddies at the APF is that they carry Antarctic Surface Water, from south of the APF, even when the eddy is in contact with the Subantarctic Front at the northern limit of the APFZ (Savchenko et al., 1978). The Subantarctic Front has not been witnessed to meander and form warm, anticyclonic eddies to populate the APFZ.

7.22 Southern Ocean circulation

The unique geography of the Southern Ocean makes it the only place where ocean currents run completely around the globe. In a narrow zone round most of the continent there is a westward-flowing coastal current, called by Deacon (1937) the *East Wind Drift* because it is attributed to the prevailing easterly winds near the coast. Farther north the remainder of the Southern Ocean is dominated by a strong, deep eastward-flowing current known as the *Antarctic Circumpolar Current* (ACC) (Fig. 7.6).

The ACC has been referred to as the "West Wind Drift" because it may be attributed to the stress of the strong westerly winds in the region, i.e. winds *from* the west causing flow *to* the east. The westerly wind in the Southern Ocean was notorious in sailing ship days and, together with the eastward current, made it difficult for such vessels to round Cape Horn from the Atlantic to the Pacific.

The surface flow of the ACC is driven primarily by the frictional stress of the westerly wind which led to the earlier name of West Wind Drift. The wind stress, combined with the Coriolis force, also contributes a northward component to the surface current resulting in the formation of fronts, i.e. convergences, within the APFZ. Below the wind-driven surface layer, the density structure appears to be in geostrophic balance with the circulation.

In its circuit round the continent the ACC is obstructed only in the narrow Drake Passage between South America and the Palmer Peninsular projecting north from Antarctica. In general, the ACC is not very fast with surface speeds of only about 4 cm/s in the Antarctic zone increasing to 15 cm/s north of the Polar Front and then decreasing again toward the Subtropical Convergence. However, the current is very deep and its volume transport was estimated by Sverdrup as up to 110 Sv. Other workers have determined transport values, at various times, from small westward flows of up to 15 Sv to as much as 290 Sv eastward (the usual direction). Peterson (1988) has presented a table of values published since the first in 1933 and has discussed the relations of the transport to the wind and tides. The earlier estimates were from geostrophic calculations made from the density distribution obtained from measurements of water temperature and salinity made during periods of only a few days at a time and with the usual uncertainties in converting the relative velocity profiles to absolute values. Some had limited current measurements to assist this conversion but most relied on "depth of no motion" assumptions. Nowlin *et al.* (1977) used both geostrophic and current meter measurements and obtained transport during a period of 3 weeks ranging from 110 to 138 Sv and averaging 124 Sv. From year-long current meter records, Bryden and Pillsbury (1977) obtained an average value of 139 Sv but during the year the transport varied from 28 to 290 Sv. Typical time and space scales for variations were 2 weeks and less than 80 km (compared with the 800 km width of the Drake Passage). Many of these variations are associated with the strong zonal jets at the major fronts. In these jets, current speeds of 50 to 100 cm/s have been observed.

Most of the transport measurements have been made in the Drake Passage because here the current is clearly limited north and south and the width is at its minimum value. As part of the International Southern Ocean Studies (ISOS) a dense array of current meters was deployed for a year, from which it is hoped to obtain an adequate description of the short- and long-term variations of the ACC which, at its maximum, is clearly the mightiest current in the oceans. It shows some variations in direction as it flows round the continent and there is evidence that some of these are due to the effects of the submarine topography. Some of the current branches off and flows north between Australia and New Zealand at subsurface levels and some, including surface water, flows north as the Peru Current up the west coast of South America into the South Pacific (Fig. 7.6), making significant contributions to the circulation and water masses of that ocean. There is some flow from the ACC northward into the Atlantic between South America and the Falkland Islands as the Falkland Current (Fig.7.6).

Likewise ACC flow is lost in other sectors where the north-south boundaries narrow, restricting the eastward flowing ACC. Another good example is west of the Drake Passage where a good portion of the ACC turns north along the coast of Chile to become part of the northward flowing Peru Current. There is also a northward turn of a portion of the ACC to the east of Australia and New Zealand where, due to an interaction with the bottom topography, the current turns abruptly to the north. Some of the ACC also flows northward when the current encounters the African continent but the passage here is fairly wide and does little to constrict the current flow. In addition, the bottom topography in this sector is quite simple and does not interact with the flow to cause any significant variations in flow path.

PLATE 1 (top).   C.S.S. *Vector*, a 40-m Canadian oceanographic research vessel.

PLATE 2 (bottom left).   Medium-duty hydrographic winch, hydraulic drive.

PLATE 3 (bottom right).   Meter wheel with remote indicator.

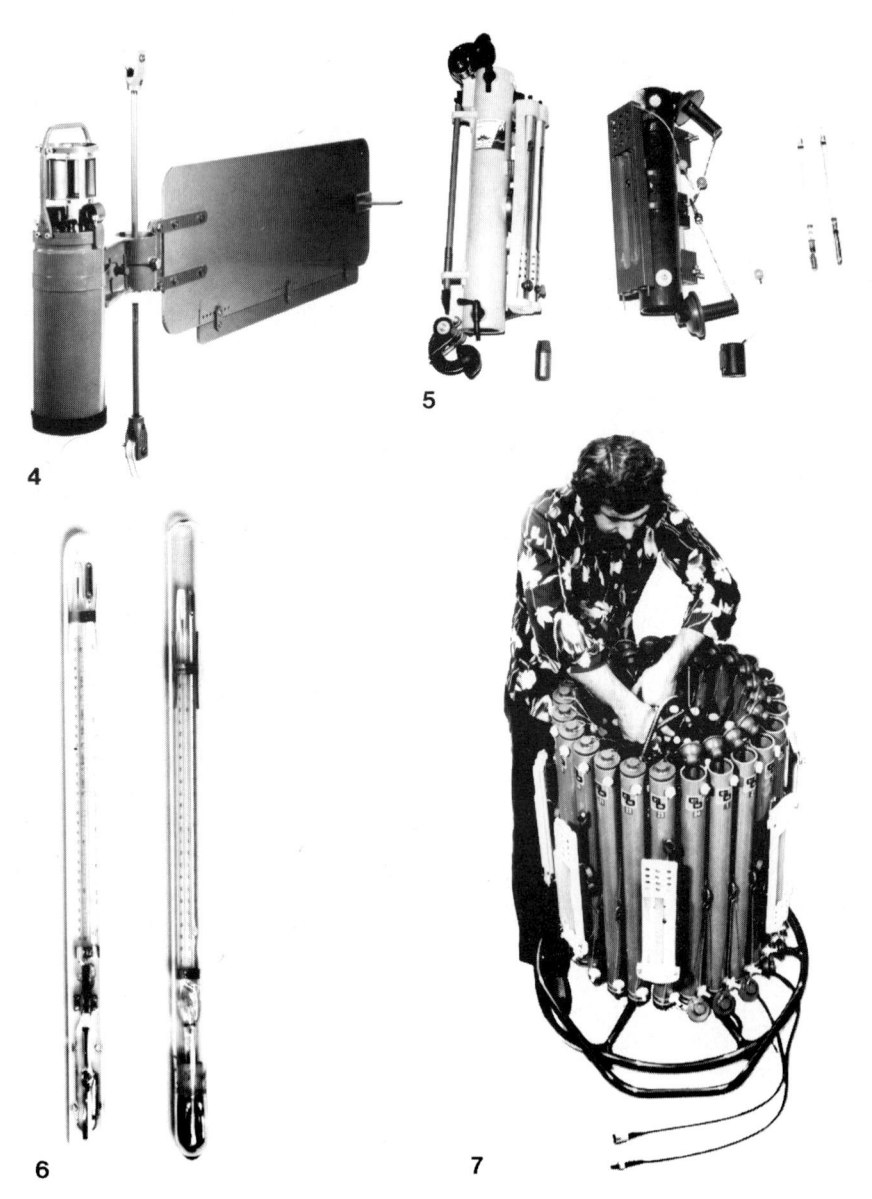

PLATE 4 (top left).   Aanderaa current meter; water property sensors just below current rotor on top of electronics casing.

PLATE 5 (top right).   Water-sampling bottles (N.I.O. and Niskin), messengers, and protected and unprotected reversing thermometers.

PLATE 6 (bottom left).   Unprotected and protected thermometers (30 cm long).

PLATE 7 (bottom right).   Rosette water sampler.

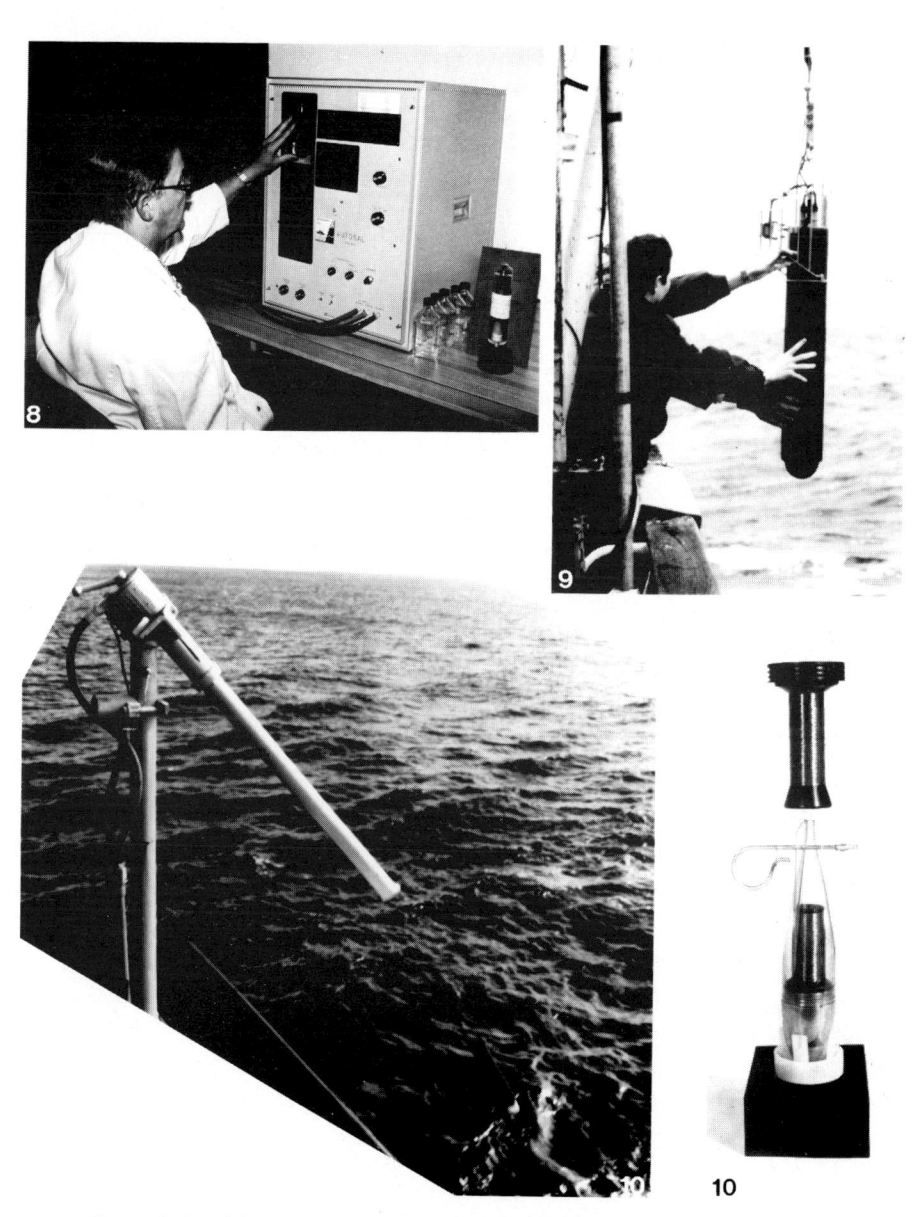

PLATE 8 (top left). "Autosal" laboratory salinometer, water-sample bottles and ampoule of Standard Sea-water.

PLATE 9 (top right). Sensor head of Guildline CTD.

PLATE 10 (bottom). (left) XBT launcher and XBT about to enter water, (right) display XBT in transparent casing to show temperature probe location at bottom and dual wire-spools.

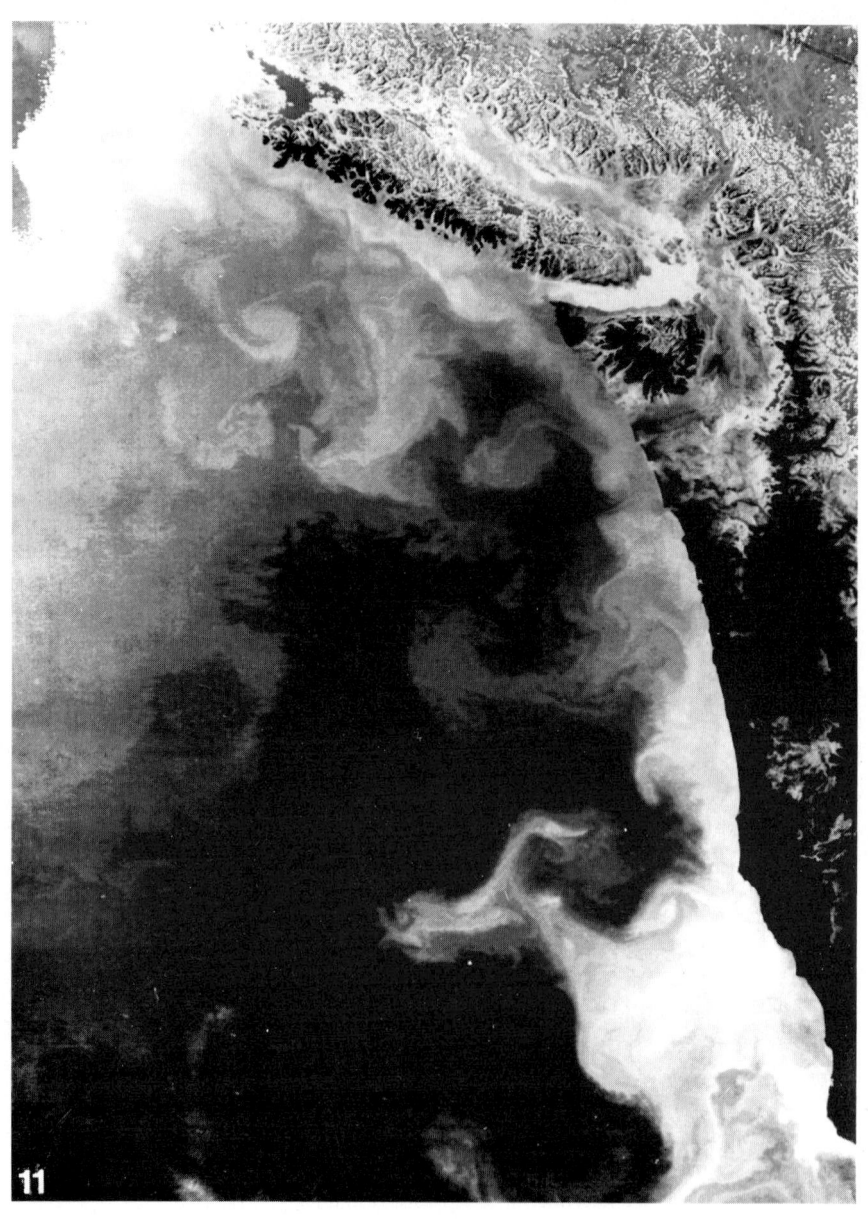

PLATE 11. Satellite infra-red image off west coast of North America from about 51° to 40°N; white is coolest water, black is warmest.

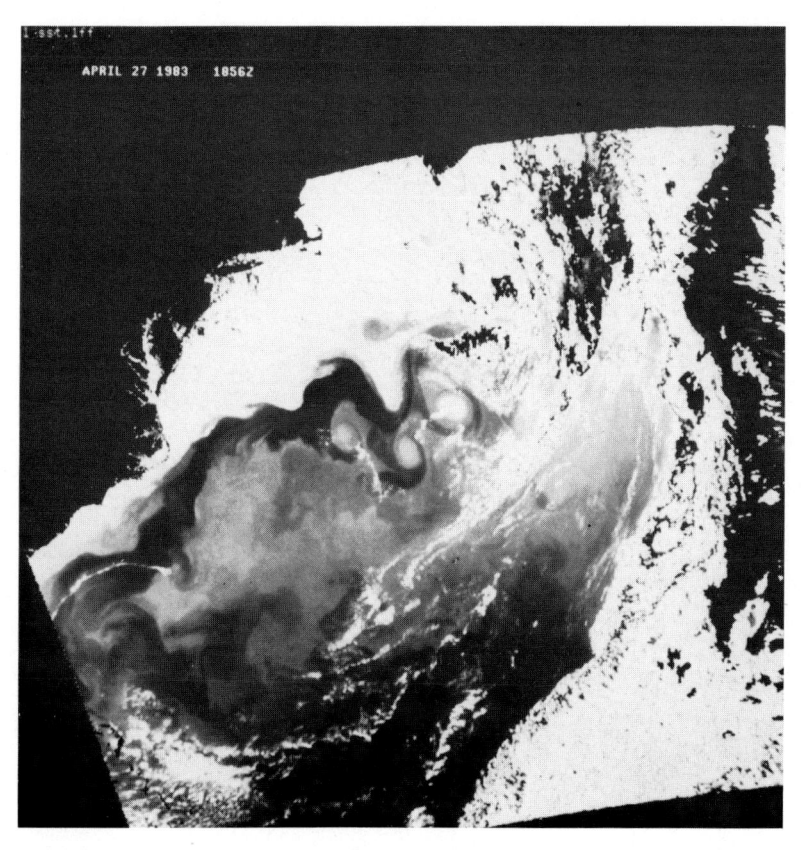

PLATE 12. Satellite infra-red image off east coast of North America, from about 33° to 44°N, white is coolest water, black is warmest.

7.23 Southern Ocean water masses

7.231 Antarctic zone

The water masses of the Southern Ocean (Fig. 7.7) have typical high latitude characteristics. South of the APF the *Antarctic Surface Water* has properties which are determined by ice melting in summer and by cooling in winter. This layer of 100- to 250-m thickness has a salinity of less than 34.5 and a low temperature ranging down to the freezing point, about $-1.9°C$ at that salinity. The temperatures are lowest in the south and increase toward the north due to the absorption of heat during the summer. It must be remembered that in the polar regions, south or north, an effect of sea-ice is to limit the range of temperature variation between winter and summer. In one direction, the freezing point of sea-water at the salinities found at high latitudes is no lower than $-1.9°C$ and limits the reduction of temperature which can occur in winter. In the other direction, the melting of ice requires a considerable proportion of the heat inflow during the summer, because of the large latent heat of melting of ice, leaving only a small part to raise the temperature of the water.

Below the surface and extending to the bottom at depths to 4000 m is the *Antarctic Circumpolar Water*. Its temperature in the Antarctic zone rises to a maximum of 1.5 to 2.5°C at 300 to 600 m and then decreases to between 0 and 0.5°C at the bottom. Its mean salinity is close to 34.7. These characteristic properties are found all round the Antarctic continent at the same depths, evidence that this water is carried around the continent by the ACC.

A very important water mass is the *Antarctic Bottom Water*. Most of it is formed in the Weddell and Ross Seas in the Antarctic continent (Gill, 1973) (Fig. 7.6). Its water is a mixture of the Antarctic Circumpolar Water with *Shelf Water* which has attained its properties on the continental shelf region of these seas. The Shelf Water has a temperature of $-2.0°C$ (the freezing point decreases with increase in pressure) and a salinity of 34.4 to 34.8. Its density (σ_t) of 27.96 is among the highest found in the southern oceans. (Values to 28.1 are found in the Ross Sea.) The formation of these dense waters has been the subject of much discussion; a review is presented by Warren (1981). The mixing of Shelf Water with Circumpolar Water reduces its density to slightly below 27.9 as it flows out from the Weddell Sea. Because of this high density it flows down the continental slope into the South Atlantic (Fig. 7.8) and also eastward through the Indian and Pacific sectors of the Southern Ocean. The eastward flow round the continent is deduced from the continuity of temperature and salinity values together with the steady decrease in oxygen content from the Atlantic sector through the Indian and Pacific sectors. If we knew the rate at which oxygen is used up in the water it would be possible to estimate the speed of flow from the rate of decrease of oxygen. Unfortunately we do not have any reliable figures for the rate of oxygen consumption in this region.

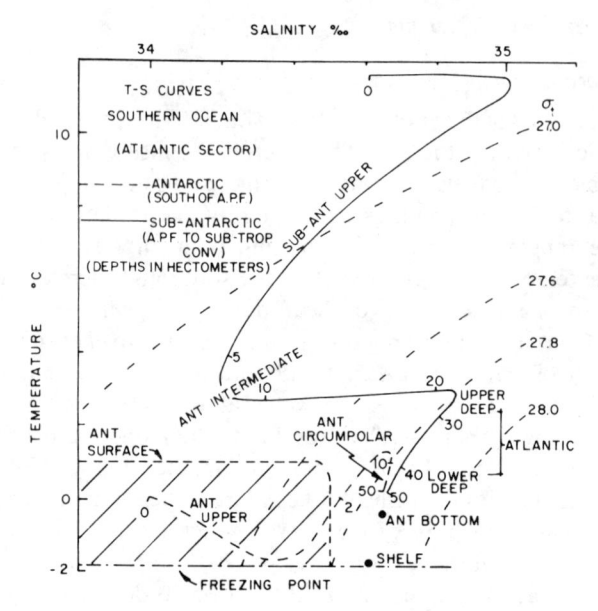

Fig. 7.7. T–S curves for Southern Ocean waters, Atlantic sector.

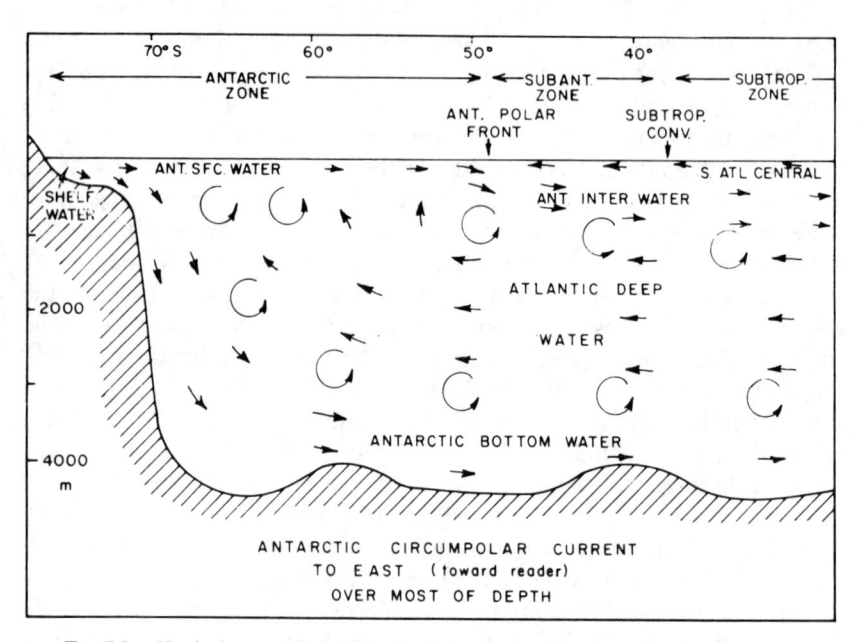

Fig. 7.8. Vertical components of the circulation in the Atlantic sector of the Southern Ocean.

7.232 Subantarctic zone

In the Subantarctic zone the water masses become more numerous (Fig. 7.7, full line). Some of them originate in the Southern Ocean but others originate outside and information which will be given later in this chapter has been used to identify them. In addition, the upper water masses show some differences between the Atlantic, Indian and Pacific sectors, in contrast to the relatively uniform character of the Antarctic surface water.

The *Subantarctic Upper Water* occupies the upper 500 m or so and has a temperature of 4 to 10°C in the winter and up to 14°C in summer, and a salinity from 33.9 to 34.9 in winter and as low as 33 in summer as ice melts. The lowest temperatures and salinities are found in the Pacific sector and the highest in the Atlantic sector. This Upper Water has a southward component of motion which accounts for its higher temperature than the water farther south. A salinity maximum is also present in all sectors at 150 to 450 m depth.

Below this water is the *Antarctic Intermediate Water* which includes surface water from the Antarctic zone and is formed by mixing below the surface in the APFZ. It continues northward with some admixture from the Upper Water (Fig. 7.8). The Intermediate Water has a thickness of about 500 m at the northern limit of the Subantarctic zone, under the Subtropical Convergence. It is more homogeneous in properties than the Upper Water, with a temperature of 2 to 3°C and a salinity of 34.2. It sinks because its density (σ_t) of 27.4 is greater than that of the water to its north. As it flows to the north it mixes with more saline water above and below and its salinity rises gradually. The Antarctic Intermediate Water flowing north forms a tongue of relatively low salinity water with its core at 800 to 1000 m depth at 40°S and to the north of this (see Fig. 7.15). The water also has a relatively high oxygen content of 5 to 7 mL/L since it has only recently left the surface.

The next water mass below is the *(Atlantic) Deep Water* which moves with a southward component of motion in addition to its eastward circulation as a part of the ACC (Fig. 7.8). It lies between 1500 and 3000 m depth in the Subantarctic zone and has a temperature of 2 to 3°C and a salinity of 34.7 to 34.9 and an oxygen content of 4 to 5 mL/L in the Atlantic sector. It is sometimes referred to as the "warm" deep water. In a vertical profile and in the T–S diagram (Figs. 7.7, 7.20(a)) the water does show a very slight temperature maximum but it is only warm relative to the colder Intermediate and Bottom Water. It was first observed in 1821 but only recognized as being of North Atlantic origin by Merz and Wüst (1922). The Upper Deep Water shows a salinity maximum in the Atlantic and Indian Ocean sectors (Fig. 7.20) while the Lower Deep Water is difficult to distinguish from the Antarctic Circumpolar Water. In the Pacific sector (Fig. 7.20) the Upper Deep Water is less easy to distinguish from the Antarctic Circumpolar Water than in the other sectors. The Deep Water has a southward component of motion, contrary to the Intermediate Water above and the Bottom Water below. There is inevitably some mixing between them, and the waters above and below are both modified by mixing with the Deep Water. In moving south from the Atlantic, the Deep Water rises toward the surface (Fig. 7.8). It is probable that it does not actually reach the surface as a water mass but diverges north and south, mixing with surface water and contributing to the Antarctic Intermediate Water and to the Bottom Water. The Deep Water comes from the North Atlantic and has been away from the surface for a long time; it has a relatively low oxygen content by Atlantic standards and high concentrations of nutrients. This continual supply of nutrients is one of the reasons for this region being prolific in phytoplankton (plant) growth and consequently in zooplankton. The latter is a source of food for larger animals in the sea and the whaling industry of the Southern Ocean was one consequence. In a vertical section of water properties (e.g. Fig. 7.15) the Deep Water is characterized by an oxygen minimum as it lies between the Intermediate and the Bottom Waters, both of which were more recently at the surface and consequently are richer in oxygen.

The deepest water in the Subantarctic zone is the Bottom Water flowing north (Fig. 7.8). Mixing with the overlying Deep Water its temperature is raised to $0.3°C$ and its salinity to 34.7 to 34.8.

Summarizing, it is possible to divide the Southern Ocean into three zones round the continent, an Antarctic zone nearest to the land and a Subantarctic zone farther north, separated by a transitional region called the Antarctic Polar Frontal Zone. In the Antarctic zone (Fig. 7.6) there is a thin surface layer of cold Antarctic Surface Water of relatively low salinity from summer melting of ice. Below this there is the large mass of the Antarctic Circumpolar Water of uniform properties all round the continent. The cold Antarctic Bottom Water formed at the continental edge flows north and east into the three main oceans. In the Subantarctic zone there are four distinct water masses. The Subantarctic Upper Water is warmer and more saline than the Antarctic Surface Water. Below this the Antarctic Intermediate Water is fed by lower salinity Antarctic Surface Water flowing north and descending beneath the Upper Water. Below this, marked by a salinity maximum, is the large body of Deep Water flowing south and rising over the Bottom Water flowing north. The Intermediate and Bottom Waters are relatively high in oxygen content from their recent contact with the surface, or with surface water, while the Deep Water is relatively low in oxygen content. The Deep Water can be divided into an Upper Deep of slightly higher temperature and salinity than the Lower Deep Water.

7.3 Atlantic Ocean

7.31 Atlantic Ocean as a whole

The *upper water circulation* of the Atlantic Ocean as a whole consists in its gross features of two great anticylonic circulations or "gyres", i.e. a counter-clockwise one in the South Atlantic and a clockwise one in the North Atlantic (Fig. 7.9). At first sight it might seem appropriate to liken these two gyres to two gear wheels revolving and meshing near the equator. This, however, gives the impression that one may be driving the other, which is not the case. Rather the two gyres are driven separately, each by the trade winds in its own hemisphere (similar though not identical to those in the Pacific, Fig. 7.31), and they are separated over part of the equatorial zone by the eastward flowing Counter Current.

7.32 South Atlantic Ocean

7.321 South Atlantic circulation

In the South Atlantic the upper-water gyre extends from the surface to a depth of about 200 m near the equator and to about 800 m at the southern limits of the gyre at the Subtropical Convergence. The different portions of this gyre have different water properties and have individual names which are given in Fig. 7.9. It is considered that the wind stress of the south-east trade winds between the equator and 10 to 15°S is the main driving force. This acts upon the sea and causes the *South Equatorial Current* to flow west toward the American side of the South Atlantic. Part of the current passes across the equator into the North Atlantic and will be discussed later. The remainder turns south along the South American continent as the *Brazil Current* which then turns east and continues across the Atlantic as part of the Antarctic Circumpolar Current and then turns north up the African side as the *Benguela Current*. The Brazil Current is warm and saline, having come from the tropic region, while the Benguela Current is cold and less saline because of the contribution of Subantarctic Water and of upwelling along the African coast. A contribution to the water in the South Atlantic comes from the *Falkland Current* flowing north from Drake Passage up the coast of South America and separating the Brazil Current from this coast to about 30°S. The South Atlantic circulation is bounded on the south by the Subtropical Convergence. For simplicity this Convergence and the Antarctic Polar Frontal Zone are shown in Figs. 7.6 and 7.9 as lines but, as described for the Polar Frontal Zone, they must be considered in reality to be zones of finite extent with seasonal variations of position.

7.322 South Atlantic volume transports

Wüst (1957) made calculations of the volume transport of water in the South Atlantic and these are interesting both as descriptive of that ocean and as examples of the application of oceanographic principles. When considering Wüst's calculations it must be remembered that there were no direct current measurements of any consequence available to him, and the volume transports had to be deduced indirectly. Calculation by the geostrophic

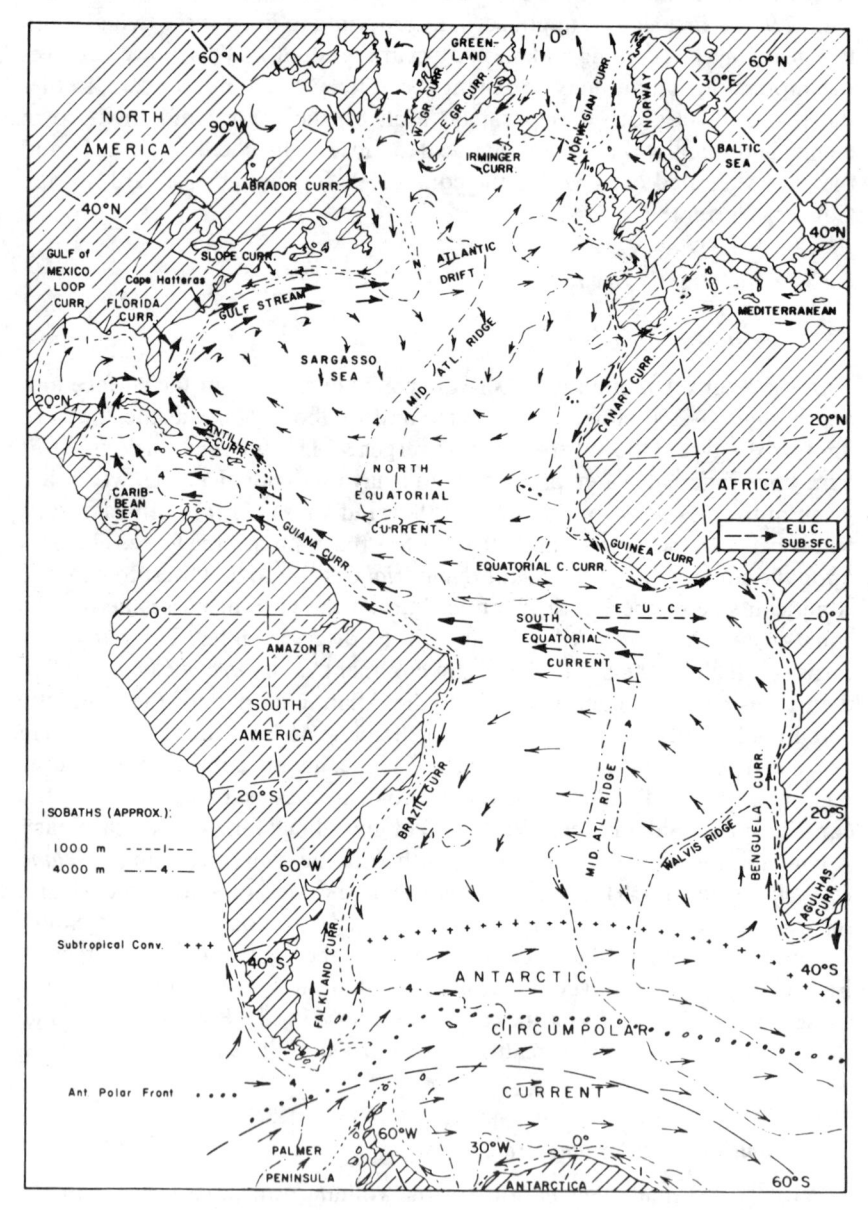

FIG. 7.9. Atlantic Ocean—bathymetry and surface circulation.

method from the density distribution gave the relative currents from top to near bottom across the ocean. To make the relative currents absolute a depth of no motion was chosen between the Antarctic Intermediate Water which must flow north and the Deep Water which must flow south. This depth varies over the ocean but is at about 1400 m for a section at 30°S from the South American to the African shore. These absolute currents could then be used to determine the flow at various depths and the positions across the section. The resulting volume transports of the different water masses had to satisfy volume continuity, i.e. there must be no net flow north or south through the transoceanic section because there is no evidence of any long-term change of sea-level north or south of the section. It also assumes that there is no net flow into or out of the North Atlantic + Arctic Sea combination. (There are certainly large exchanges of water between the Atlantic and Arctic but these are internal to the present discussion and cannot affect the mean sea-level.) The assumption of no net flow into the Atlantic + Arctic combination is not quite true because there is some flow through the Bering Strait from the Pacific into the Arctic. Recent measurements and estimates suggest that the mean transport ranges from 0.5 to 1.1 Sv, but this is small compared with the volume transports across the section in the South Atlantic and can be ignored. It should be noted that conservation of salt across the section also had to be satisfied.

Sverdrup summarized the volume transports (*The Oceans*, 1946) and these are shown schematically in Fig. 7.10. In the upper layer there is a transport to

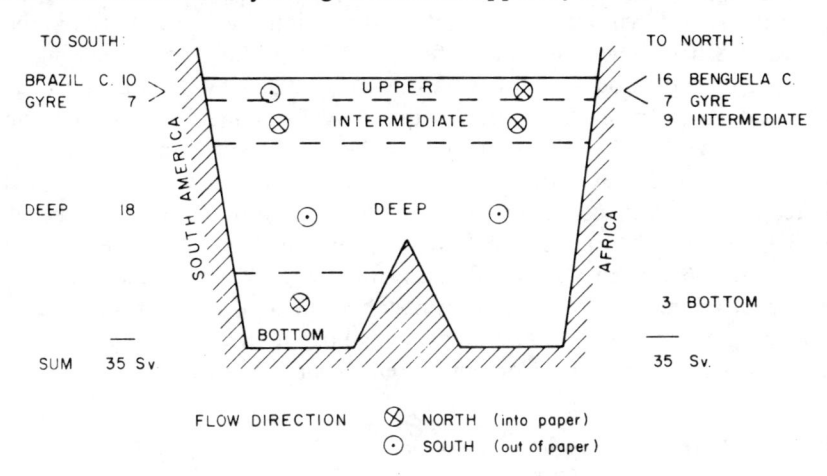

TO SOUTH:					TO NORTH:	
BRAZIL C.	10		UPPER		16	BENGUELA C.
GYRE	7				7	GYRE
			INTERMEDIATE		9	INTERMEDIATE
DEEP	18		DEEP			
					3	BOTTOM
			BOTTOM			
SUM	35 Sv				35 Sv.	

FLOW DIRECTION ⊗ NORTH (into paper)
⊙ SOUTH (out of paper)

FIG. 7.10. South Atlantic—north–south volume transports (schematic).

the north of about 23 Sv on the east side (Benguela Current and South Atlantic Gyre) as against 17 Sv to the south on the west side (Brazil Current and the Gyre). The Antarctic Intermediate Water has a transport to the north of 9 Sv, the Deep Water has a transport to the south of 18 Sv, and the Bottom

Water has a transport to the north of 3 Sv. There is therefore a net transport to the north of 18 Sv in the upper layers (from the surface to 1400 m depth) together with the Bottom Water, balanced by 18 Sv to the south in the Deep Water. This represents a fairly rapid rate of exchange of water in the north–south direction in the South Atlantic Ocean.

An example of a recent calculation of heat and fresh-water fluxes was given in Section 5.384. In that case continuity of fresh water relative to a standard salinity was used instead of conservation of salt relative to zero salinity.

7.33 Equatorial Atlantic circulation

We will describe the circulation in the equatorial Atlantic briefly, reserving a fuller description of equatorial circulation for the more extensive Pacific equatorial system (Section 7.61).

At the equator, the Atlantic extends from 10°E to 45°W, a distance of over 6000 km and the main currents are the North Equatorial CounterCurrent (NECC) flowing to the east from 8°–3°N, the South Equatorial Current (SEC) to the west from 3°N–8°S, the Equatorial Undercurrent (EUC) flowing to the east at the equator between about 50 m and 300 m, and the Brazil Coastal Current. At 30°W the eastward NECC has a maximum transport of about 16 Sv in August decreasing seasonally to about 4–10 Sv from November to May, while downstream at 10°W the seasonal variation is similar but the transport has decreased to about 6 Sv in September and 1–2 Sv for November to April due to downwelling and southward flow into the EUC. At 10°W, the westward SEC reaches a maximum flow of 16 Sv in May/June decreasing to 4–9 Sv from September to April, while downstream at 30°W the transport has increased to 27 Sv in June due to upwelling from the EUC, decreasing seasonally to 12–15 Sv from September to April. The SEC contributes to the Brazil Coastal Current which increases between 5°S and 2.5°S and then decreases as mass is lost to the EUC by 2.5°S and to the NECC by 8°N. At 30°W, the EUC has a transport which peaks in August at 25 Sv falling seasonally to 6–8 Sv in November to May, but by 10°W the peak is only 18 Sv in August falling to 5–9 Sv in November–April, the decrease being due to upwelling into the SEC. In addition to the above upper-layer currents there are deep North and South CounterCurrents near 5°N and 5°S.

7.34 North Atlantic circulation: general

The North Atlantic has been subjected to intensive study, particularly in recent years and, as a consequence, some new ideas about its circulation have developed. The classical pattern will be presented first and then new interpretations of the data will be described with some criticisms.

The circulation of the North Atlantic shown in Fig. 7.9 follows that presented by Sverdrup *et al.*, in *The Oceans* (1946) and by other authors earlier (e.g. Iselin, 1936). The clockwise gyre (Fig. 7.9) may be considered to start with the *North Equatorial Current* driven by the north-east trade winds. This current flows to the west and is joined from the south by that part of the South Equatorial Current which has turned across the equator into the North Atlantic. Part of this combined flow goes north-west as the *Antilles Current* east of the West Indies, and part goes between these islands and through the Caribbean and the Yucatan Channel into the Gulf of Mexico. In its passage through the Caribbean the flow is driven by the east winds in this region and the water piles up in the Gulf of Mexico. From there it escapes between Florida and Cuba into the North Atlantic as the *Florida Current*. The characteristics of this current indicate that its source is chiefly the North and South Equatorial Current waters which traverse the Caribbean. Little of the water from the Gulf of Mexico itself appears to be carried out in the Florida Current. Studies have shown, however, that the flow within the Gulf of Mexico sometimes forms a large loop. This loop often spawns an anticyclonic eddy which moves westward in the Gulf. Off the coast of Florida the Current is joined by the Antilles Current and from about Cape Hatteras, where the combination breaks away from the North American shore, it is called the *Gulf Stream*. The Florida Current water is distinguished by a salinity minimum due to the Antarctic Intermediate Water (Fig. 7.15) brought in by the South Equatorial Current. The Antilles Current is composed mostly of North Atlantic Water and the salinity minimum is much less evident in it. The Gulf Stream flows north-east to the Grand Banks of Newfoundland at about 40°N, 50°W. From there, the flow which continues east and north is called the *North Atlantic Current*. This divides and part turns north-east between Scotland and Iceland and contributes to the circulation of the Norwegian, Greenland and Arctic Seas which will be described later. The remainder of the North Atlantic Current turns south past Spain and North Africa to complete the North Atlantic Gyre and to feed into the North Equatorial Current. The southward flow covers the greater part of the North Atlantic as is shown in Fig. 7.9 and is sometimes referred as covering the Sargasso Sea. The flow is so slow and diffuse that it is hard to distinguish specific currents, although the *Canary Current* is recognized flowing to the south off the coast of North Africa.

7.341 Gulf Stream system

A more recent interpretation of the circulation, presented by Worthington (1976), questions several aspects of the classical pattern. His argument is based chiefly on a study of water mass characteristics, and the circulation which he proposes differs from the earlier pattern in two major respects. He argues that the circulation must be confined almost entirely to the north-western half of the North Atlantic, i.e. north-west of a line from Iceland to the mouth of the Amazon River, rather than extending over the whole ocean as in the previously described circulation pattern; he also suggests that the western circulation consists of two gyres, not one. The main reason for the limitation to the north-west half of the ocean is that (diluted) Mediterranean outflow water persists as a saline wedge extending from the coast of Europe and North Africa westward to Bermuda or further west (e.g. Fig. 7.18(b)) and from depths of a few hundred metres to more than 2000 m and that a southward flow through this mass is extremely unlikely. It is not suggested that this water itself forms a barrier to flow but that, in view of the small rate of supply of Mediterranean water (about 1 Sv), its continued existence is evidence that there can be no significant flow from north to south (of water of different characteristics) at the depths occupied by this water mass in the eastern North Atlantic. The *Gulf Stream System*, according to Worthington, then consists of a clockwise gyre of which the northern edge is the Florida Current and the Gulf Stream flowing north-east or east with transports of up to 150 Sv. Most of this recirculates to the south and west back into the Gulf Stream within the western half of the ocean. The main part of this gyre lies between 30° to 40°N and 40° to 80°W with about 30 Sv of the recirculation penetrating further south, through the Antilles and the Caribbean, to form the Florida Current. The second and smaller circulation, the "Northern Gyre", lies between about 38° to 52°N and 34° to 47°W, i.e. to the north-east of the Gulf Stream System, and has a maximum transport on its west side of about 74 Sv. This means that any flow to the east across the North Atlantic is not a direct continuation of the Gulf Stream itself. Worthington uses the term "North Atlantic Current" for the strong flow on the west side of the Northern Gyre. He mentions the possibility of irregular exchange of water between the two gyres.

It is admitted that some surface layer flow may occur in the eastern North Atlantic above the Mediterranean Water but Worthington finds little evidence for a North Equatorial Current as a significant volume transport contributing to the Gulf Stream System although he states that there are very few direct current measurements in the eastern tropical North Atlantic to confirm this.

The circulation paths are based on a study of the water mass characteristics for the whole, while the transport values are based on both geostrophic calculations and direct current measurements in the north-western North Atlantic. One of the results is a series of estimates of vertical flows by means of a so-called "box model" in which the ocean is divided from bottom to surface into five layers by isothermal surfaces at 4°, 7°, 12° and 17°C. Circulation patterns and horizontal transports are estimated for each of these layers and the vertical transports are determined by assuming continuity of volume and of water characteristics. This box model leads to an average residence time for water in the North Atlantic of 241 years, which makes it one of the youngest of the oceans. (Note that as defined in Section 5.221, "residence time" is not necessarily the same as "age".)

It must be pointed out that Worthington's interpretation is regarded as controversial. Specifically, Clarke, Hill, Reiniger and Warren (1980), from a multi-ship study off Newfoundland, have presented arguments against

Worthington's two main points of difference from the classical circulation pattern. They point out that Worthington's two-gyre hypothesis involves a rejection of the basic physics of geostrophic circulation and that lateral mixing can account for the water property differences which Worthington explains by his two gyres. Clarke et al. do recognize an eddy to the east of the Grand Banks (also observed earlier by Mann, 1967) but consider this to be a minor feature and that the North Atlantic Current is definitely a continuation of the Gulf Stream and that there is only one main gyre in the North Atlantic. They do agree that if the Mediterranean influenced water mass in the North Atlantic were maintained by *advection* (flow), Worthington's argument about no southward flow in the eastern North Atlantic would be valid. However, they point out that the Mediterranean outflow is small and that several studies have shown that the saline water mass in the North Atlantic could be maintained by lateral *diffusion* to the west from the high salinity Mediterranean outflow water concentrated along the eastern shore of the Atlantic, even in the presence of an ocean-wide gyre with slow southward flow in the east as envisaged by the classical picture of the North Atlantic circulation.

It is by such advancement of hypotheses or models, followed by critical study which may either support them or suggest revision, that our understanding of the ocean circulation advances. Accounts of the development of ideas about the Gulf Stream have been presented by Stommel (1965) and by Fofonoff (1981).

The salient feature of the Gulf Stream System in either the classical pattern or that of Worthington is the fast, concentrated north-eastward flow of the Florida Current and the Gulf Stream on the west side (westward intensification) contrasting with the broad and less well-defined south and westward flow over the rest of the ocean. The Gulf Stream is sometimes misrepresented as an individual current of warm water distinct from its surroundings, like a "river in the ocean". It is better thought of as the rapidly moving western edge of the Sargasso Sea, the 700- to 800-m deep pool of warm water which forms the upper layer of most of the North Atlantic Ocean. The Florida Current, which precedes the Gulf Stream, flows over the continental slope but the Gulf Stream itself is over deep water. The speeds of up to 250 cm/s (9 km/h) in the Gulf Stream are among the highest found in ocean currents. Radar altimeter measurements from *Seasat* for 3 months in 1978 (Cheney and Marsh, 1981) gave an average speed (calculated geostrophically from sea surface slope) of 150 cm/s, average width of 115 km and average change of surface elevation across the Stream of 140 cm, corresponding to a surface slope of 1.2×10^{-5}. The Gulf Stream is usually sharply defined on its north-west side, but much less so on the south-east toward the centre of the gyre. Due to the small scale of Fig. 7.9 the Gulf Stream is shown as a series of straight arrows directed to the north-east. However, when examined in detail it can perhaps best be described as consisting of a series of filaments of current which are usually sinuous or meandering. While one can indicate on a chart the general region where these filaments which comprise the Stream are likely to be found, it is not possible to predict where they will be individually at any particular time. Some of these meanders were recorded in detail during an unusual (at that time) oceanographic survey in 1950, called "Operation Cabot" (Fugilister and Worthington, 1951), when six oceanographic ships investigated a part of the Gulf Stream region simultaneously for 3 weeks. In particular, one meander developed to the extent that it broke off to form an individual eddy while the filament of current closed up behind it. More will be said about such eddies and similar features in the ocean in Section 7.344.

The North Atlantic Current is sometimes described as consisting of several branches or filaments of current forming a continuation of the Gulf Stream. Fugilister (1955) took data for the region between 75°W and 25°W for several cruises within a short period of time in 1953 and showed three significantly different interpretations of the same data. These interpretations varied from a single rather tortuous stream to a large number of fairly straight filaments. This exercise demonstrated the deficiencies in our observational knowledge of this region. The density of observations which it is practical to make from ships is still low and such observations are not sufficiently simultaneous to give an adequate picture of a complex region. However, observations from aircraft and from satellites can in many cases delineate temperature boundaries in the Stream and neighbouring waters and are helping to improve our knowledge of the Stream's location and of its changes with time.

Using such a technique, Cornillon et al. (1987) analysed 30 months of infra-red satellite images to locate the position of the Gulf Stream, identified as a marked separation between the warmer and colder surface water. This surface thermal position of the shoreward boundary, or "cold wall", of the Gulf Stream effectively reduced the two-dimensional data of the series of images to a single line per image. Taken together in Fig. 7.11(a) these lines present a time-series of the shape and position of the Gulf Stream over a period of two and one-half years. This particular description was based on a line for every two days between Cape Hatteras (35.5°N) to about the latitude of New York (40°N). These lines were computed relative to a straight line connecting the positions (36°N, 74°W) and (39°N, 60°W). In this time-series presentation individual features can be tracked as they move, or propagate, to the east. Some examples are connected with heavier solid lines in Fig. 7.11(a). It should be noted that solid lines sloping upward to the west indicate meanders or eddies that are propagating to the west, rather than to the east, against the mean flow of the Gulf Stream.

Collected into different time periods (Fig. 7.11(b)) this same data set clearly indicates the seasonal and interannual variations of the Gulf Stream. April to December, 1982, shows a general bias to the northern side of the Gulf Stream while the longer time-series (up to September, 1984) appears much more

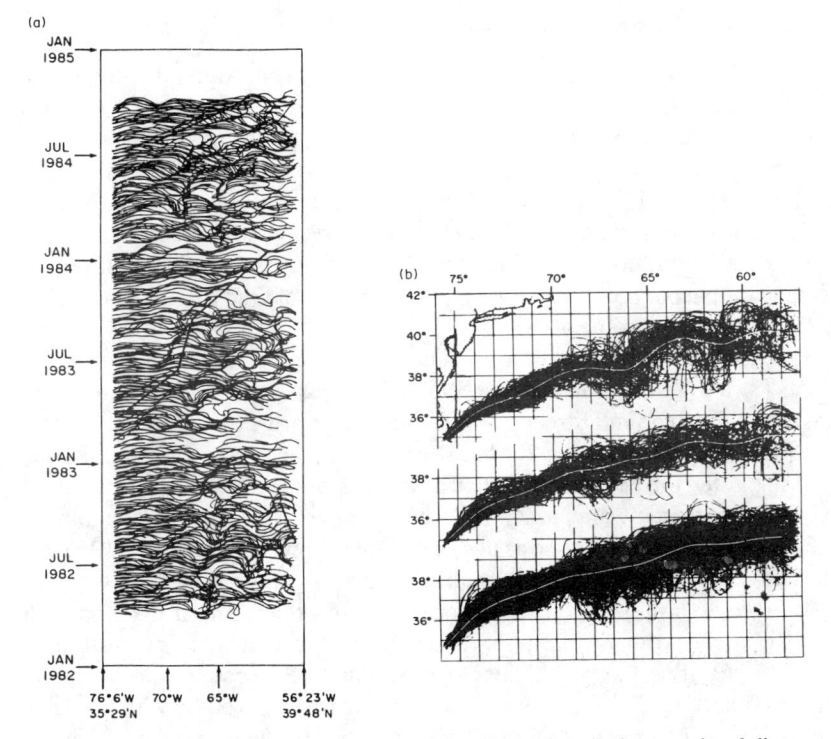

FIG. 7.11. (a) Positions of the Gulf Stream northern edge relative to a rhumb line from 36°N, 74°W to 39°N, 60°W from 1982 to 1984. One realization plotted for every two days. (b) Gulf Stream northern edges from (a) but superimposed on one another for (top) April to December 1982, (middle) all of 1983 and (bottom) April, 1982 to September 1984. The white lines represent the mean tracks.

symmetrical around a mean line. The middle set from 1983 is clearly different from the 1982 data in the top set of curves. It should be noted that most of this temporal variability is in the eastern half of these curves while to the west there is fairly good agreement in Gulf Stream mean position between the various data sets. Occasionally there is a sharp northward meander which deviates from the overall envelope of Gulf Stream patterns.

Between the Gulf Stream and the shore of North America there is a southwestward-flowing coastal current with an elongated counter-clockwise gyre between it and the Stream. The exact mechanism which maintains this gyre has been much discussed as a problem in dynamical oceanography. The coastal current is partly supplied from the Labrador Current which flows south out of the Labrador Sea (Fig. 7.26) and round Newfoundland. The presence of this southward-flowing, cold, low-salinity Labrador Sea water in proximity to the warm, saline waters of the Gulf Stream gives rise to a very complicated oceanographic situation in this western part of the North Atlantic.

7.342 Gulf Stream volume transport

According to the 1976 review of the North Atlantic circulation by Worthington, the average transport of the Florida Current (off Florida) is 30 Sv (with variations from 15 to 38 Sv according to other studies using the electromagnetic method). The transport increases north-east to 85 Sv off Cape Hatteras and to 150 Sv by 65°W and then decreases eastward to 37 Sv at 40°W. The increase west of 65°W is supplied by recirculation to the south-west from east of this longitude including water from the Antilles Current. A significant part of the increase may be in the form of Gulf Stream rings which could account for some 40 Sv.

One feature of the circulation which was observed directly with Swallow floats (Swallow and Worthington, 1957, 1961) was that underneath the north-eastward-flowing Gulf Stream was a south-westward-flowing countercurrent with speeds of 9 to 18 cm/s. Earlier studies using the geostrophic relationship had suggested the possibility of such a current but the different interpretations were not consistent and to many oceanographers the idea of a significant current close to the bottom was not acceptable. However, Stommel (1965) had developed a theory of the thermohaline circulation from which there was good reason to expect such a southward deep current. The 1957 and 1961 measurements with Swallow floats demonstrated that in the region about 33°N, 75°W it certainly did exist at those times. Subsequent measurements (see Fofonoff, 1981) have shown flows varying from 2 to 50 Sv with a mean of 16 Sv. Wüst computed that a similar southward flow must occur in the western South Atlantic. There is some indication of a southward flow below the Kuroshio on the west side of the North Pacific but the measurements are few and of short duration and do not really provide convincing evidence.

7.343 Gulf Stream temperature and salinity distributions

Figure 7.12 shows the distribution of temperature and salinity in a section across the Gulf Stream. The Stream itself is associated for dynamic reasons with the steeply sloping isotherms and isohalines in a relatively narrow band of about 120 km width as shown. The fact that the water to the left of the current is cold compared with that at the same depths to the right, and the steepness of the isotherms in such a section, gave rise to the term "cold wall" to describe the water to the left of the Stream. Although the term is striking, it must be remembered that the real slope of the isotherms in the sea here is only of the order of 1 in 200. This is certainly very steep for the slope of a property surface in the sea but it is not much for a wall! It is clear from the sections that the Gulf Stream is not an individual flow of warm water and is hardly distinguishable from the Sargasso Sea to its east. One other feature is the wide spacing (*thermostad*) between the 16° and 20° isotherms over most of the section. A body of so-called *18° Water* has been a permanent feature of the

Circulation and Water Masses of the Oceans 191

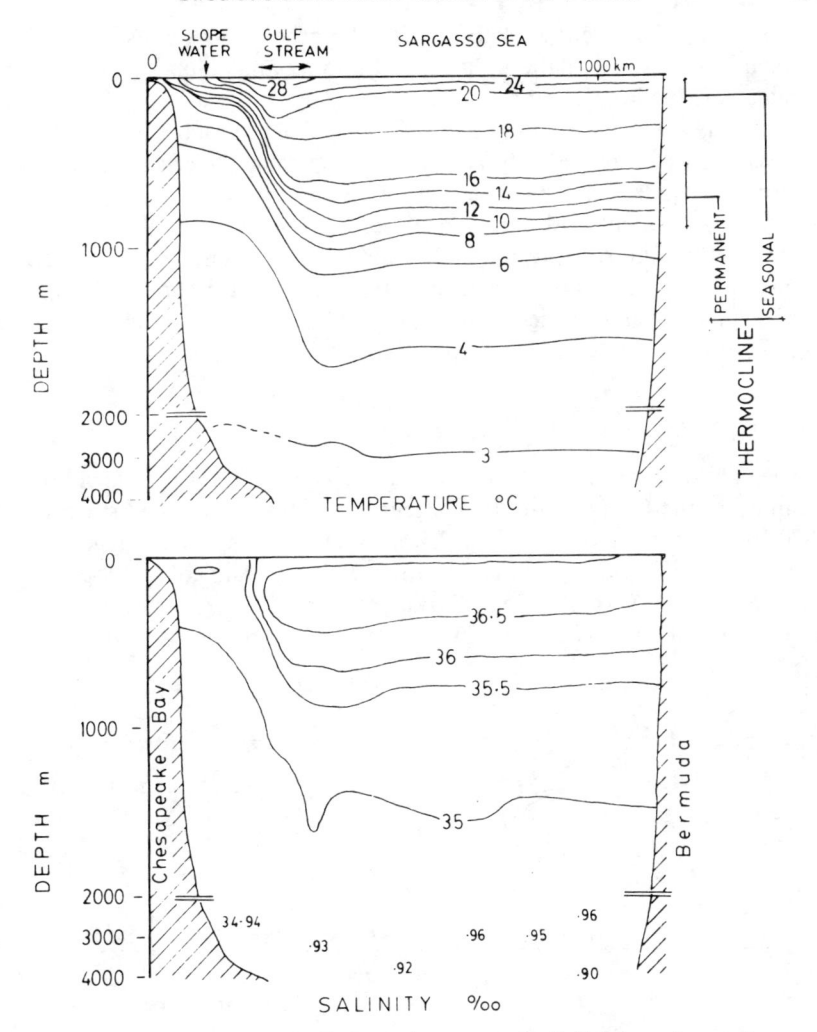

FIG. 7.12. Temperature and salinity sections across the Gulf Stream, August–September 1932.

west side of the Sargasso Sea since it was first observed from the *Challenger* in 1873, and its properties are very consistent within a few tenths of a degree on either side of 18°C and with a salinity of 36.4 to 36.6. If it is winter-formed water the remarkable feature is the consistency of its properties, particularly temperature considering the variation of air temperatures in winter. Estimates of the annual rate of formation of this water vary from about 50% of its average volume to zero. The rate of formation is calculated from the heat loss expected in winter and the differences in rate of formation are due to the use of different eddy diffusion coefficients (Section 5.35), particularly for evaporation, by different workers. As Worthington (1976) points out, "Of the problems in ocean circulation, that of the amount and manner of water-mass formation is the least tractable".

For many years, information on the Gulf Stream area of the North Atlantic has been collected by the U.S. Naval Oceanographic Office and later by the National Oceanic and Atmospheric Administration from a variety of sources such as government and commercial ships, from aircraft using radiation thermometers, photography and radar, and from satellites. Since 1966 a digest of this information has been published monthly with sea temperatures, Gulf Stream position, and articles on special features of interest in this region ("*gulfstream*", N.O.A.A., U.S. Department of Commerce).

7.344 Gulf Stream rings and ocean eddies

The cold or cyclonic eddy observed in "Operation Cabot" to form from a meander of the Gulf Stream drew attention to this process and since about 1970 much more information has been obtained. Fugilister suggested the name *Gulf Stream Rings* for these features which entrain colder Slope Water into the warm Sargasso Sea. Parker (1971) illustrated this process of formation as shown in Fig. 7.13. In this example, the result is an anticlockwise

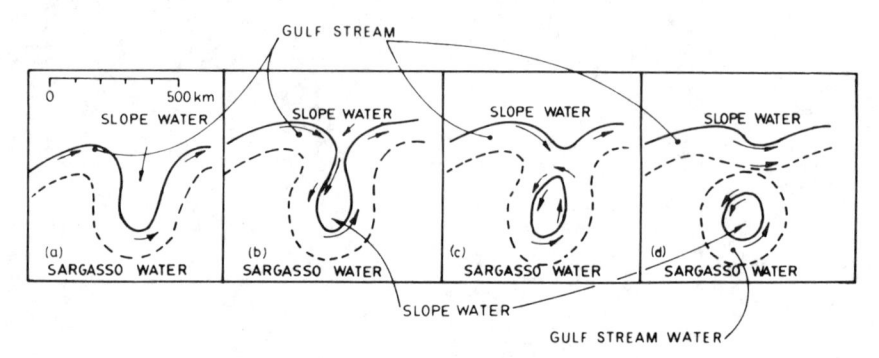

FIG. 7.13. Diagram of Gulf Stream ring formation.

(cyclonic) rotating ring of cool, Slope Water in the warm Sargasso Sea. Alternatively, a meander may start on the north side of the Gulf Stream and result in a clockwise rotating ring of warm Sargasso Water in the cool Slope Water. Although first observed in the Gulf Stream region such rings are now known to be general features of the ocean circulation, the sense of rotation for the same type of ring being opposite in the two hemispheres to satisfy the geostrophic relationship. A cold ring is the oceanic equivalent of an atmospheric low-pressure system, i.e. a cyclone, while a warm ring is the equivalent of a high-pressure system, i.e. an anticyclone. For this reason, the sense of rotation of a cold ring is often referred to as *cyclonic* (anticlockwise in

Circulation and Water Masses of the Oceans   193

the northern hemisphere, clockwise in the southern) and for a warm ring as *anti-cyclonic* (clockwise in the northern hemisphere, anticlockwise in the southern). These rings are 150 to 300 km in diameter and some 3000 m in vertical dimension and have lifetimes of up to 2 years. At any time there may be three or more warm-core rings north of the Stream and eight to fifteen cold-core rings south of it in the area west of 50°W and north of about 30°N. After formation the rings move in a south or south-westerly direction at speeds of a few kilometres per day and eventually merge with the Gulf Stream again (e.g. see Richardson, 1980). In a newly formed cold ring the thermocline is about 500 m higher than in the surrounding water and, in addition, it carries both the chemical characteristics and the biological populations of the Slope Water into the Sargasso Sea area (Ring Group, 1981).

In Fig. 7.14(a) are shown, from a study by Richardson, Cheney and Worthington (1978), the positions of nine cold and three warm rings in the spring of 1973, obtained from four XBT and CTD surveys together with other data centre information, including satellite observations (infra-red). There was also a meander at 62°W which had formed when a warm ring coalesced with the Gulf Stream; it later formed a new warm ring. At 57°W there was a meander/ring which continued during the period of study. In Fig. 7.14(a) are shown two lines, A and B, which show the positions of two vertical temperature sections reproduced in Fig. 7.14(b) and (c). In section A, the Gulf Stream appears at the left where the isotherms slope down steeply to the east and two cold rings are shown at 8 and 9 where lower temperature isotherms rise up toward the surface underneath the thermocline. In section B, the warm ring 1 appears at the left where the isotherms dip down to a maximum depth at the 100 km distance, while the Gulf Stream is at the right.

Other features seen in section A in Fig. 7.14(b) are the 18° water, and evidence of the Gulf Stream recirculation to the west revealed by the isotherms (which approximate isopycnals here) sloping up to the east at 300 km distance. The slope of the isotherms for ring 8 shows the cyclonic direction of rotation, to the south at the west side of 8 and to the north at its east side.

Seasat altimeter measurements in 1978 (Cheney and Marsh, 1981) indicated that the surface level at the centre of cold rings was up to 95 cm deeper than at the edge while in warm rings it was up to 75 cm higher at the centre than at the edge. Infra-red satellite images also clearly reveal the position of the Gulf Stream and the eddy structure associated with it. In Plate 12 we present a single image of the Gulf Stream region which not only shows the core of the Stream by the significantly warmer (darker) water but also shows the presence of the strong cyclonic, cold rings to the south of the Stream. A single warm-eddy expression can be seen directly to the north of the dominant cyclonic meander in the centre of the satellite image. As discussed earlier it is this type of meander which is primarily responsible for the formation of such cold eddies which then propagate to the west. Likewise warm rings (or eddies) form through northward excursions (called anticyclonic meanders) of the mean

194 Descriptive Physical Oceanography

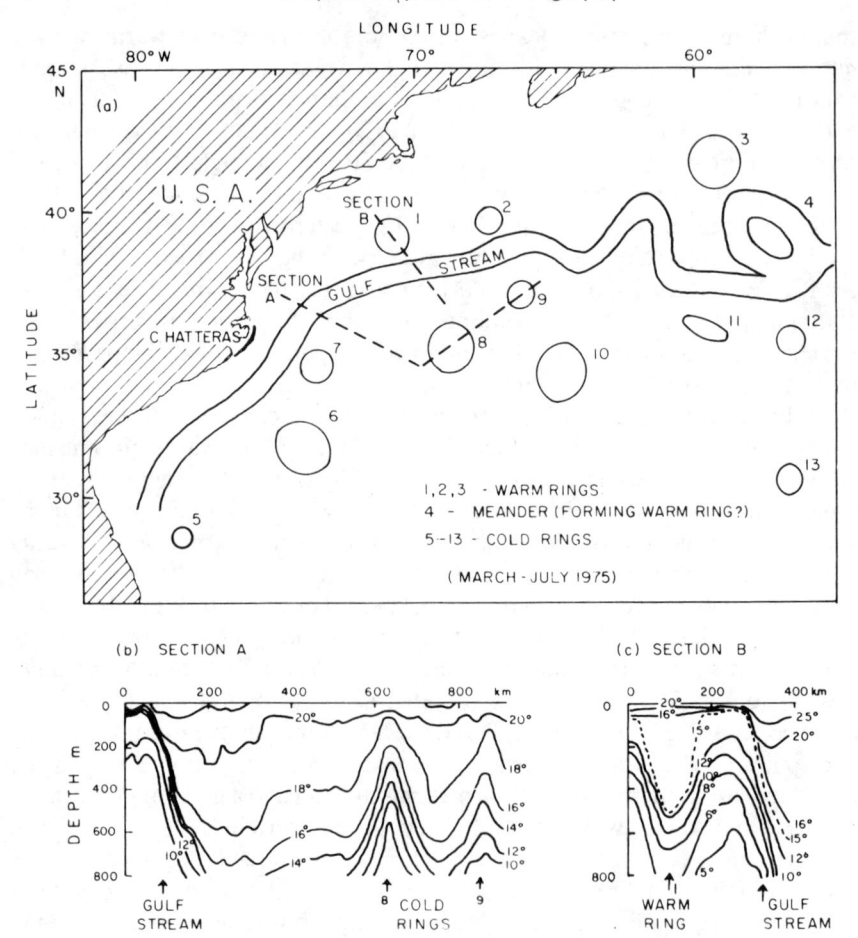

FIG. 7.14. (a) Locations of Gulf Stream and of warm and cold rings, March–July 1975, (b, c) Vertical temperature sections along lines A and B in (a) showing Gulf Stream and cold and warm ring structures.

Gulf Stream which then separate and are thereafter more quickly erased in terms of their surface signature.

Evans et al. (1985) intensively studied a single warm-core Gulf Stream ring with both satellite and in situ data and found that the ring interacted significantly with the surrounding waters by exchanging both angular momentum and water properties. These authors did note a problem in following the satellite infra-red surface signatures of the warm-core Gulf Stream rings which does not appear to be the case as strongly with the cold cyclonic rings. Examining in situ data from this same ring, Olson et al. (1985) and Schmitt and Olson (1985) determined that warm-core ring decay rates were similar to those of other rings and that decay processes progressed

Circulation and Water Masses of the Oceans 195

according to linear theory. It was discovered that the reabsorption of such warm rings back into the Gulf Stream system provided a source of high-salinity, high-oxygen anomalies in the main thermocline of the Sargasso Sea making the reabsorption of these rings an important thermocline ventilation mechanism for the Sargasso Sea.

A now accepted feature of the open ocean circulation about which there have been hints since the 1930s or even earlier (see the article by Wunsch, 1981) is that embedded in the general large-scale circulation there are large numbers of eddies, some circular, some elongated. They have been studied in the North Atlantic since about 1960, most intensively since about 1970, e.g. in the Soviet POLYGON-70 study and the U.S. MODE-1 study in 1973 (MODE Group, 1978) which were precursors to the 1978 international POLYMODE study. The eddies are typically tens to hundreds of kilometres across and often extend from the surface to the bottom. The mode of formation of the Gulf Stream rings and others near major currents is clear (although the reasons for the meanders starting are not obvious) but the origin of the mid-ocean eddies is not yet understood. The ocean eddies have been likened to the storms and weather systems of the atmosphere, the smaller size and speed of motion of ocean eddies being related to the greater density of water than air. In spite of these characteristics, ocean eddies have been observed to persist over periods of many months. An excellent account of Ocean Eddies is presented in the Spring 1976 edition of *Oceanus* (MODE Group, 1976).

7.35 Atlantic Ocean water masses

When studying the water mass characteristics of the Atlantic Ocean we are fortunate in having a considerable amount of information and in particular the data from two extensive expeditions, both of which covered the greater part of the area systematically and in a relatively short time. The two expeditions were the German *Meteor* expedition during 1925 to 1927, and the International Geophysical Year studies of 1957–58 carried out by Woods Hole Oceanographic Institution of the United States with co-operation from the National Institute of Oceanography of Great Britain (Fugilister, 1960). The second of these studies was deliberately arranged to cover the South and Equatorial Atlantic along the same lines of stations as those occupied by the *Meteor* in order to obtain a direct comparison of the distribution of water properties after the interval of 30 years. One of the immediate results of the comparison of the two sets of data was the recognition that the distributions of temperature, salinity and dissolved oxygen were almost identical. This result was very comforting to oceanographers who have made a practice of assuming that the state of affairs which they observe in the ocean is a reasonably steady one, at least when averaged over a year to eliminate seasonal changes.

It should be noted that the spacing between stations in the above surveys was too large to resolve the mesoscale eddies mentioned earlier. These eddies are a source of shorter time-scale changes which, when averaged over a sufficient time, do not alter the mean large-scale circulation studied by the surveys.

7.351 Atlantic Ocean upper waters

In the surface layer of the Atlantic, the salient temperature characteristic is that it is high at low latitudes and decreases to higher latitudes (Figs. 4.1, 4.2). Between the equator and about 20°N the temperature ranges between 25° and 28°C with little seasonal change (cf. the Pacific Ocean, Fig. 4.7). In the South Atlantic, south of the equator to 40°S, there is a seasonal change of temperature of about 5K between winter and summer; at higher latitudes the seasonal range of temperature decreases. In the North Atlantic the difference between winter and summer temperatures rises to 10K at 40°N and then decreases at high latitudes. The salinity does not show any significant seasonal change (except close to ice). There is a minimum value of 35 just north of the equator and then maxima of 37.3 just north of the equator and then maxima of 37.3 in the tropics at about 20°N and S. From here the salinity decreases to 34 or less at high latitudes.

There is a marked difference in the surface water characteristics between the west (American) side and east (European–African) sides. In the North Atlantic there is a difference of about 25K in the sea-surface temperature between Florida and Labrador, compared with only about 10K between the same latitudes on the east side (North Africa to Scotland). In salinity there is a south–north difference of about 3 on the west compared with 1.5 on the east. These differences between west and east are clearly associated with the differences in currents. In the west, there is the contrast between the warm, saline Florida Current in the south and the cold, low salinity Labrador Current in the north whereas on the east side there is just the slow, diffuse southward flow of the North Atlantic gyre with no contrasting water masses.

In the South Atlantic, the effect of the Brazil Current is present near the equator in contrast to the Falkland Current in the south but the differences in properties are not as marked as in the North Atlantic. On the east side of the ocean, low temperatures immediately off the coast of South Africa are due to upwelling of subsurface waters. Low salinities in this region are also characteristic of this water but the low salinities farther north, off tropical Africa, are due to river runoff.

7.352 Atlantic Ocean deep-water masses and circulation

The main features of the deep-water characteristics and circulation of the Atlantic are well shown in Fig. 7.15 which is based on the GEOSECS data for the Atlantic collected in 1972–73 (Bainbridge, 1976). The four sections show the distributions of potential temperature, salinity, potential density and

Circulation and Water Masses of the Oceans

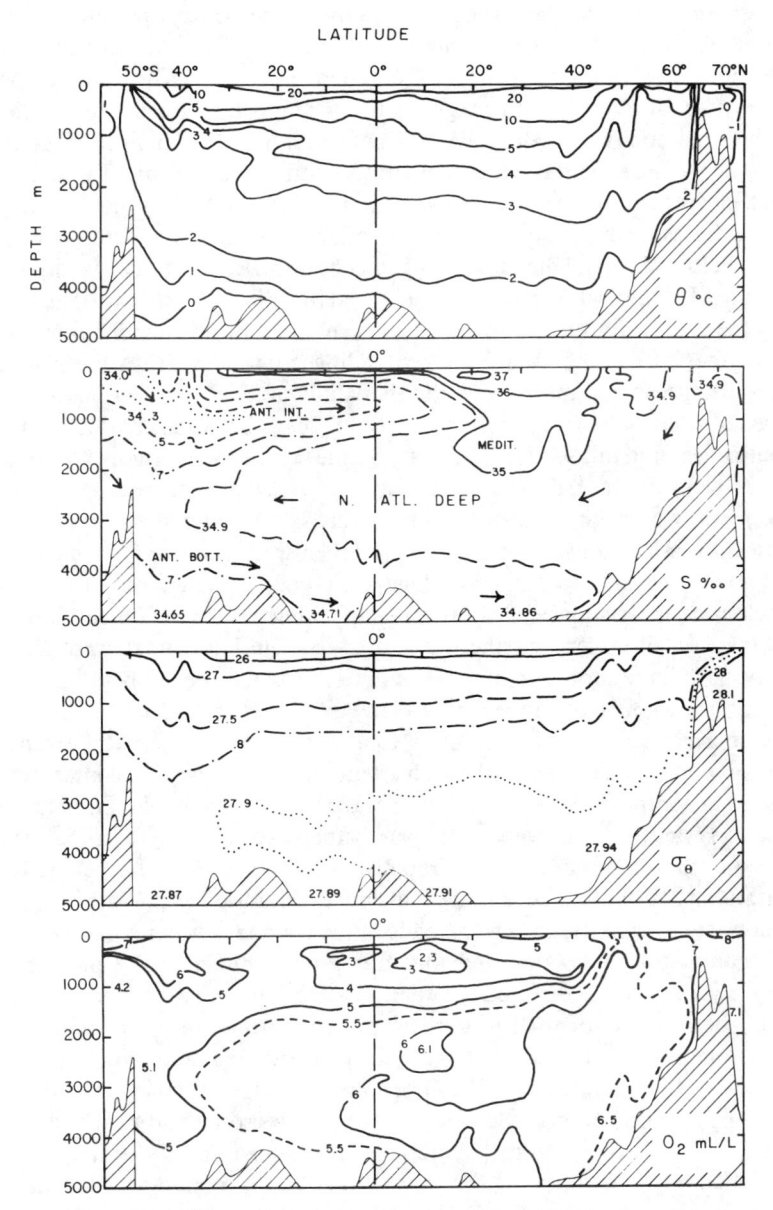

FIG. 7.15. Atlantic Ocean—south–north vertical sections of water properties along the western trough (data from GEOSECS Atlas, 1976).

dissolved oxygen along the western trough of the Atlantic Ocean, i.e. between the American continent and the Mid-Atlantic Ridge. The temperature section shows clearly that the largest variations both horizontally and vertically are in the upper layer, and emphasizes the statement made earlier that vertical property gradients in the sea are generally much greater than horizontal ones.

The Antarctic Intermediate Water from south of the Polar Frontal Zone is evident to some extent as a low-temperature tongue centred at 1000 m, but is much clearer as a low salinity tongue extending to 25°N. It is also apparent, although less clearly, by an oxygen maximum at about 800 m depth to about 10°S. Below this and flowing along the bottom to the north is the cold Antarctic Bottom Water which can be traced by its low salinity to about 45°N.

Between the Intermediate Water above and the Bottom Water below there is the great bulk of the Deep Water extending from the North Atlantic to the Southern Ocean and most evident in the salinity and oxygen sections. (It also appears to be shown by the σ_θ section but this is an artefact which will be discussed later in this section.) For a long time the accepted origin of this Deep Water was in the Labrador and Irminger Seas off the southern tip of Greenland where winter cooling of relatively saline water was considered to cause it to sink to considerable depths and then spread south. This suggestion was made by Nansen (1912), and Dietrich (1969) presented data from 1958 which apparently showed this process happening. In the region south of the southern tip of Greenland the water properties in late winter were almost uniform from the surface to near bottom. For instance, the temperatures from the surface to 3000 m ranged only between 3° and 3.25°C in an area over 150 km across. From this column, a tongue of 3° to 3.25°C water extended south at depths of 2000 to 2500 m. This water had a high oxygen content; even at 2000 m depth it was still 90% of that at the surface. The explanation for this situation was that the deep water was actually forming at the time, being cooled at the surface and sinking rapidly to 2000 to 3000 m depth, and then continuing south at its density level as North Atlantic Deep Water.

However, no calculations were made of the volume of water so formed and, as accumulating data indicated that this process did not take place every winter, attention was directed to other possible sources or mechanisms. As a result both of studies of water property distributions and of current measurements, it is now considered that the major source of North Atlantic Deep Water (up to 80%) is in the Greenland Sea where deep water is formed, by the open-ocean process (Section 7.11), and flows south and west and mixes with North Atlantic (Central) Water above it to form the *Northeast Atlantic Deep Water* (2.5°C and 35.03 salinity). Some of this continues south in the eastern basin of the North Atlantic over bottom water of Antarctic origin. The remainder of the Northeast Atlantic Deep Water flows west across the Mid-Atlantic Ridge into the western basin, below the Labrador Sea water in the north (3.4°C and 34.89) but above the *Northwest Atlantic Bottom Water* (1°C and 34.91). The North Atlantic Deep Water which flows south across the equator therefore has two components, the Northeast Atlantic Deep Water and the Northwest Atlantic Bottom Water (see Warren, 1981 for a fuller review). Another feature of the NADW is that evidence gathered in 1981 showed wide-spread reduction in salinity, in the 20 years since 1962, by 0.2 in the deeper water and more in the shallower parts, suggesting a relatively rapid response to climate variations.

One other point to mention is that both current meter measurements and water property distributions suggest that the overflows between Scotland, Iceland and Greenland are not continuous but probably occur in pulses—the dynamic reasons are not known.

The density characteristic plotted in Fig. 7.1 is σ_θ for the following reason. At shallow depths, the difference between t and θ, and so between σ_t and σ_θ, is small and either may be used to describe the density distribution. In deeper water, the difference between t and θ becomes significant and σ_θ is preferred because the effect of pressure on the temperature of the water has been removed. In the upper 1000 m or so, σ_θ follows θ rather than S, e.g. the effect of the low salinity of the Antarctic Intermediate Water is not apparent in the σ_θ section. However, in the deeper water, the effect of salinity is seen in the relation of σ_θ to the North Atlantic Deep salinity maximum combined with decreasing temperature (Lynn and Reid, 1968; Reid and Lynn, 1971).

The decrease of σ_θ with increasing depth below about 4000 m was earlier thought to indicate that the bottom layer was gravitationally unstable because it appeared that denser water was above less dense. This is not really the case—it is an artifact of using σ_θ, i.e. $\sigma_{\theta,S,0}$, the value for pressure = 0. Because the thermal expansion and compressibility of sea-water vary with temperature (in particular, colder water is more compressible), the vertical rate of change of σ_θ is not a good measure of stability when considered over large depth ranges. A better estimate is obtained by using a value of σ for a higher pressure, e.g. $\sigma_4 = \sigma_{\theta,S,4}$ for a pressure of 40,000 kPa (= 4000 dbar) which is close to the pressure at 4000 m depth. If this is done for the deep Atlantic, the value of σ_4 increases to the bottom, as shown in Fig. 7.16.

The effect of the difference in compressibility can be seen by the following "thought experiment". Imagine two equal volumes of sea-water at, say, 5000 m depth, the upper one more saline but warmer and the lower one less saline but cooler so that it was *slightly* more dense than the upper. The two would then be in stable equilibrium. Now, if the two volumes were raised adiabatically to the surface, the lower (cooler one) would expand more than the upper (warmer) one and could become less dense than the upper one—the situation would be statically unstable (less dense water under more dense). The imagined action of raising the two water samples from depth to the surface is, in effect, what we do when we calculate σ_θ, i.e. calculate the effect of taking a sample from high pressure (depth of 5000 m in our thought experiment) through a large pressure reduction to the surface (water pressure zero) as in Fig. 7.15 (σ_θ). The less drastic procedure of calculating σ_4 is equivalent to a

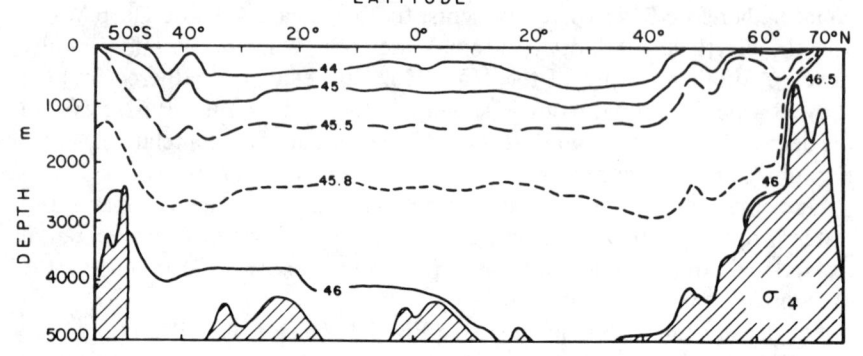

FIG. 7.16. Atlantic Ocean—south–north vertical section of σ_4 along the western trough (data from GEOSECS Atlas; 1976).

much smaller pressure reduction (5000 m to 4000 m depth) and the expansion of the cooler sample could be insufficient to make it less dense than the warmer one above it and the pair would still be in static equilibrium at 4000 m. This is the situation in the deep water of the South Atlantic (Fig. 7.16) that we described above.

The bottom waters in the western and eastern basins of the Atlantic have noticeably different properties in the South Atlantic because of the barrier of the Mid-Atlantic Ridge. Figure 7.17(a) shows transverse sections of temperature and salinity across the ocean at 16°S to illustrate this. On the west side the Antarctic Bottom Water has temperatures down to 0.4°C whereas on the east side the minimum is 2.4°C. The Antarctic Bottom Water is prevented from flowing directly into the eastern basin by the Walfish Ridge with a sill depth of less than 3500 m extending from Africa to the Mid-Atlantic Ridge. The salinity difference between the two sides is less remarkable, 34.7 on the west compared to 34.9 on the east. It is interesting to note that the densities on either side of the Ridge are identical, the temperature and salinity differences being in such directions are to cancel their effects on density. Other water masses evident in the figure are the Antarctic Intermediate Water with its low salinity core at about 800 m and the more saline North Atlantic Deep Water with its higher salinity core at about 2500 m.

Farther north at 36°N (Fig. 7.17(b)) the temperature difference has fallen to only 0.3K and the salinities are almost identical. The influence of the outflow of the Mediterranean Water is evident in the salinity section with its core at about 1000 m, but is less evident in the temperature section because the water has settled to a level determined by its density which is determined mainly by temperature at this depth. Figure 7.18 shows horizontal sections of temperature (*t*) and salinity (*S*) at 1000 m depth for the North Atlantic (Wüst and Defant, 1936) showing how this Mediterranean water spreads west and north across most of this region. At the west (Fig. 7.17(b)) the steep

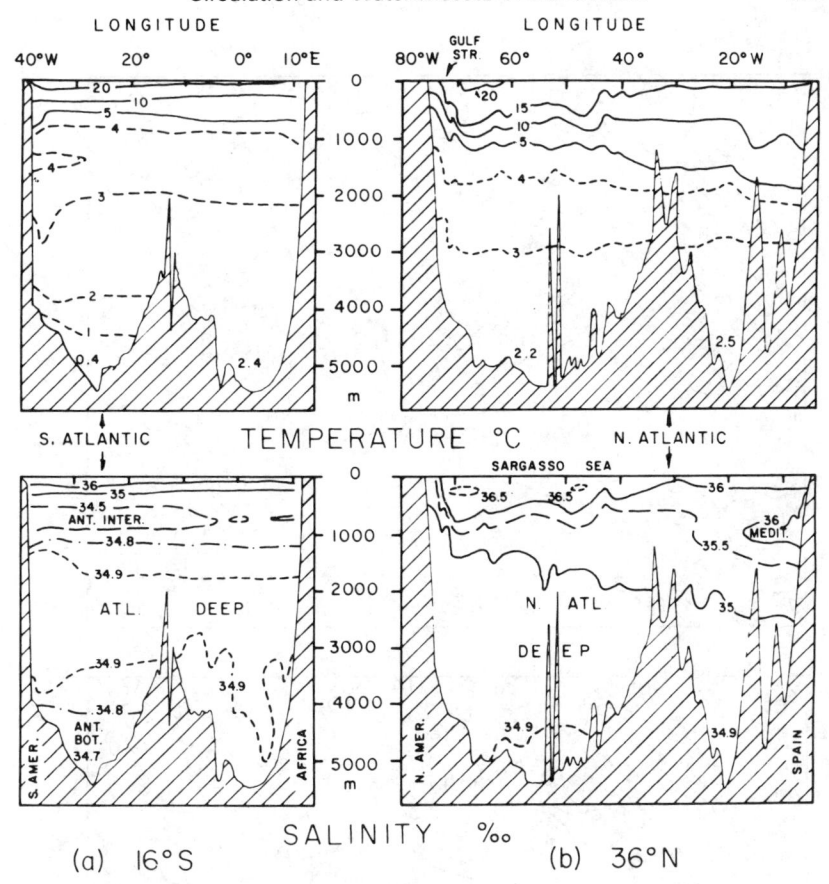

Fig. 7.17. West–east sections for temperature and salinity across the Atlantic Ocean at 16°S and 36°N.

slopes of the isotherms and isohalines associated with the Gulf Stream are also evident in the upper water.

7.353 T–S characteristics of subsurface waters of the Atlantic Ocean

The average T–S characteristics of the waters below the surface layers are shown in Fig. 7.20. The upper waters below the surface layer appear on this diagram as the *Atlantic Central Waters, North* and *South*, extending to depths of 300 m on either side of the equator and deepening to 600 to 900 m at mid-latitudes and then getting somewhat shallower at high latitudes. Both of the Central Water masses appear on the T–S diagram as straight lines extending from high temperature and salinity to lower temperature and salinity. At first sight, these Central Water Masses might appear to be examples of water masses produced by the vertical mixing of water types represented by the

202   Descriptive Physical Oceanography

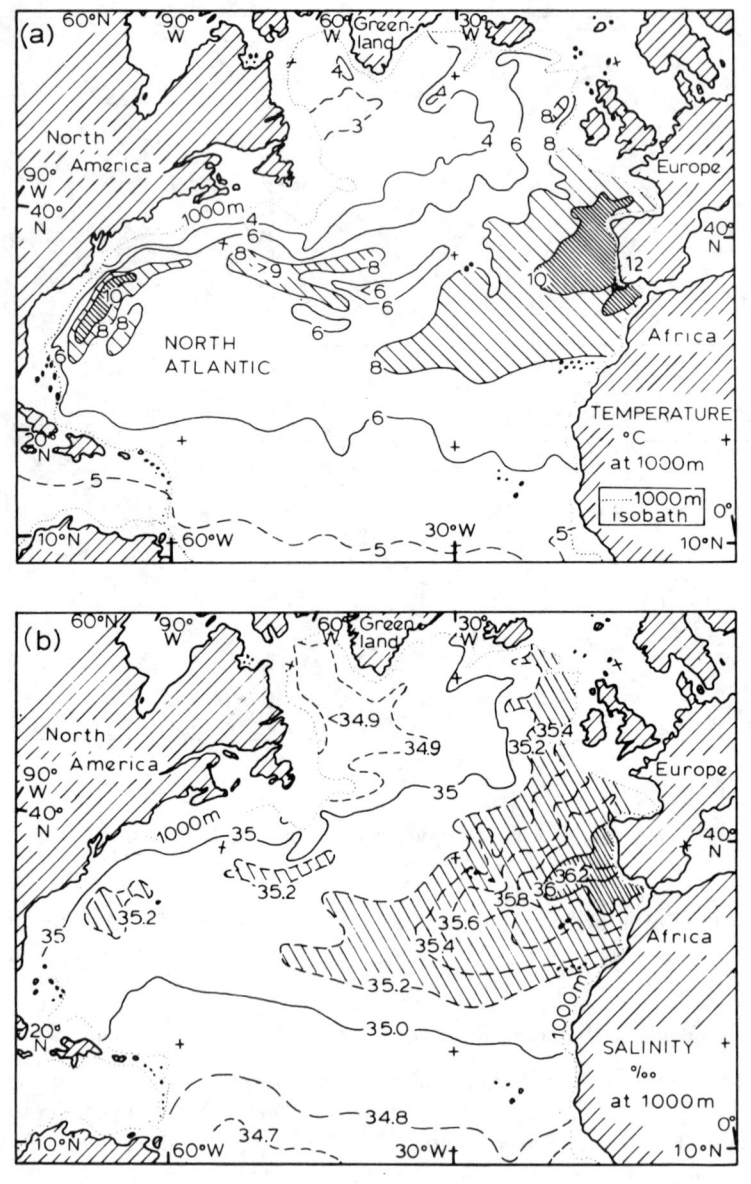

FIG. 7.18. North Atlantic: horizontal sections at 1000 m depth of (a) temperature, (b) salinity, to show the spreading of Mediterranean Water.

characteristics at the end of the lines (on the diagram). This, however, is believed not to be the case. Iselin (1939) pointed out that if one examines the winter T–S characteristics in the north–south horizontal direction at the surface over a range of latitudes near the subtropical convergences one finds the same T–S characteristics that one finds in the vertical below the surface in the Central Water Masses at lower latitudes. He then suggested that these water masses originate by sinking on the equatorward side of the subtropical convergences. The cooler, less saline water sinks at higher latitudes while the warmer, more saline water sinks at lower latitudes and flows equatorward above the cooler water. In the South Atlantic, the Central Water terminates at depth where it merges into the well-defined Antarctic Intermediate Water. In the North Atlantic, the Arctic Intermediate Water is a much less significant water body and the Central Water merges into this at high latitudes and into the Mediterranean Water at lower latitudes.

The effect of the Mediterranean Water can be seen in the T–S diagram (Fig. 7.20(a)) as a salinity maximum above the Deep Water maximum. The Mediterranean maximum is very marked for waters of the eastern North Atlantic but is less conspicuous in the west and in the south as it gradually loses its characteristic properties by mixing with the waters above and below it. However, by careful analysis using the core method, the Mediterranean Water can be traced down through the South Atlantic.

Three points must be mentioned about Fig. 7.19 and 7.20. The first is that the lines indicate mean values for large areas; the individual points scatter over a band whose width is about one-third of the separation of the two lines which represent the mean values for the South and the North Atlantic, i.e. a band of width equivalent to about 0.1 in salinity on the diagram. (This is also true for the other T–S curves in Figs. 7.19 and 7.20.) To better illustrate this scatter we present, as an example, Fig. 7.21 which shows all available data points for a 10° square, i.e. 10° latitude by 10° longitude, for the tropical Pacific south of Hawaii. In this figure, the T–S relationship is expressed by three lines. The centre line represents the mean T–S curve for that square while the bracketing lines represent one standard deviation in salinity on either side of the mean for all data within that square. Thus we can examine not only the distribution of the mean property-to-property relationship but also the degree of variability in this relationship. This variability is usually greatest at the upper, or warm, parts of the curves, e.g. the Tropical Upper Water in Fig. 7.21. It is caused by changes at the sea surface which alter both the temperature and salinity. In some regions, lateral meanders of boundaries between water masses also cause variability even below the surface as in Fig. 7.21 where both the North Pacific Intermediate and the Pacific Equatorial Waters are found within the same square.

The second point is that Figs. 7.19 and 7.20 present only the salient features of the T–S curves averaged over large oceanic areas. When one looks at the T–S curves for local regions, considerable variations from the mean curves are

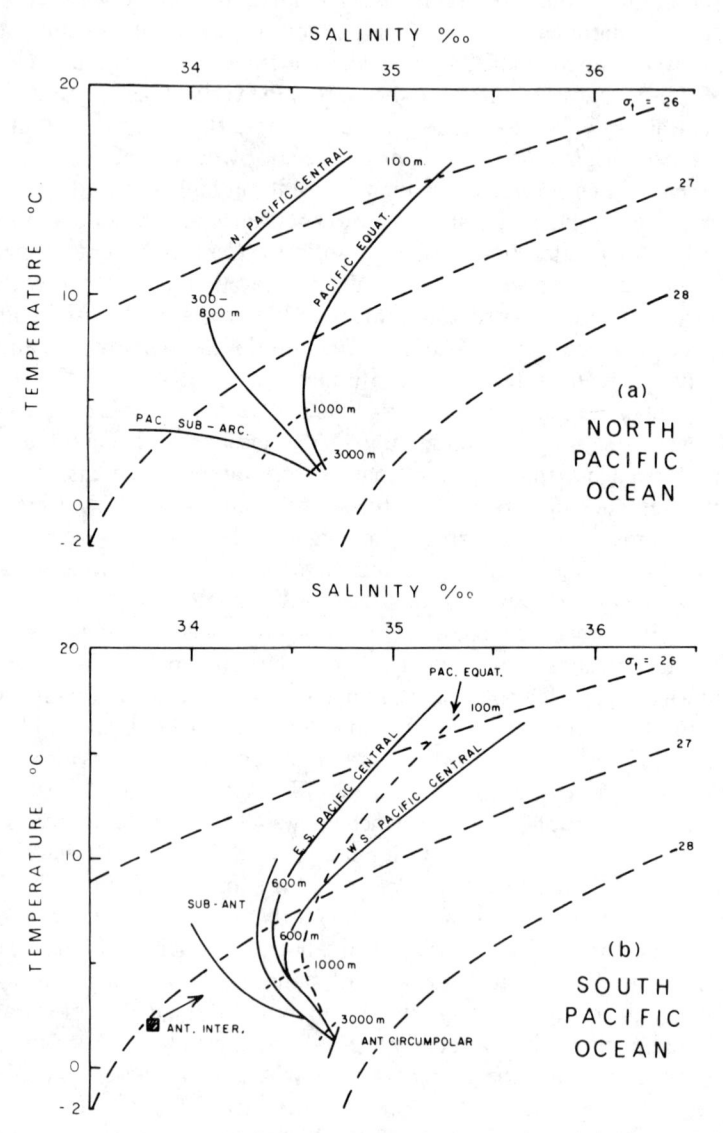

FIG. 7.19. Average temperature–salinity (T–S) diagrams for the main water masses of the Pacific Ocean.

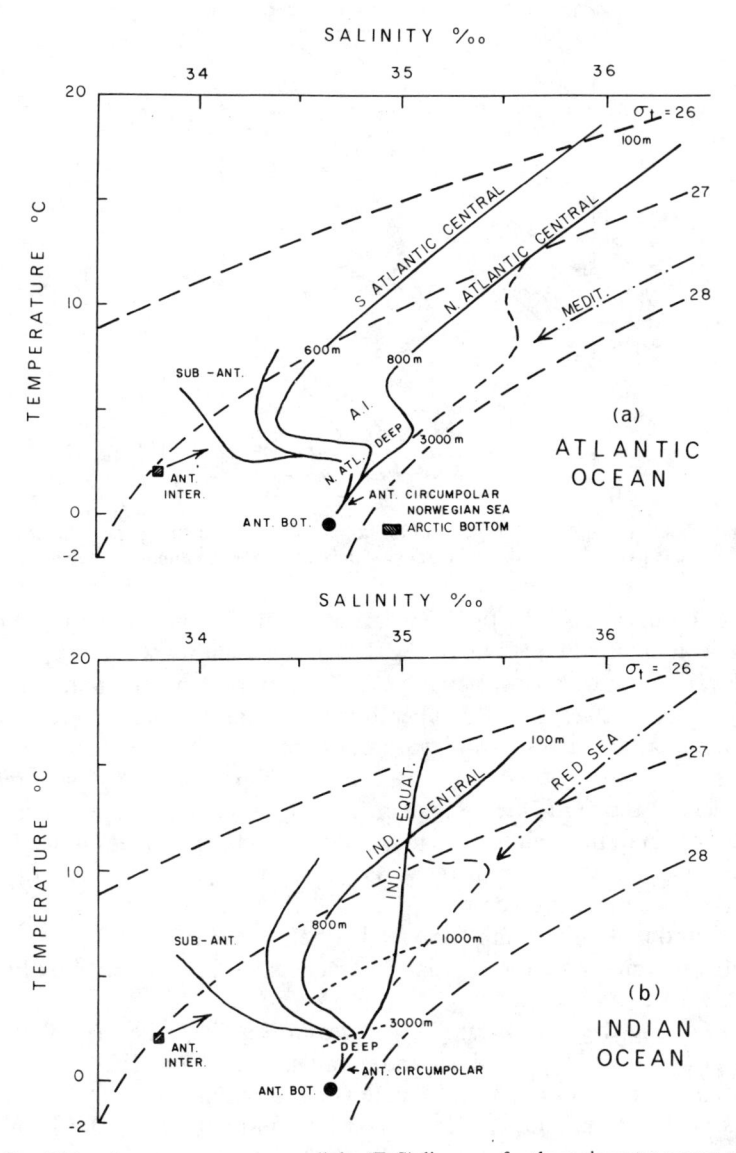

FIG. 7.20. Average temperature–salinity (T–S) diagrams for the main water masses of the Atlantic and Indian Oceans.

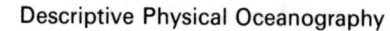

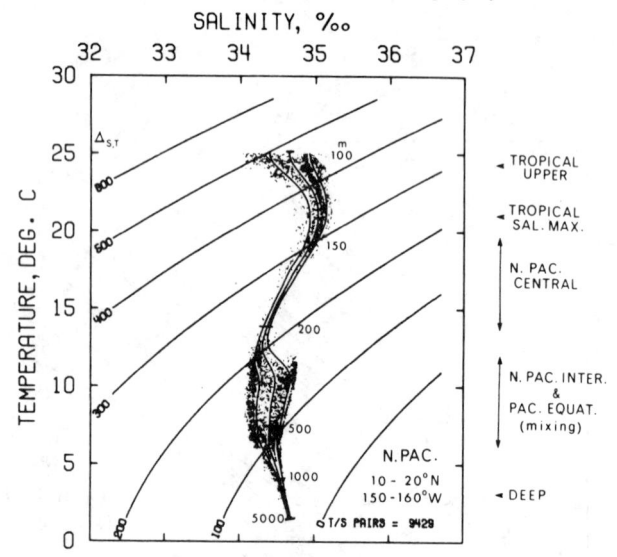

FIG. 7.21. Example of T–S "scatter plot" for all data within a 10° square with mean T–S curve (centre line) and curves for one standard deviation in salinity on either side.

found. To illustrate this, the T–S curves for all 5° squares for the North Atlantic are plotted on a map in Fig. 7.22 (Emery and Dewar, 1982). In this display, the considerable variety of local T–S curves is apparent and one can easily trace the characteristic temperature and salinity values associated with individual water masses. Strong variations are seen over the entire temperature range at 40° to 50°N off the east coast of North America. These reflect meanders in the Gulf Stream system, carrying very different water masses north and south through the 5° squares. At the higher latitudes (above 50°N) in the western North Atlantic, the variability is largest at low temperatures as here the sea surface is coldest.

The third point is to remind the reader that, as described in Section 6.542, updated information on water mass characteristics was presented by Emery and Meincke (1986).

Considering the local mean T–S curves in Fig. 7.22 one can follow the transitions of various maxima and minima. At about 2°C the salinity minimum of the Antarctic Intermediate Water is well marked in the southern latitudes. This minimum erodes to the north, losing its character by about 30°N. In the south, an almost straight line links the Antarctic Intermediate Water with the salinity maximum (~ 36) of the upper waters. To the north, these maximum values are found at the warmest temperatures.

In northern latitudes, the temperature range is limited to values below 15°C. As has been mentioned, in these regions the coldest temperature are frequently at the top of the water column, emphasizing the significant role of salinity in determining density in high latitudes. These regions are marked by a cold

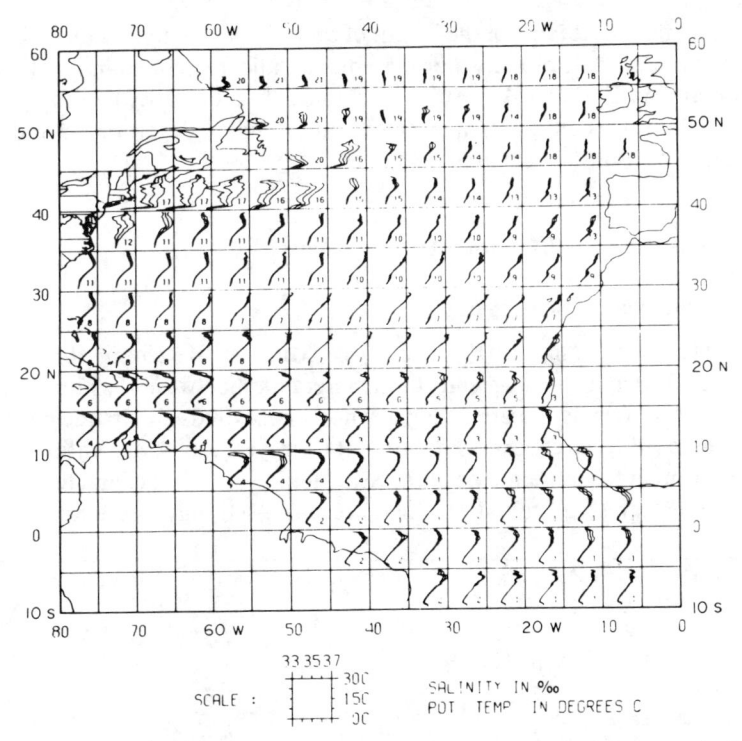

Fig. 7.22. Atlantic Ocean: mean T–S curves and one standard deviation curves by 5° squares.

salinity maximum. In contrast, the waters in the central far North Atlantic appear almost isohaline over the entire temperature range.

At mid-latitudes, off the Strait of Gibraltar, the saline outflow from the Mediterranean leads to a salinity maximum at mid-depth at about 10°C. From the sharp bend in the T–S curves one can trace this maximum as it spreads north, west and south. As has been discussed, this water mixes with those above and below as it spreads, eroding the salinity maximum.

The great variety in T–S curve shapes shown in Fig. 7.22 (prepared in 1980) compared with those in Fig. 7.20 (prepared before 1942) is partly because the latter were intended primarily as a synopsis of the main water mass features for large areas but is chiefly because there are so many more data available now, making it possible to divide the ocean into relatively small areas (Fig. 7.22) but still have sufficient data within each for the T–S curves to be significant. Also, computer processing makes possible the examination of these large quantities of data which it would be impractical to analyse with "paper and pencil" methods.

It is left to the reader to study Fig. 7.22 further to identify other features of the North Atlantic water mass characteristics and their spatial distribution, bearing in mind that most of the length of each T–S curve represents only the upper 500 to 1000 m of the water column as indicated, for example, by the depth marks in Fig. 7.21. It may also be noted that the paper by Emery and Dewar (1981) also contains maps of mean and standard deviation t, z and S, z profiles for the same 5° squares.

7.4 North Atlantic Adjacent Seas

7.41 Mediterranean Sea

The Mediterranean Sea (Fig. 7.23) (see Miller in Ketchum, 1983) is of interest in displaying some results of the interaction between the atmosphere and the ocean. Due to the prevailing dry northwest winds and frequent sunny days, there is a large excess (about 100 cm/year) of evaporation over precipitation in the eastern part of the Sea. Here the water is characterized by high temperatures and salinities which are surpassed only in the Red Sea.

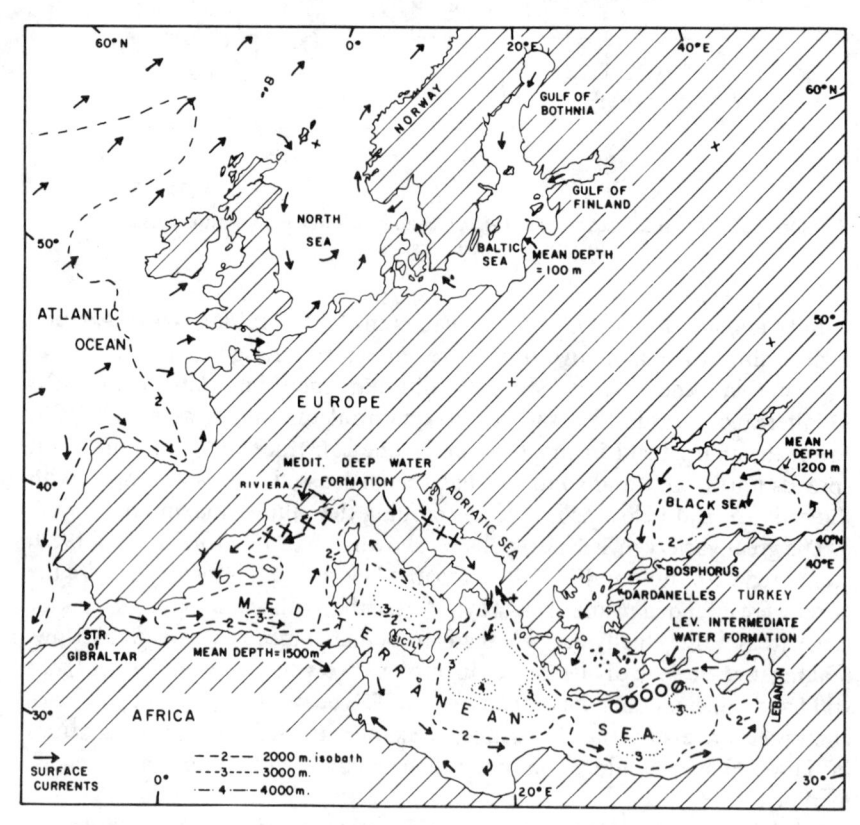

FIG. 7.23. Mediterranean, Black and Baltic Seas: surface circulation and regions of Intermediate and Deep Water formation for Mediterranean.

Circulation and Water Masses of the Oceans 209

The Mediterranean is divided essentially into a western and an eastern basin by a sill of depth about 400 m extending from Sicily to the North African coast. The maximum depths are about 3400 m in the western basin and 4200 m in the eastern. The tidal range is small, decreasing from 0.8 m at Gibraltar in the west to 0.4 m at Port Said in the east and to as low as 0.2 m along the French coast in the north. Sea level decreases in a north-easterly direction by about 0.7 m from the African coast to the Aegean Sea.

The two characteristic Mediterranean subsurface water masses are the Intermediate and the Deep waters. The former, called by Wüst (1961) the *Levantine Intermediate Water* (LIW), is formed in the winter off the south coast of Turkey (circles in Fig. 7.23) with a temperature of 15°C and a salinity of 39.1. It flows west at 200 to 600 m depth (Fig. 7.24) along the North African coast and out as a bottom layer through the Strait of Gibraltar into the Atlantic (Figs. 7.23, 7.24). By the time that it starts to flow down the continental slope its properties have been modified to 13°C and $S = 37.3$ by mixing with overlying Atlantic Water during its passage to the west, particularly through the Strait. The outflow in the bottom layer is replaced by an upper-layer inflow from the Atlantic (Fig. 5.2(a)). The flows through the Strait of Gibraltar have been studied by numerous expeditions, the most recent being the "1985–86 Gibraltar Experiment" (Kinder and Bryden, 1987). The real situation is considerably more complicated than assumed in Section 5.221 for the first-order estimate of the flushing time for the Mediterranean due to this exchange. The Strait is relatively deep (to 850 m) in the Tarifa Narrows which extend westward for about 25 km from the east end, south of Gibraltar, and the two bottom sills which are about 300 m deep are actually west of these Narrows. The hydraulic control on the Atlantic Water inflow occurs at the lateral constriction of the Narrows while the two sills control the Mediterranean Water outflow.

The boundary between the Atlantic Water and the Mediterranean Water can be considered to be in the middle of the pycnocline which, in the mean, rises from about 700 m depth about 80 km west of the Narrows to 50 m or less in the Narrows itself, with mean inflow above and mean outflow below the pycnocline. The instantaneous flow is complex and much influenced by the tides. During the maximum flood, there is strong inflow in the upper layer and weak outflow in the lower; during the maximum ebb there is outflow in both layers. At certain stages of the tide, hydraulic jumps occur near the sills and large amplitude internal waves can develop. A paper by Farmer and Armi (1989) discusses the hydraulics of the flow in detail and contains numerous references to earlier work in the region.

It was earlier thought that the bottom layer outflow to the west was entirely LIW but recent measurements have shown the presence (possibly episodic) of *Western Mediterranean Deep Water* (WMDW) from the Alboran Sea (just east of Gibraltar) in the outflow in the Atlantic outside the 300 m sill. The

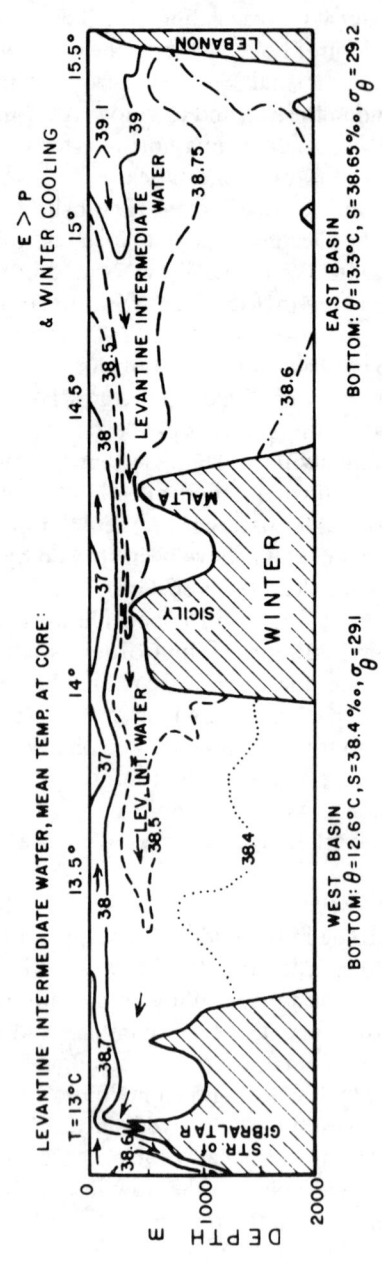

FIG. 7.24. Mediterranean Sea: longitudinal salinity section in winter to show the Levantine Intermediate Water.

WMDW is evident at the south side of the outflow below 300 m while the LIW is to the north and above the WMDW.

The *Deep and Bottom Waters* are formed at the northern edges of the basins in winter, chiefly off the Riviera in the western basin and in the southern Adriatic Sea (crosses in Fig. 7.23). In the western basin the deep-water potential temperature is 12.6°C and the salinity is 38.4, in the eastern basin the properties are 13.3°C and 38.65 with very high σ_θ values of up to 29.2. These values may be contrasted with those at 4000 m in the Atlantic of 2.4°C and 34.9 and a σ_θ of 27.8. The oxygen content of the deep water in these basins is fairly high, up to 4.7 mL/L in the western and 5.0 mL/L in the eastern, suggesting frequent replenishment by the considerable vertical convection ("open-ocean sinking", see Section 7.11) which carries the winter-formed water down to the depths of the basins as described in the next paragraph.

In January to March 1969, a six-ship operation off Toulon in the south of France observed this process in action for the first time. It took place in three phases. The preconditioning phase was when the winds from the north (Mistral) cooled the surface water until it was only just stable. Then a violent mixing phase occurred when stronger, dry winds blew for 6 days during which the water column was further cooled and was mixed to uniformity down to 1400 m depth. After this the water sank to depths of at least 2400 m and spread out, displacing upward the older deep water. (The maximum depth in this north-western basin is less than 2900 m.) A very interesting feature is that the horizontal extent of the formation area was only about 40 km across but nevertheless the volume of deep water formed (in a few days) was estimated to be equal to the outflow through the Strait of Gibraltar for about 6 weeks. (For a fuller account see Sankey, 1973.)

7.42 Black Sea

The oceanographic characteristics of the Black Sea (Fig. 7.23) (see Sorokin in Ketchum, 1983) which connects with the eastern Mediterranean are in direct contrast to those described above. In particular, it is a region where river runoff and precipitation exceed evaporation. The surface circulation is in the form of an anticlockwise gyre in each of the west and east basins. The Black Sea has a maximum depth of over 2200 m and connects with the Mediterranean through the narrow passages of the Bosporus and the Dardanelles, which have depths of only 40 to 92 m. In the Black Sea, a sharp pycnocline centred at 50 m due to the salinity distribution separates the upper low-salinity, oxygenated (*oxic*) water from the deeper oxygen-free (*anoxic*) water. The Deep Water in the Black Sea has a salinity of 22.3 and potential temperature of 8.9°C and there is now no dissolved oxygen below about 95 m; it has been replaced by hydrogen sulphide whose concentration increases with depth. In fact, the Black Sea is the archetypical anoxic basin.

There is evidence that over geological time, the water in the Black Sea has varied from being fresh (as recently as 7000 years ago) with no hydrogen sulphide, to moderately saline. It is not know whether the present deep water salinity represents a steady state or not, although observations which we will describe shortly suggest that changes do take place on a decadal time scale. There is considerable river runoff into the Sea, although this has decreased by 15% in the last 20 years due to the diversion of the river water for agricultural purposes, and a very limited influx of salt water from the Mediterranean, which flow is expected to increase with a decrease in fresh water input. The currents through the Bosporus and Dardanelles consist of a surface outflow of low-salinity water, carrying the precipitation and river runoff, and a subsurface inflow of more saline water. The narrowness and shallowness of the passage result in considerable current speeds and current shear and the consequent turbulence causes vertical mixing. The result is that the surface water which leaves the Black Sea with a salinity of about 18 reaches the Mediterranean with its salinity increased to about 30, while the incoming subsurface water from the Mediterranean with a salinity of 38.5 has this reduced to about 34 by the time that it reaches the Black Sea. There is some difference of opinion as to how much saline water actually flows into the Black Sea. There is no doubt that there is a considerable flow from the Mediterranean into the Dardanelles but much of this saline water is mixed upward and carried out by the surface flow. This is an example of the estuarine type of circulation discussed in Section 8.42. It is evident from the stagnant condition of the deep water of the Black Sea (Section 5.222) that the volume of well-oxygenated, saline water which penetrates through the Bosporus to sink in the Black Sea must be small.

The 1988 Black Sea Expedition (Murray *et al.*, 1989) carried out an intensive study of the physical and chemical characteristics of the southern (Turkish) part of the Black Sea and revealed a number of features, particularly that distributions of physical/chemical properties in the upper 200/300 m had changed significantly in the 20 year interval since the last major study in 1969. Figure 7.25 shows profiles of water properties in the upper 200 m (adapted from the above reference) for a station in the centre of the western basin and compares values for temperature and salinity in 1988 with the 1969 observations. The difference in upper layer temperature was because the 1969 observations were made in April but the 1988 ones in June. Salinity was higher by about 0.1 in 1988 than in 1969. The Cold Intermediate Layer (CIL) ($<8°C$) is a feature of the temperature distribution. One possibility is that this cool water originates from the northwest shelf region of the Sea but other suggestions have been made. The most conspicuous feature of the Black Sea, the sharp decrease in dissolved oxygen to near zero values occurred at about 55 m in 1988 compared with about 125 m in 1969 and significant hydrogen sulphide concentrations started at about 90 m compared with 125 m previously. Between 55 m and 95 m in 1988 there was a *suboxic* zone where the dissolved oxygen content was less than 5 μM/kg (~ 0.02 mL/L). In previous

Circulation and Water Masses of the Oceans 213

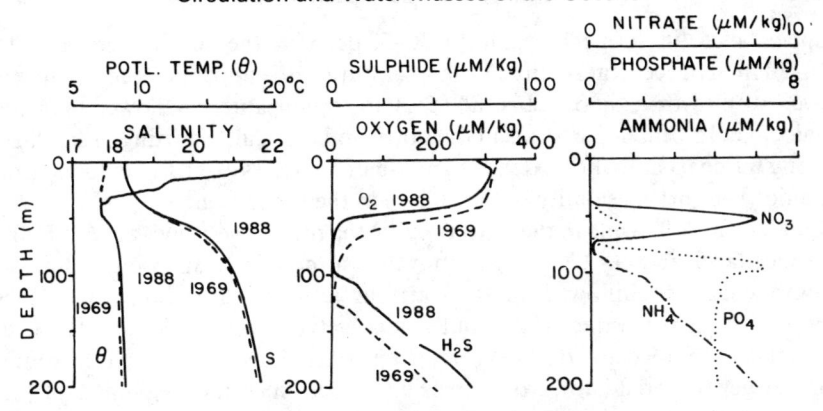

FIG. 7.25. Water properties in the upper 200 m in the Black Sea, 1988. (Adapted from Murray et al., 1989.)

studies, hydrogen sulphide was also found in the suboxic zone at concentrations of $<15\ \mu M/kg$ but there was no such overlap in 1988. Hydrogen sulphide, ammonia and phosphate continued to increase rapidly to about 900 m and then more slowly to the bottom. A conspicuous nitrate peak occurred in the suboxic zone, and ammonia increased rapidly below this. The phosphate concentration showed a small maximum in the suboxic zone and then increased rapidly in the deep water.

7.43 Baltic Sea

The Baltic Sea (Fig. 7.23) (see Kullenberg in Ketchum, 1983), including the Gulf of Bothnia to the north and the Gulf of Finland to the east is the largest area of brackish water in the ocean system. It has an irregular bottom topography with a mean depth of 57 m and a number of basins of which the deepest is 459 m deep. The Baltic is connected with the Atlantic at its southwest end through a complex of passages with a sill depth of 18 m leading to the Kattegat and the North Sea.

Evaporation and precipitation are estimated to be nearly equal at about 47 cm/year and therefore cancel. The annual river runoff is equivalent to 130 cm depth of water over the whole area of the Sea, with the bulk flowing into the Gulf of Finland, equivalent to 400 cm/year over that gulf, and 170 cm/year to the Gulf of Bothnia, but there are significant year-to-year variations. The general circulation is very weak, speeds of ~ 1 cm/s, but at the connection with the North Sea there is generally an estuarine circulation with the upper-layer outflow in the Kattegat having a salinity of 20 and the bottom-layer inflow having a salinity of 30–34. Large-scale meteorological conditions can overide the estuarine circulation and result in full depth inflow or outflow at times.

The water in the Baltic is basically a two-layer system with a well-mixed upper-layer (in terms of salinity) 30–50 deep in the south increasing to 60–70 m in the central Baltic. The temperature of the upper layer in summer is over 10°C with a thermocline at 15–20 m, significantly shallower than the halocline. In winter, ice forms in the north and east gulfs, starting in January, along the coast and often extends to mid-gulf but is less extensive in the central Baltic. The surface salinity in the gulfs is in the range from 2–7 and the deep values from 3–9, while in the central Baltic the respective values are 6–8 in the surface layer and 10–13 in the deep waters although it can rise to 20 in the south when large inflows from the Kattegat occur. Dissolved oxygen reaches saturation in the surface layers but is relatively low in the deep water, with variations on a decadal time-scale related to variations of inflows at the south, and a general trend toward decreasing values since the beginning of this century. Anoxic conditions occur in many of the deep basins in which the residence times are several years.

7.44 Norwegian and Greenland Seas

Bordering the North Atlantic are two adjacent seas of some significance, the Norwegian and Greenland Seas to the east of Greenland and the Labrador Sea and Baffin Bay area to the west (Fig. 7.26). The *Norwegian Current* is a continuation of part of the North Atlantic Current which turns north and passes over the Wyville–Thompson Ridge between the Shetland and Faroe Islands into the Norwegian and Greenland Seas. Along the Greenland coast there is the south-westward-flowing *East Greenland Current* which is composed of the outflow from the Arctic Sea and some water from the Norwegian Current. The speeds in these two currents are up to 30 cm/s. They are upper-layer currents and the submarine ridges which extend from Greenland to Scotland with maximum sill depths of less than 1000 m prevent deeper Atlantic water from entering the Norwegian Sea and hence the Arctic. Between the two currents are gyral circulations in the Norwegian and Greenland Seas.

A rather curious feature is that apparently the subsurface water which enters the Arctic Sea from this area comes from the gyre of the Norwegian Sea to the south rather than from the Greenland Sea which is north of it and closer to the Arctic. This was shown by Metcalf (1960) from data obtained in the winters of 1951 to 1955. Characteristically, the water of the Greenland Gyre above 1500 m has properties of -1.1 to $-1.7°C$ and $S=34.86$ to 34.90, while below this depth the water is almost isohaline at 34.92 with a temperature of $-1.1°C$ or colder. The deep water of the Norwegian Gyre has the same salinity but its temperature is $-0.95°C$ or warmer, properties which are similar to those of the Bottom Water of the Arctic Basin. This Norwegian Sea water is also found to the east and north of the Greenland Sea and apparently in some way forms a barrier to the passage of the colder Greenland Gyre water into the Arctic.

The deep water of both the Norwegian and Greenland Seas have high

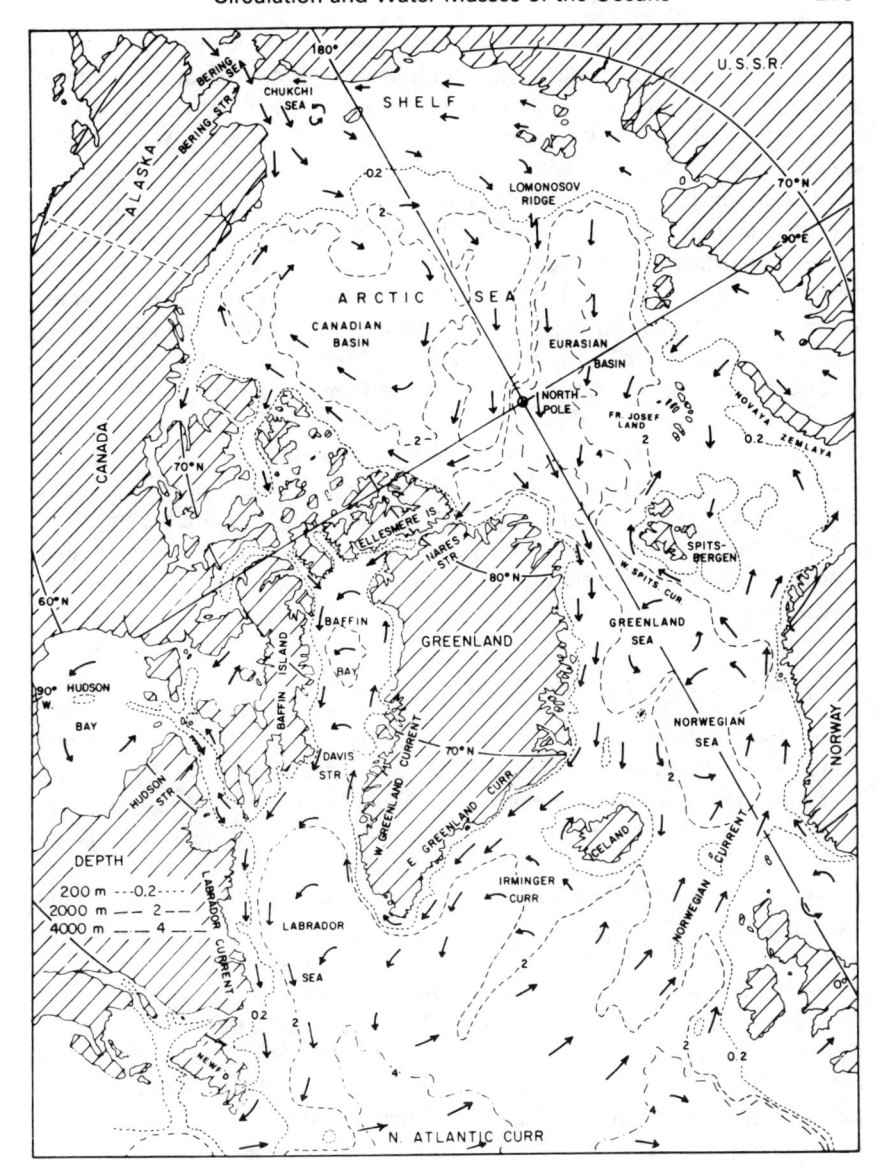

Fig. 7.26. Arctic Sea and North Atlantic adjacent seas: bathymetry and surface currents.

oxygen contents of 6 to 7.5 mL/L indicating frequent formation. It had been assumed that winter cooling at the surface resulted in overturning which mixed the water from surface to bottom. However, Carmack and Aagaard (1973) have pointed out that although there is considerable evidence that the formation of *Greenland Sea Deep Water* (GSDW) is affected by the severity of the particular winter, when good winter measurements in this area became available there was no evidence of a homogeneous water column. Therefore, unless the formation takes place in a very small area, some other mechanism must be responsible for the formation of Greenland Sea Deep Water. They suggest that the deeper Atlantic water (relatively warm and saline) loses heat by conduction to the Greenland Gyre Water above it (cool and less saline) at a greater rate than it loses salt. The result is that the bottom of the Gyre Water, being heated from below, becomes unstable and convects upward carrying heat to the surface and discharging it to the atmosphere. The Atlantic Water which is cooled by this process at its contact with the Gyre Water, without losing salt, becomes more dense and sinks to form the GSDW. The exchange across the interface between the Atlantic Water and the Gyre Water is identified as a double diffusion process. However, there is some question about this since a typical double diffusion process is a short-range one (decimetres to metres) whereas the proposed Greenland Sea process requires heat transfer over 50 to 200 m. Note that if it is an example of double diffusion, it is of the second variety described in Section 7.14 in which there is no mixing between layers—there is simply convection within each layer.

7.45 Labrador Sea, Baffin Bay and Hudson Bay

The East Greenland Current (Fig. 7.26) carries much ice from the Arctic down the coast of Greenland, maintaining the low temperatures and rendering access to the east coast of Greenland difficult. The Current flows round the southern tip of Greenland into the Labrador Sea, having picked up some Atlantic water south-west of Iceland. It continues north up the west coast as the *West Greenland Current* from which water branches off to the west until the Current eventually peters out in Baffin Bay. The inflow to this area is balanced by the southward flow, along the west side of Baffin Bay, of the *Baffin Land Current* which continues south as the *Labrador Current* down the west side of the Labrador Sea back into the Atlantic. In this region the differences between the properties of the in- and outflowing currents are not as great as to the east of Greenland. The West Greenland Current has temperatures around 2°C and salinities of 31 to 34, while the Labrador Current water is at 0°C or less and 30 to 34. The frequently quoted calculations of Smith, Soule and Mosby (1937) for the Labrador Sea indicate that above 1500 m depth the total inflow is 7.5 Sv, comprised of the East Greenland Current from the south (5 Sv), the Baffin Land Current from the north (2 Sv) and the inflow from Hudson Bay (0.5 Sv). The outflow totalling 5.6 Sv consists of 1 Sv northward along the west Greenland coast to Baffin Bay and 4.6 Sv south in the Labrador Current. They concluded that the balance of 1.9 Sv must flow out as deep water from the Labrador Sea into the Atlantic. Other estimates have indicated that considerable variations in the net flow may occur but the consensus is that in

Circulation and Water Masses of the Oceans   217

the long-term average there is a net inflow in the upper layers and an outflow of deep water.

The Labrador Sea is another region where it has been assumed that deep, convective overturn as a result of winter cooling was the mechanism for deep-water formation. As some winter cruises did not show the vertically homogeneous water mass which would be expected in this process, a winter cruise in C.S.S. *Hudson* in 1966 (Lazier, 1973) was designed to study the region with closely spaced stations in case the formation area was very small in horizontal extent. No evidence of large-scale convective overturn was obtained and a more plausible explanation of the data was that surface cooling resulted in a flow downward along moderately sloping σ_θ surfaces rather than vertically. However, in 1967 data from Weather Station BRAVO in the centre of the Labrador Sea did show evidence of deep convective overturn to 1500 m and it is possible that both vertical overturn and flow along σ_θ surfaces may at different times play their part in the formation of deep water there.

Some contribution to the Labrador Current comes from Hudson Bay (Fig. 7.26). This is an extensive body of water averaging only about 90 m in depth with maximum values of about 200 m. The Bay is usually covered with ice in the winter but free for a time in summer. As a result, most of the oceanographic data is for the summer, mainly from cruises in 1930, 1960 and 1961. There is considerable seasonal river runoff into the Bay from the south and east sides, giving rise to a marked horizontal stratification and an estuarine-type circulation. In summer, the upper water properties range from 1° to 9°C and $S = 25$ to 32 while the deeper water is from -1.6 to 0°C and 32 to 33.4. The low salinities are generally in the south and east, near the main sources of runoff and consistent with a general anticlockwise circulation in the upper layer. A few observations taken in winter through the ice indicate upper salinities from 28 in the south-east to 33 in the north, with temperatures everywhere at the freezing point appropriate to the salinity. The implication is that the waters are vertically mixed each year; the high dissolved oxygen values of 4.5 to 8 mL/L in the deep water are consistent with this.

The Labrador Sea is open to the Atlantic to depths of over 4000 m but Baffin Bay, with a maximum depth of 2200 m, is separated from the Labrador Sea, and hence from the Atlantic, by the sill in the Davis Strait of depth only about 600 m. The deep waters in Baffin Bay below sill depth are markedly different from those at the same depths outside. The temperature is between 0° and -0.5°C and the salinity 34.5 with relatively low oxygen values of 4 mL/L. Sverdrup interpreted the *Baffin Deep Water* as Labrador Sea water which had moved north below the surface, mixed with cool, low salinity surface water, and then had its density increased by freezing to sink to fill the basin. Since Baffin Bay is ice-covered during the winter there is little information for this season but there are indications that water having the properties of the Deep Baffin Water does not occur at the surface. An alternative suggestion by Bailey (1957) is that the source of the Baffin Deep Water is probably inflow from the

Arctic through Nares Strait, the passage between Greenland and Ellesmere Island. Water of the same salinity and temperature as the Baffin Deep Water certainly occurs at the appropriate depth in the Arctic Sea. The annual inflow to Baffin Bay is presumably relatively small so that the water in the basin has a long residence time there and the oxygen content gets depleted.

7.46 Adjacent seas; inflow and outflow characteristics

Sverdrup contrasted the horizontally separated inflow and outflow of the northern seas with the vertically separated flows into and out of the Mediterranean but, as will be demonstrated below, the distinction is not clear cut.

Figure 7.27 shows schematically some of the flow characteristics for seas or other bodies of water adjacent to a main ocean. Three examples of seas with vertically separated exchange (type A) are illustrated. The European Mediterranean, and the Red Sea to be described later, have upper layer inflow and deeper outflow over their sills. In both cases, the excess of evaporation over fresh-water input increases the salinity of the upper waters. From the Mediterranean, the outflow is chiefly of the Levantine Intermediate Water from depths of a few hundred metres only, the deeper water being stratified. In the Red Sea, the deep waters are much less stratified but the outflow is still from a shallow intermediate layer. The Baltic Sea provides an example of estuarine circulation (Section 8.42) with surface outflow driven by an excess of fresh-water input over loss by evaporation, but with significant vertical mixing in the basin. Most fjords follow this pattern. The Black Sea is an extreme example of estuarine circulation in which the upper layer is of low density and high stability to below sill depth, so that little vertical circulation takes place. A few fjords also possess these properties.

A clear example of the second type (B, Fig. 7.27) is the American Mediterranean (the Caribbean Sea) into which inflow is through the Antilles Islands in the south-east and outflow from the north-west through the Straits of Florida. Less strict examples of type B are the northern seas adjacent to the North Atlantic. For the Norwegian/Greenland Seas, the water of the North Atlantic Current which becomes the Norwegian Current has temperatures from 4° to 13°C and salinities from 34.9 to 35.3, whereas the outflowing East Greenland Current water is at −1.5° to 2°C and 31 to 34. On a T–S diagram the upper waters of the inflowing Norwegian Current are represented by a line roughly parallel to the temperature axis, with the density mainly determined by temperature, whereas the upper-layer water of the outflowing East Greenland Current is represented by a line roughly parallel to the salinity axis with the density determined almost entirely by salinity as is characteristic of high latitude waters. Another example of a type B basin is the Labrador Sea/Baffin Bay region in the north-west.

The reason why these cannot be considered as strict type B (horizontally

Circulation and Water Masses of the Oceans

ADJACENT SEAS, TYPE CIRCULATIONS

TYPE

A. ENTRY FLOWS SEPARATED VERTICALLY:

A.1- MEDITERRANEAN SEA
RED SEA   E > P + R & WINTER COOLING

OCEAN IN / OUT SEA

↓ = WINTER SINKING
REFRESHES DEEP WATER

REL. HIGH O_2

A.2- BALTIC SEA
HUDSON BAY   P + R ≫ E

OUT / IN SEA MIXING
HIGH O_2

SHALLOW with VERTICAL MIXING
= LITTLE STAGNATION

(A.2 & A.3 HAVE "ESTUARINE CIRCULATION")

A.3- BLACK SEA   P + R ≫ E

OUT / IN SEA MIXING

STABLE UPPER LAYERS
NO DEEP SINKING,
NO REFRESHMENT
OF DEEP WATER.
STAGNANT
ZERO O_2

B. ENTRY FLOWS SEPARATED HORIZONTALLY:

OCEAN OUT / IN   SEA   POSSIBLE GAIN or LOSS of HEAT & F.W.

e.g.

B.1- ARCTIC SEA (heat loss)
     CARIBBEAN

B.2- (with deep outflow also)
     NORWEGIAN/GREENLAND SEAS
     LABRADOR SEA/BAFFIN BAY

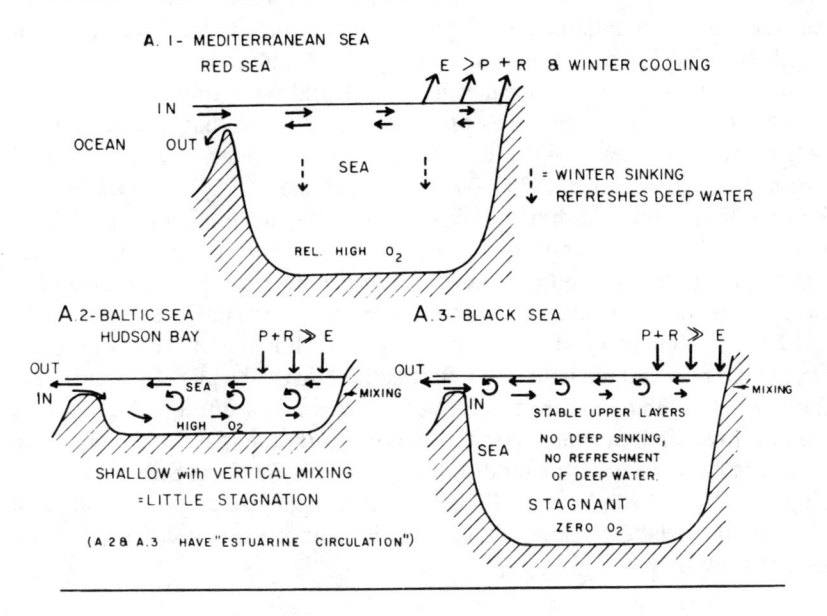

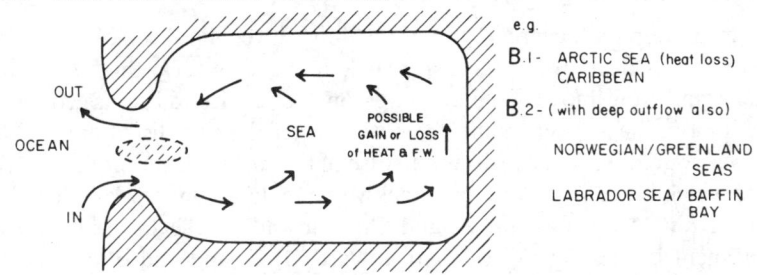

FIG. 7.27. Schematic diagram of various types of circulation in seas adjacent to oceans.

separated) flows is that in both cases there are now known to be large deep-water outflows as well as the horizontally separated upper-layer flows.

7.5 Arctic Sea

Our knowledge of the Arctic Sea has developed considerably since the mid-1950s. Numerous soundings and oceanographic stations have been taken from ships, as well as through the ice from semi-permanent camps on ice islands or ice floes or from temporary camps established by aircraft transportation. In particular it has been demonstrated that the Arctic Sea is divided into two basins by the Lomonosov Ridge which extends from Greenland past the North Pole to Siberia (Fig. 7.26). These two basins have been named the Canadian Basin (depth about 3800 m) and the Eurasian Basin (depth about 4200 m). Soundings along the Lomonosov Ridge are not numerous enough to determine its sill depth with any certainty but comparisons of water properties on either side suggest that it is at 1200 to 1400 m below sea level. A feature of the bottom topography is the broad continental shelf off Siberia, 200 to 800 km wide and occupying about 36% of the area of the Arctic Sea but containing only 2% of the total volume of water in the Sea. The main connection with the other oceans is with the Atlantic through the gap between Greenland and Spitsbergen with a main sill depth there of 2600 m while the sill depth between Spitsbergen, Franz Josef Land and Novaya Zemlaya is only about 200 m (Coachman and Aagaard, 1974). The Bering Strait connection to the Bering Sea and the Pacific Ocean has a sill depth of about 45 m and is narrow, but the water flow into the Arctic is not insignificant. There are also connections from the Arctic through the Canadian Archipelago by several channels, principally Nares Strait (sill depth 250 m) and Lancaster Sound (sill depth 130 m) which lead to Baffin Bay and thence to the Atlantic. These passages are difficult of access because of ice and are not fully charted.

7.51 Arctic Sea: upper-layer circulation

Information on the circulation of the upper layers has been obtained from the records of the movements of ships held in the ice, such as the *Fram* and the *Sedov*, and from movements of camps on the ice. In addition, geostrophic calculations have been made from the water-density distribution. These various sources yield a consistent picture of the surface-layer movement (Fig. 7.26) which is best described as a clockwise circulation in the Canadian Basin leading out to the East Greenland Current and, in the Eurasian Basin, a movement by the most direct path toward Greenland and out in the East Greenland Current. The speeds are of the order of 1 to 4 cm/s or, perhaps, more meaningfully stated as 300 to 1200 km/year when considered in relation to the size of the Arctic Sea which is about 4000 km across. The speed and distance may be compared to the 3 years taken by the *Fram* to drift from north of the Bering Strait to Spitsbergen, and the $2\frac{1}{2}$ years for the *Sedov* to drift about 3000 km. The movement is not by any means steady but has frequent variations of speed and direction which average out to the figures quoted.

7.52 Arctic Sea water masses

Three main water masses are recognized (Coachman and Aagaard, 1974). These are (Table 7.1 and Fig. 7.28(a)) the surface or *Arctic Water* from the sea

Circulation and Water Masses of the Oceans

TABLE 7.1
Arctic Sea water masses

Water mass		Properties	
Name (circulation direction)	Boundary depth		Seasonal variation
ARCTIC SURFACE	Surface	T: Close to F.P., i.e. −1.5 to −1.9°C S: 28 to 33.5	ΔT: 0.1K ΔS: 2
	25 to 50 m		
ARCTIC SUBSURFACE		T: Canadian Basin −1 to −1.5°C Eurasian Basin −1.6° to 100 m, then increase S: Both basins 31.5 to 34	Small
	100 to 150 M		
ARCTIC LOWER		Intermediate between Subsurface and Atlantic	
	200 m		
(all above masses circulate clockwise)			
ATLANTIC (anticlockwise)		T: Above 0°C (to 3°C) S: 34.85 to 35	Negligible
	900 m		
DEEP or BOTTOM (uncertain, small)		T: Canadian Basin 2000 m Bottom −0.4° −0.2°C Eurasian Basin −0.8° −0.6°C S: Both Basins 34.90 to 34.99	(rise adiabatic)
	Bottom		

surface to 200 m depth, the *Atlantic Water* from 200 to 900 m, and the *Deep* or *Bottom Water* from 900 m to the bottom. One of the features of the water structure is that its density is determined largely by the salinity (which is why the marked temperature maximum of the Atlantic Water can exist).

7.521 Arctic Water

The Arctic Water (0 to 200 m) can be divided into three layers which will be called the *Surface Arctic*, the *Sub-surface Arctic* and the *Lower Arctic Waters*.

Surface Water (Table 7.1) is much the same across the whole Arctic and extends from the surface to between 25 and 50 m depth. The salinity is strongly influenced by the freezing or melting of ice and has a wide range from 28 to 33.5. The temperature is also controlled by the melting or freezing which involves considerable heat transfer at constant temperature (the freezing

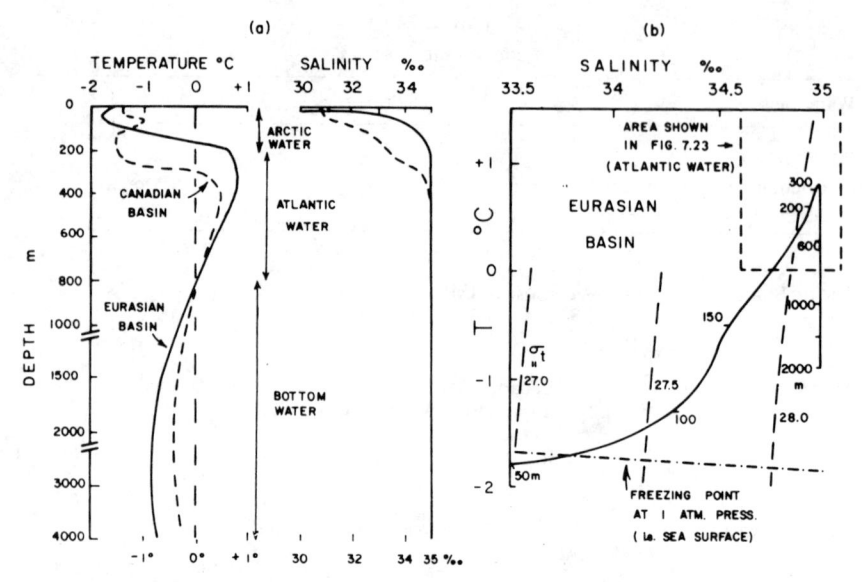

FIG. 7.28. Arctic Sea: (a) typical temperature and salinity profiles for the two basins, (b) T–S diagram for the Eurasian Basin water.

point). In consequence the temperature remains close to the freezing point of sea-water which varies only from $-1.5°C$ at a salinity of 28 to $-1.8°C$ at a salinity of 33.5. Seasonal variations in water properties are limited to this layer and range up to 2 in salinity and 0.2K in temperature.

The *Subsurface* layer (25/50 m to 100 m) in the Eurasian Basin is isothermal to 100 m (Fig. 7.29(b) and then increases but there is a strong halocline between 25 and 100 m (Fig. 7.29(a)). Below 100 m the temperature increases markedly but the salinity only increases slowly. The fact that the Subsurface water is isothermal (limited to the freezing point) but not isohaline indicates that its structure cannot be due to vertical mixing between the Surface layer and the deeper layers. It is probable that the Subsurface Water is maintained by horizontal advection (flow) from the Eurasian Shelf. The mechanism suggested by Coachman is that the saline Atlantic Water which enters near Spitsbergen continues below the surface along the Eurasian continental slope which is indented by several deep submarine canyons. At the same time the considerable river runoff from northern Asia flows north over the shelf as a cold, low-salinity surface layer. It mixes at its subsurface contact with the warmer, more saline Atlantic Water to form the Subsurface Water which is close to its freezing point. The Subsurface Water continues out into the Arctic Sea to maintain the layer there between 25 and 100 m. The canyons are necessary to feed the saline Atlantic Water into the shelf area, and the vertical mixing process is similar to that which occurs in an estuary where fresh river water flows over saline sea-water as described in Section 8.4.

Circulation and Water Masses of the Oceans 223

The Subsurface Water in the Canadian Basin also shows a halocline from 25 m to 100 m but its temperature structure is different from that in the Eurasian Basin. There is a characteristic temperature maximum at 75 to 100 m depth (Fig. 7.29(c)) with a consequent temperature minimum of $-1.5°C$ at

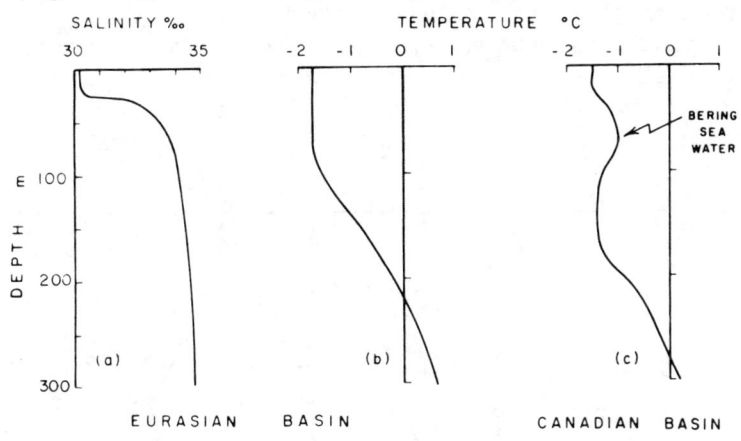

FIG. 7.29. Salinity and temperature profiles for the Arctic Water Mass.

about 150 m and then an increase to the deeper water values. The temperature maximum is attributed to Bering Sea water coming into the Arctic through the Bering Strait. This water is warmer than the Arctic Surface layer but slightly denser because of its salinity. It presents one of the relatively few examples of a subsurface temperature maximum occurring in the open ocean. The reason that it occurs here is because the water is close to its freezing point and the effect of salinity preponderates over that of temperature in determining density. The temperature maximum is found to be most prominent in the Chukchi Sea north of the Bering Strait, and it diminishes around the clockwise circulation of the Canadian Basin.

The *Arctic Lower Water* is essentially a mixing layer with properties intermediate between the Subsurface Arctic Water above and the Atlantic Water below.

7.522 Atlantic Water

The second water mass, the *Atlantic Water* (Fig. 7.28(a) vertical profiles, (b) T–S diagram), is recognized chiefly by having a higher temperature than the water above or below it. Where it enters as the *West Spitsbergen Current*, on the east side of the Greenland–Spitsbergen gap, its temperature is up to 3°C and its salinity 34.8 to 35.1. In the Arctic Sea its temperature decreases gradually to 0.4°C and its salinity is within the limiting range from 34.85 to 35. Its movement has been traced by the core method along the Eurasian continental slope, with some water branching off to the north and out as part of the East Greenland Current. The remainder flows across the Lomonosov Ridge into the Canadian Basin.

The application of the core method is as follows. Figure 7.28(b) shows a T–S diagram for the water column in the Eurasian Basin with the marked temperature maximum of the Atlantic Water near the salinity of 35. (Note that the temperature maximum is emphasized by the magnified temperature scale in this diagram, 2K equivalent on paper to 1 in salinity, compared with the other T–S diagrams shown previously, e.g. Fig. 7.18 where 10K was equivalent to 1.) Then Fig. 7.30(a) shows the temperature maximum part of

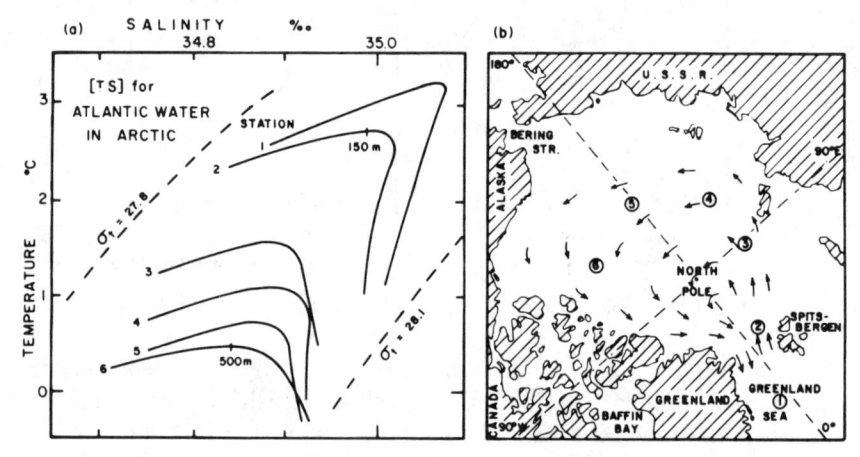

Fig. 7.30. Arctic Sea: (a) temperature maximum part of T–S diagram for core method analysis of flow direction for Atlantic Water, (b) circulation inferred from successive erosion of core shown in (a) for stations 1 to 6.

the T–S diagram (outlined in Fig. 7.28(b), the core of the Atlantic Water, for a selection of stations from the Greenland Sea and around the Arctic basins (Fig. 7.30(b)) to show how the core is gradually eroded away by mixing during its circuit. The water mass itself appears to remain in the depth range from 200 to 900 m, but the depth of the temperature maximum increases from about 150 m at the top of the Atlantic Water just before entry into the Arctic near Spitsbergen to nearly 500 m in the Canadian Basin. The reason is that the temperature gradient in the upward direction is greater than in the downward direction (Fig. 7.28(a)) with the result that more heat is lost upward from the layer than downward and the temperature maximum increases in depth. (There may also be some descent of the Atlantic Water.) The circulation of the Atlantic Water is then basically counter-clockwise around the Arctic Sea, the opposite direction to that of the Arctic Water above it.

7.523 Deep or Bottom Water

The *Arctic Deep or Bottom Water* (Fig. 7.28; Table 7.1) extends from the lower 0°C isotherm, about 800 m depth, to the bottom and comprises about

Circulation and Water Masses of the Oceans   225

60% of the total water volume of the Arctic Sea. Its salinity range through the whole volume is very small, from 34.90 to 34.99, and in any particular area the change in the vertical direction is generally smaller than this. There is a tendency for the salinity to increase very slightly with depth. The *in situ* temperature varies over a range of 0.2K in the vertical column. In the Eurasian Basin the temperature reaches a minimum of $-0.8°C$ at 2500 m while in the Canadian Basin the minimum is $0.4°C$ at 2000 m (Fig. 7.28(a)). Below the minimum the temperature rises by about 0.2K to the bottom, the rate of increase being equal to the adiabatic rate, i.e. it can be attributed entirely to compression of the water as it sinks.

Aagaard *et al.* (1985) distinguished four components of the Deep Water as: cold, relatively fresh Greenland Sea Deep Water (GSDW), warmer and more saline Norwegian Sea Deep Water (NSDW), (both named from their sources), still warmer and more saline *Eurasian Basin Deep Water* (EBDW) and, the most saline and warmest, *Canadian Basin Deep Water* (CBDW), (these two named for their locations). Through Fram Strait (between Greenland and Spitsbergen) there is inflow of GSDW below outflow from the Arctic. The authors also suggested that a source for the higher salinity of the CBDW than of the EBDW is brine formed during freezing on the continental shelf and flowing down the slope to increase the salinity of the deep water, i.e. a nearshore boundary process. (Note that this will also contribute to ventilation of the deep water in the basin.) It was pointed out that the brine formation must also occur along the extensive Eurasian shelf and Aagaard *et al.* suggested that the lower salinity and silicate content of the EBDW than those of the CBDW may be accounted for by the inflow through Fram Strait into the Eurasian Basin of low salinity, low silicate GSDW which does not pass into the Canadian Basin below the level of the Lomonsov Ridge sill.

7.53 *Arctic Sea budgets*

The Arctic Sea has always attracted the attention of oceanographers wishing to exercise their talents by investigating the water budget, and many sets of calculations have been made. The results obtained vary somewhat in detail but the main features are substantially the same. A recent calculation by Aagaard and Greisman (1975) has been chosen for demonstration because it also includes heat and salt budgets (Table 7.2). This does not represent an absolute budget in the sense that all the flow terms and water properties are known exactly; in fact, information on most of the quantities is sparse and, in any case, variations with time are likely to occur. Nevertheless, it is worth making even trial budgets in order to determine which terms appear to be important and where one's observational efforts should be directed in order to improve the budgeting.

From the budget in Table 7.2, it can be seen clearly that the main volume fluxes are through the Greenland–Spitsbergen gap but that the Bering Strait

TABLE 7.2

Annual mean water, heat and salt budgets for the Arctic Ocean (Adapted from Aagard and Greisman, 1975)

	Mean water properties		Transports		
	Temp. (°C)	Sal.	Volume (Sv)	Heat (10^{12} W)	Salt (10^6 kg)
Bering Strait, Water	0.5	32.4	1.5	4	49
Ice	−10	3	negl.	−2	negl.
Arctic Archipelago	−0.7	34.2	−2.1	5	−72
East Greenland Current					
Polar Water	−1.2*	34.0	−1.8	8	−61*
Atlantic Water	0.5*	34.9	−5.3	−13	−185*
Ice	−10	3	−0.1	34	
West Spitsbergen Current	2.2	35.0	7.1	68	249
Spitsbergen–Fr. Josef Land	2.7	34.9	−0.1	−1	−4
Fr. Josef Land–Nov. Zemlaya	0.9*	34.7	0.7	3	24
Runoff	5.0	0	0.1	2	0
Total Inflow/heat gain	1.8	34.6	9.4	124	322
Total outflow/heat loss	−0.1	34.6	−9.4	−16	−322
Net exchange			0	108	0

Notes: (1) Positive values represent inflow or heat gain, negative values represent outflow or heat loss.
(2) Heat transport is calculated relative to 0.1°C, the mean outflow temperature.
(3) Items marked with an asterisk (*) have been adjusted, within observed data limits, to satisfy continuity.

and Arctic Archipelago flows are not negligible. The main heat flux (larger than in previous budgets) is carried by the West Spitsbergen Current and overall there is a net advective inflow (Q_v positive) to the Arctic; this must be lost through the sea surface. (Note that because of its high latent heat of fusion an outflow of ice is equivalent to an inflow of heat and vice versa.)

Calculations of heat budgets are done for other reasons than for displaying virtuosity in deduction and calculation. From the water budget, in relation to the volume of the basin itself, one can obtain an idea of the rate at which the water in the basin is exchanged. This may be important for determining the rate of replenishment of nutrients in an area important for fisheries or for the rate of removal of sewage or industrial effluent. For the Arctic it is estimated that the surface water is substantially all replaced in a period (residence time) of 3 to 10 years, the deep water in 20 to 25 years (from the above budget) and the bottom water in about 150 years.

7.54 Ice in the sea

Ice in the sea has two origins, the freezing of sea-water and the breaking of ice from glaciers. The majority of ice comes from the first of these sources and will be referred to as *sea-ice*; the glaciers supply *icebergs*. The importance of sea-ice is that it drastically alters the momentum transfer from atmosphere to ocean, reduces the surface albedo, is a thermal insulator, damps surface waves, changes the temperature and salinity structure in the upper layer by melting and freezing and is a major hindrance to navigation.

7.541 Ice physics

When water loses sufficient heat (by radiation, conduction to the atmosphere, convection or evaporation) it freezes to ice, i.e. changes to the solid state. Initial freezing occurs at the surface and then the ice thickens by freezing at its lower surface as heat is conducted from the underlying water through the ice to the air.

The initial freezing process is different for fresh and low-salinity water than for more saline water because the temperature at which water reaches its maximum density varies with salinity. Table 7.3 gives the values of the freezing

TABLE 7.3
Temperatures of the freezing point (t_f) and of maximum density ($t_{\rho max}$) for fresh and salt water.

S	0	10	20	24.7	30	35
t_f	0	−0.5	−1.08	−1.33	−1.63	−1.91°C
$t_{\rho max}$	+3.98	+1.83	−0.32	−1.33	—	—°C

point and temperature of maximum density for water of various salinities. (Note that the above values are for freezing, etc. at atmospheric pressure—increased pressure lowers the freezing point which decreases by about 0.08K per 100 m increase in depth in the sea.)

To contrast the freezing process for fresh- and sea-water, first imagine a fresh-water lake where the temperature initially decreases from about 10°C at the surface to about 5°C at about 30 m depth. As heat is lost through the surface, the density of the water there increases and vertical convective mixing (overturn) occurs with the temperature of the surface water-layer gradually decreasing. This continues until the upper mixed-layer cools to 3.98°C and then further cooling of the surface water causes its density to decrease and it stays near the top. A result is rapid loss of heat from a thin surface layer which soon freezes. For sea-water of salinity = 35 of the same initial temperature distribution, surface cooling first results in density increase and vertical mixing by convention currents occurs through a gradually increasing depth, but it is not until the whole column reaches −1.91°C that freezing commences. As a much greater volume of water has to be cooled through a greater temperature range than in the fresh water case, it takes longer for freezing to start in salt water than in fresh water. A simple calculation for a column of fresh water of

100 cm depth and 1 cm^2 cross-section initially at 10°C shows that it takes a heat loss of 163 joules to freeze the top 1 cm layer, whereas for a similar column of sea-water of $S = 35$ it takes a loss of 305 joules to freeze the top 1 cm because the whole column has to be cooled to -1.91°C, rather than just the top 1 cm to 0°C for the fresh water.

Note that as sea-water of salinity less than 24.7 has a higher temperature of maximum density than its freezing point, it will behave in a similar manner to fresh water, although with a lower freezing point. For sea-water of salinity > 24.7, in high latitudes the salinity generally increases with depth and the stability of the water column usually limits to 30–50 m the depth to which the convection currents will sink. Therefore ice starts to form at the surface before the deep water reaches the freezing point.

Generally, sea-ice forms first in shallow water near the coast, particularly where the salinity is reduced by river runoff and where currents are minimal. The first process is the formation of needle-like crystals of pure ice which impart an "oily" appearance to the sea surface (*frazil ice*). The crystals increase in number and form a *slush* which thickens and breaks up into *pancakes* of a metre or so across. With continued cooling, these grow in thickness and lateral extent, eventually forming a continuous sheet of *floe* or *sheet* ice.

Once ice has formed at the sea surface, when the air is colder than the water below, freezing continues at the lower surface of the ice and the rate of increase of ice thickness depends on the rate of heat loss upward through the ice (and any snow cover). This loss is directly proportional to the temperature difference between top and bottom surfaces and inversely proportional to the thickness of the ice and snow cover.

With very cold air, a sheet of sea-ice of up to 10 cm in thickness can form in 24 h, the rate of growth then decreasing with increased ice thickness. Snow on the top surface insulates it and reduces the heat loss markedly, depending on its degree of compaction. For instance, 5 cm of new powder snow may have an insulation equivalent to 250–350 cm of ice, 5 cm of settled snow can be equivalent to only 60–100 cm of ice while 5 cm of hard-packed snow can be equivalent to only 20–30 cm of ice.

As an example of the annual cycle of the development of an ice sheet at a location in the Canadian Arctic, the ice was observed to start to form in September, was about 0.5 m thick in October, 1 m in December, 1.5 m in February and reached its maximum thickness of 2 m in May, after which it started to melt.

In the initial stage of ice-crystal formation, salt is rejected and increases the density of the neighbouring sea-water, some of which then tends to sink and some of which is trapped among the ice crystals, forming pockets called "brine cells". The faster the freezing, the more brine is trapped. Sea-ice in bulk is therefore not pure water-ice but has a salinity of as much as 15 for new ice (and less for old ice as gravity causes the brine cells to migrate downward in time). With continued freezing, more ice freezes out within the brine cells leaving the brine more saline. Some of the salts may even crystallize out. The salinity of first-year ice is generally 4 to 10, for second-year ice (ice which has remained frozen beyond the first year) it decreases to 1–3 and for multi-year ice may be less than 1. If sea-ice is lifted above sea-level, as happens when ice becomes thicker or rafting occur, the brine gradually trickles down through it and eventually leaves almost saltless, clear old ice. Such ice may be melted and used for drinking whereas melted new ice is not potable. Sea-ice must therefore be considered to be a material of variable composition and properties, which depend very much on its previous history. (For more detail see Doronin and Kheisin, 1975.)

The density of pure water at 0°C is 999.9 kg/m^3 and that of pure ice is 916.8 kg/m^3. However the density of sea-ice may be greater than this last figure (if brine is trapped among the ice crystals) or less (if the brine has escaped and gas bubbles are present.) Values from 924 to 857 kg/m^3 were recorded on the Norwegian *Maud* Expedition (Malmgren, 1927).

Malmgren gave values for the specific heat of sea-ice of as much as 67 kJ/kg K at −2°C and S = 15. This surprisingly high figure arises because the measurement of specific heat requires heating through a finite range of temperature and while so doing there will be some melting of ice crystals into the brine. The 67 kJ/kgK then includes some latent heat of melting and the high values are not true specific heats in the sense that the term is used for pure substances. The latent heat of melting decreases from 335 kJ/kg (80 cal/g) at 0°C and S = 0 to only 63 kJ/kg at −1°C and S = 15.

The amount of heat required to melt sea-ice varies considerably with its salinity. For S = 0 (fresh water ice) it requires 19.3 kJ/kg from −2°C and 21.4 kJ/kg from −20°C, while for sea-ice of S = 15, it requires only 11.2 kJ/kg from −2°C but 20.0 kJ/kg from −20°C. The small difference of heat (2.1 kJ/kg) needed to raise the temperature of fresh-water ice from −20°C to −2°C is because no melting takes place, i.e. it is a true measure of the specific heat of pure ice. However, for sea-ice of S = 15, it requires more heat (8.8 kJ/kg) to raise its temperature through the same range because some ice near brine cells melts and thus requires latent heat of melting as well as heat to raise its temperature. Note also that less heat is needed to melt new ice (S = 15) than old ice which has a lower salinity.

Because of the spongy nature of first year sea-ice (crystals + brine cells) it has much less strength than fresh-water ice. Also, as fast freezing results in more brine cells, the strength of ice so formed is less than when freezing occurs slowly, i.e. sea-ice formed in very cold weather is initially weaker than ice formed in less cold weather. As the temperature of ice decreases, its hardness and strength increase, and ice becomes stronger with age as the brine cells migrate downward. When ice forms in calm water, the crystals tend to line up in a pattern and such ice tends to fracture along cleavage planes more easily than ice formed in rough water where the crystals are more randomly arranged and cleavage planes are not formed.

The mechanical behaviour of sea-ice when temperature changes is complex. As the ice temperature decreases below its freezing point, the ice expands initially, reaches a maximum expansion and then contracts. For instance, an ice floe of S = 4 will expand by 1 m per 1 km length between $-2°C$ and $-3°C$, reaches its maximum expansion at $-10°C$ and thereafter contracts slightly. Ice of S = 10 expands 4 m per 1 km length from $-2°C$ to $-3°C$, and reaches its maximum expansion at $-18°C$. The expansion on cooling can cause an ice sheet to buckle and "*pressure ridges*" to form, while contraction on further cooling after maximum expansion results in cracks, sometimes wide, in the ice sheet.

Pressure ridges can also develop as a result of wind stress on the surface driving ice sheets together. The ridges on top are accompanied by a thickening of the lower surface of the ice by four to five times the height of the surface ridges. Sea-ice generally floats with about five-sixths of its thickness below the surface and one-sixth above and so relatively small surface ridges can be accompanied by deep ridges underneath—depths of 25 to 50 m below the sea surface have been recorded. Thickening of an ice sheet may also result from *rafting* when wind or tide forces one ice sheet on top of another or when two sheets, in compression, crumble and pile up ice at their contact. Old ridges, including piled up snow, are referred to as *hummocks*. As they are less saline than newer pressure ridges, they are stronger and more of an impediment to surface travel than the younger ridges.

The motion of pack ice is determined by several forces:

(a) Wind stress at the top surface, the magnitude depending on the wind speed and the roughness of the ice surface, so that ridges, etc. increase the wind stress. Typical ice speeds are 1 to 2% of the wind speed.

(b) Frictional drag on the bottom of an ice sheet moving over still water will tend to slow it down, while water currents (ocean and tidal) will exert a force on the bottom of the ice in the direction of the current. Because current speeds generally decrease with increase in depth, the net force on deep ice and icebergs will be less than on thin ice, and pack ice will move past icebergs when there is significant wind stress.

(c) In both of the above cases, the effect of the Coriolis force will be to divert the ice motion by 15° to 20° to the right of the wind or current stress in the northern hemisphere (to the left in the southern). (It was the observation of the relation between wind direction and ice movement by Nansen, and communicated by him to Ekman, which caused the latter to develop his well-known theory of wind-driven currents.) It is convenient to note that as surface friction causes the surface wind to blow at about 15° to the left of the surface isobars, the direction of the latter is approximately that in which the ice is likely to drift (northern hemisphere).

(d) If the ice sheet is not continuous, collisions between individual floes may occur with a transfer of momentum (i.e. decrease of speed of the faster floe and increase of speed of the slower). Energy may go into ice deformation and

Circulation and Water Masses of the Oceans 231

building up of ridges at impact—this is referred to as internal ice resistance. This will increase with ice concentration, i.e. proportion of area covered by ice. The effect of upper surface roughness (R on a scale of 1 to 9) and ice concentration (C on a scale of 1 to 9) on the speed of the ice V (expressed as a percentage of the wind speed) is given by: $V = R(1 - 0.08\,C)$ (to be taken to only one place of decimals), so that the speed of the ice increases with roughness but decreases with increased ice concentration. It should be noted that for very close pack ice, stresses of wind or current will be integrated over quite large areas and the local motion may not relate well to the local wind.

Ice break-up is caused by wave action, tidal currents and melting. Melting of ice occurs when it gains enough heat by absorption of solar radiation and by conduction from the air and from sea-water nearby to raise its temperature above the melting point. The absorption of radiation depends very much on the albedo of the surface (proportion of radiation reflected) which varies considerably, e.g. for sea-water the albedo is from 0.05 to 0.10 (i.e. it is a very good absorber of radiation), for snow-free sea-ice it is from 0.3 to 0.4, while for fresh snow it is 0.8 to 0.9. Dark materials, like dirt and dust, have a low albedo of 0.1 to 0.25 and so absorb radiation well, and such material on ice can form a centre for the absorption of radiation and consequent melting of ice around it, so that puddles can form. These can absorb heat because of the low albedo of water and may even melt right through an ice sheet. When any open water forms, this absorbs heat and causes rapid melting of ice floating in it.

7.542 Distribution of sea-ice

In the Arctic Sea, the sea-ice may be divided into three categories. The most extensive is the *Polar Cap Ice* which is always present and covers about 70% of the Arctic Sea, extending from the pole approximately to the 1000 m isobath. It is very hummocky and, on the average, is several years old. Some of this Cap Ice melts in the summer and the average thickness decreases to about 2.5 m. Open water spaces, *polynyas*, may form. In the autumn these freeze over and the ice in them gets squeezed into ridges, or rafted. In the winter, the average ice thickness is 3 to 3.5 m but hummocks increase the height locally up to 10 m above sea level. The occasional ice islands, which have fairly uniform ice thickness considerably greater than the regular Cap Ice, originate from glaciers in northern Ellesmere Island.

Although this Polar Cap Ice is always present it is not always the same ice. Up to one-third of the total Cap and Pack Ice is carried away in the East Greenland Current each year while other ice is added from the Pack Ice described below. The Polar Cap Ice is only penetrable by the heaviest icebreakers.

The *Pack Ice* lies outside the Polar Cap and covers about 25% of the Arctic area. It is lighter than the Cap Ice and its area varies somewhat from year to year but extends inshore of the 1000 m isobath. It penetrates farther south in the East Greenland and the Baffin Land/Labrador Currents by which it is carried. Its areal extent is least in September and greatest in May. Some of it melts in summer and some gets added to the Cap by rafting. The ice which impedes navigation in the northern parts of the Canadian Archipelago, along the east coast of Greenland, in Baffin Bay and the Labrador Sea area and in the Bering Sea is the Pack Ice which is of a few metres thickness. It can be penetrated by ice-breakers.

The general circulation of the Cap and Pack Ice is similar to that of the upper water-layer in the Arctic. In the Beaufort Sea (Canadian Basin) there is a clockwise circulation and, in the Eurasian basin, a direct flow at about 3 km/day toward Fram Strait (between Greenland and Spitsbergen) and down the east coast of Greenland. It has been estimated that the volume of fresh water, as ice, exported there is approximately equal to the total continental runoff into the Arctic basin.

In recent years, much attention has been directed to studies near the edge of the Pack Ice (Marginal Ice Zone Experiments). In MIZEX, (MIZEX Group, 1986) along the Greenland Sea ice edge, and in MIZEX WEST in the Bering Sea, it has been shown that the break-up of the Pack is partly due to surface wave energy arriving. MIZEX results also showed that upwelling, eddies and jets are features in the sea along the ice edge, and that higher levels of biological productivity than in surrounding waters are found in the MIZ. LIMEX (Labrador Ice Margin Experiment) started in 1987 and is expected to be repeated at two-year intervals with particular reference to oil exploration and exploitation off Newfoundland—much of this study is being carried out by remote sensing from aircraft.

Lastly, *Fast Ice* is that which forms from the shore out to the Pack. This ice is "fast" or anchored to the shore and extends to about the 20 m isobath. In the winter it develops to a thickness of 1–2 m but breaks up and melts completely in summer. When it breaks away from the shore it may have beach material frozen into it and this may be carried some distance before being dropped as the ice melts, giving rise to "erratic" material in the bottom deposits.

7.543 Build-up and break-up of sea-ice in several regions

To give some idea of the variation in ice conditions with latitude, we present brief accounts of the build-up and break-up of sea-ice from about 48°N to about 80°N in the Canadian north.

In the Gulf of St. Lawrence (46°–51°N), there is only first-year ice. Ice forms first in the inner area (river), then along the north shore, becomes a hazard to shipping in the main Gulf by January, covering most of the area by the end of February with ice to 0.6 m thickness. Break-up starts in mid-March and ships can move freely by mid-April along mid-Gulf over the deep Laurentian Channel and all the ice melts by the summer. In severe winters, build-up and break-up can be two months earlier/later respectively.

Circulation and Water Masses of the Oceans 233

In Baffin Bay/Davis Strait (63°–78°N), there is mostly first-year ice of 1.5–2 m thickness with some older ice up to 3 m thick entering at the north through Smith Sound from the Arctic. This is a region where ice cover is more common than open water. Baffin Bay and the west side of Davis Strait are largely covered by ice until mid-May, mostly clear by mid-August except off Baffin Island and ice then starts to develop from the north by late October. Interannual variations are considerable with some areas clearing by mid-June in a good year but freeze-up starting as early as the end of August in a bad year. A feature of the north end of Baffin Bay is the "North Water", a large "recurrent" polynya which is generally clear of persistent ice throughout the winter.

In the Canadian Archipelago (north of 75°N, from Baffin Bay to about 120°W) ice cover may break up but floes remain present throughout the year and ice-breakers are needed for surface supply to northern outposts there. First-year ice develops to 2.4 m thickness and multi-year ice to 4.5 m. Some clearing does take place in Lancaster Sound (74°N, leading west from Baffin Bay) and in passages further west by July, but floes continue to be present.

The Western Arctic (120°W to the Bering Strait) is largely an open sea area north from the Canada/Alaska coast which is at about 70°N, with only a small tidal range to 0.7 m and a slow current to the east in the open. Multi-year ice (Arctic Pack) of up to 4.5 m thickness is general over the open sea south to 72°N while fast ice develops to 2 m thickness along the coast. Open water is usually found near the coast from mid-August to mid-September and can even extend to 73°N but in extreme years the Arctic Pack may extend to the coast in August. Ship movements along the coast are generally limited to September.

In the open North Pacific ice does not occur, but it is formed in the adjacent seas to the north and west, i.e. the Bering Sea, the Sea of Okhotsk and the north of the Sea of Japan. In the Bering Sea, Pack Ice extends in winter to about 58°N but clears completely in the summer, retreating north through the Bering Strait to 70–72°N.

7.544 Icebergs

Icebergs differ from sea-ice in that they originate on land, have no salt content, have a density of about 900 kg/m^3 (which is less than that for pure ice because of gas bubbles in bergs) and have much greater vertical dimensions and are a more serious hazard to shipping because of their large mass. In the North Atlantic, the chief source of icebergs is calving from the glaciers of West Greenland with a much smaller number from the west coast of Baffin Bay, the total number formed each year being estimated at as many as 40,000. Icebergs vary considerably in dimensions (height above sea level/length) from 1.5 m/5 m for *growlers*, 5–15 m/15–60 m for *small bergs*, 50–100 m/120–220 m for *large bergs*, while bigger ones are called *very large bergs*. The ratio of volume below sea level to that above is close to 7 to 1 but the ratio of maximum depth below sea level to height above it is less than this, depending on the shape of the iceberg. It varies from 5 to 1 for *blocky bergs* to only 1 to 1 for very irregular bergs (*horned* or *winged*). North Atlantic bergs are generally in the *pinnacled bergs* category with ratio 2 to 1.

The drift of icebergs is determined chiefly by water currents because of their greater draft than floe or pack ice whose movement is more determined by wind stress. Icebergs have an average life of 2–3 years and may travel up to 4000 km from their origin in Greenland whence they move north in the West Greenland Current, east across Baffin Bay and then south-east in the Baffin Island and Labrador Currents at about 15 km/d, many becoming grounded on the shelf. A small proportion pass into the North Atlantic off Newfoundland where they are usually a few tens of metres high, the highest recorded being 80 m high and the longest about 500 m. The main season for icebergs in the Grand Banks region is from March to July.

In the Southern Ocean, the immediately available information on the distribution of ice is not great, although in recent years a considerable amount of material has been gathered using satellite observations in conjunction with ship reports. On the basis of available data it appears that icebergs may be found between 50° and 40°S while pack ice extends only to 65° to 60°S. The relatively zonal distribution is probably due to the zonal character of the currents in the Southern Ocean. An outstanding feature of the Antarctic is the Ross Ice Barrier with a sea front of about 700 km to the Pacific and a height of 35 to 90 m above sea level with corresponding depth below. This *shelf ice* represents the extension of glaciers from the Antarctic continent out on to the sea where they float until bergs break off. These *tabular bergs* may be up to 80 to 100 km long and tens of kilometres wide. In late 1987, the biggest berg on record broke off from the Ross Ice Shelf, 208 km long, 53 km wide and 250 m thick—claimed to provide enough fresh water if melted to satisfy the needs of Los Angeles or New Zealand for 1000 years. At that time it was drifting north and west at 5 km/day, (in the East Wind Drift), and it was estimated that it would be a year before it got far enough north to enter the West Wind Drift.

Polynyas also occur in the Antarctic and much information about them has been obtained from satellite observations in conjunction with observations from ships (Gordon and Comiso, 1987). These polynas are of two types. *Coastal polynyas* are a result of offshore winds driving the ice away from the shore leaving open water there. On this, new ice soon forms as latent heat is discharged to the atmosphere at a rate of as much as 300 W/m^2. The salt rejected as the ice forms contributes to the formation of Antarctic Bottom Water at a rate estimated at up to 10 Sv. Away from the coast, *open ocean polynyas* result from relatively warm Circumpolar Atlantic Deep Water coming to the surface (probably deflected upward over rises in the bottom topography) and melting the ice there. One very large polynya (350 × 1000 km) formed in the Cosmonaut Sea area (43°E, 66°S) of the Weddell Sea in 1974 and recurred in 1975 and 1976. This one had a measurable effect on the water characteristics in its vicinity. A polynya was also observed in this area in 1980, 1982 and 1986, and another open ocean polynya has been observed over the Maud Rise at 2°E, 64°S.

7.6 Pacific Ocean

Our oceanographic knowledge of the Pacific Ocean is less complete in some respects than that of the Atlantic. The North Pacific can be described well, and our information on the disposition of currents in the Equatorial Pacific appears to be good and quantitative information on their volume transports is increasing. Data on the South Pacific are sparse and it is not possible to describe it as adequately as the South Atlantic.

7.61 Pacific Ocean circulation

The circulation of the upper waters of the Pacific (Fig. 7.31) as a whole is very similar in its main features to that of the Atlantic. There is a clockwise gyre in the North Pacific and a counterclockwise one in the South Pacific, with an equatorial current system between them. The whole equatorial current system is well and clearly developed in the Pacific and it will be described first, and details of the two gyres given later.

7.611 Pacific Ocean: equatorial circulation

The Pacific equatorial current system (Fig. 7.31) is now recognized to include at least four major currents, three of which extend to the surface and one of which is below the surface. The three major upper-layer currents (evident at the surface) are the westward-flowing *North Equatorial Current* (NEC) between about 20° to 8°N (in mid-Pacific at 160°W), the westward *South Equatorial Current* (SEC) from about 3°N to 10°S and the narrower *North Equatorial Countercurrent* (NECC) flowing to the east between them. The fourth current, the *Equatorial Undercurrent* (EUC), flows to the east below the surface straddling the equator from about 2°N to 2°S. This current was first recognized as recently as 1952 but the surface currents must have been known to Pacific islanders long before Europeans arrived in the Pacific. This equatorial system can be traced almost from the Philippines in the west to the Gulf of Panama in the east, a distance of some 15,000 km.

The surface current system is driven by the trade winds and is asymmetrical about the equator because the trade-wind system is asymmetrical as is shown in Fig. 7.32, a situation which is related to the asymmetrical distribution of land and sea about the equator. Although the trade wind system is present throughout the year it undergoes some seasonal variations which will be described. (In doing so we will identify the seasons by month names when considering areas which include both hemispheres to avoid the ambiguity of

236 Descriptive Physical Oceanography

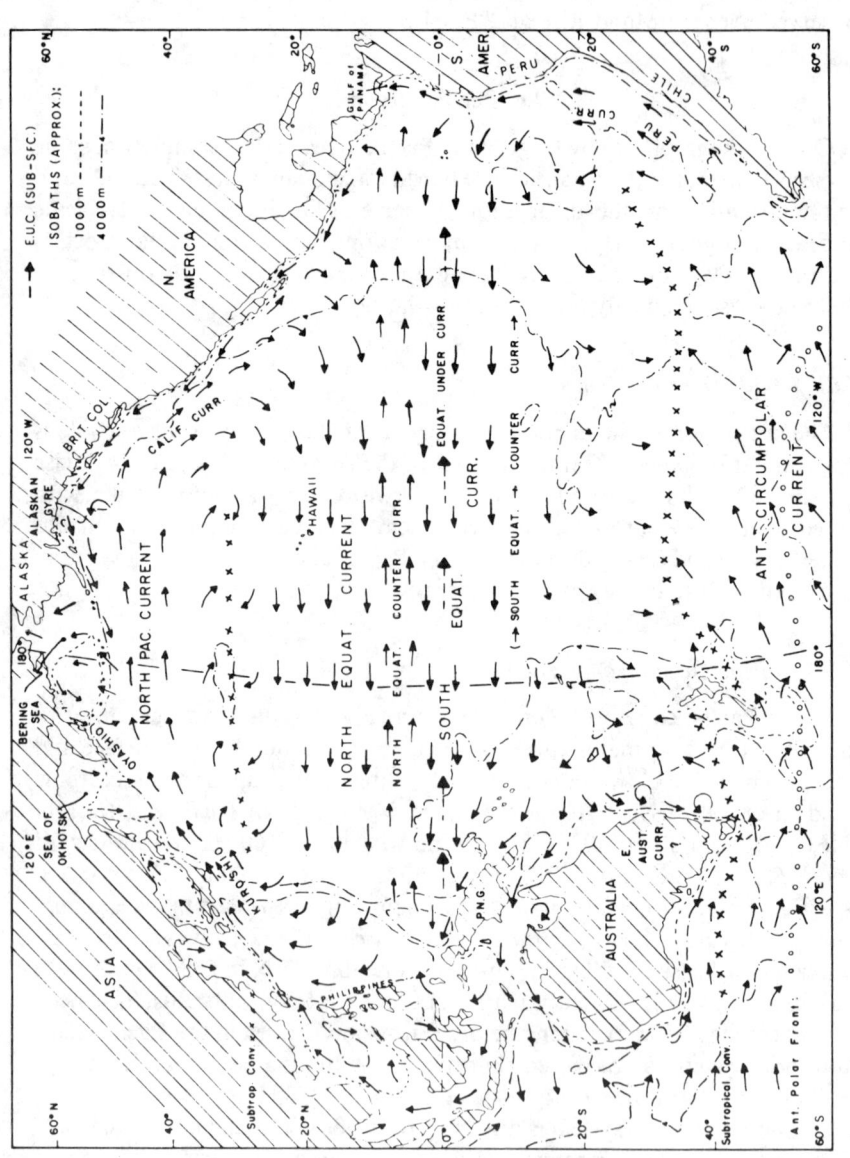

Fig. 7.31. Pacific Ocean—surface circulation.

Circulation and Water Masses of the Oceans   237

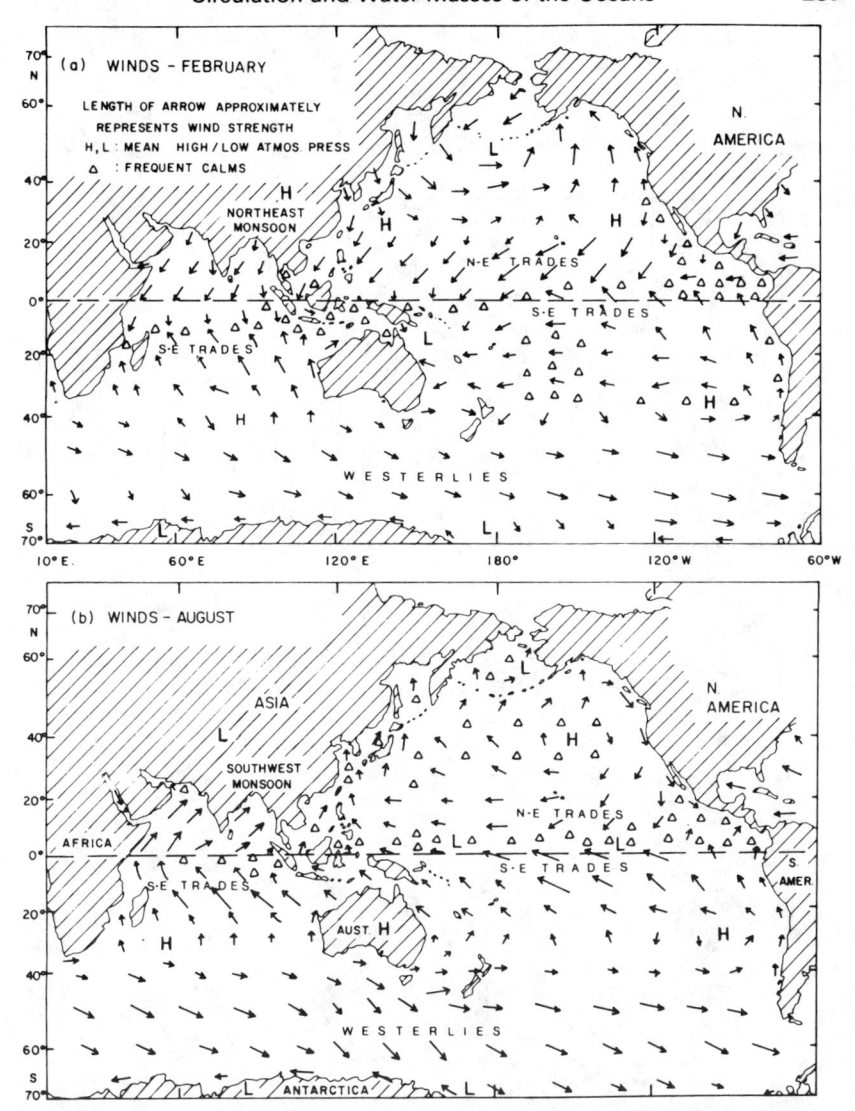

FIG. 7.32. Indian and Pacific Oceans—winds and mean atmospheric pressure highs and lows: (a) February (north–east monsoon, (b) August (south–west monsoon).

the terms "winter" and "summer".) In February, the monsoon season, the north-east Trade winds stop short of the equator at about 5°N in the east but extend to the equator in the west and even have an eastward component there for short periods. At this time, the south-east Trades extend northward across the equator in the east. In August, the southern limit of the north-east Trades is at 10°N across the whole Pacific. (The seasonal change in the equatorial wind system is more marked in the Indian Ocean and will be discussed in

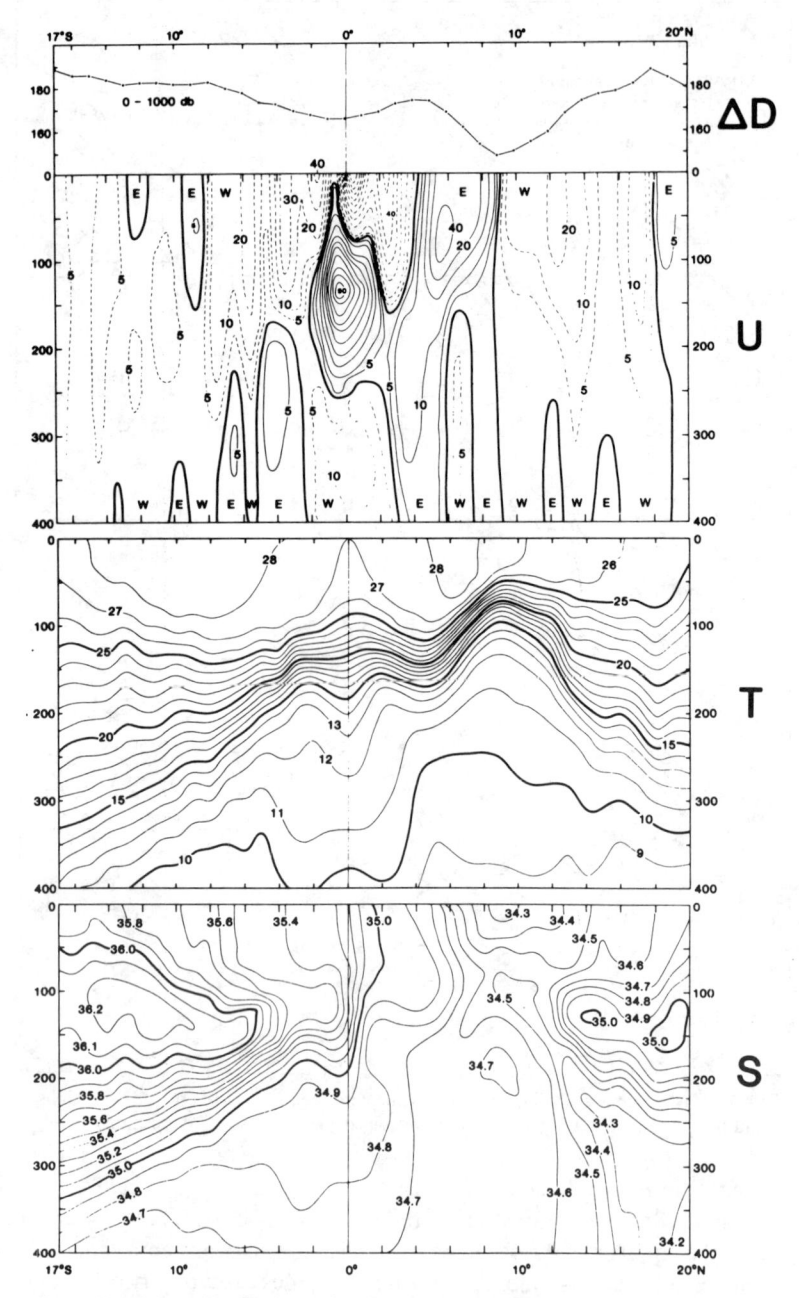

FIG. 7.33. Mean distributions of surface dynamic height (ΔD dyn. cms.) relative to 1000 db (dyn. cm.) and vertical meridional sections of zonal geostrophic flow (U cm/s), temperature (t°C) and salinity (S) between Hawaii and Tahiti, for 12 months from April 1979. (Wyrtki and Kilonsky, 1984.)

Circulation and Water Masses of the Oceans   239

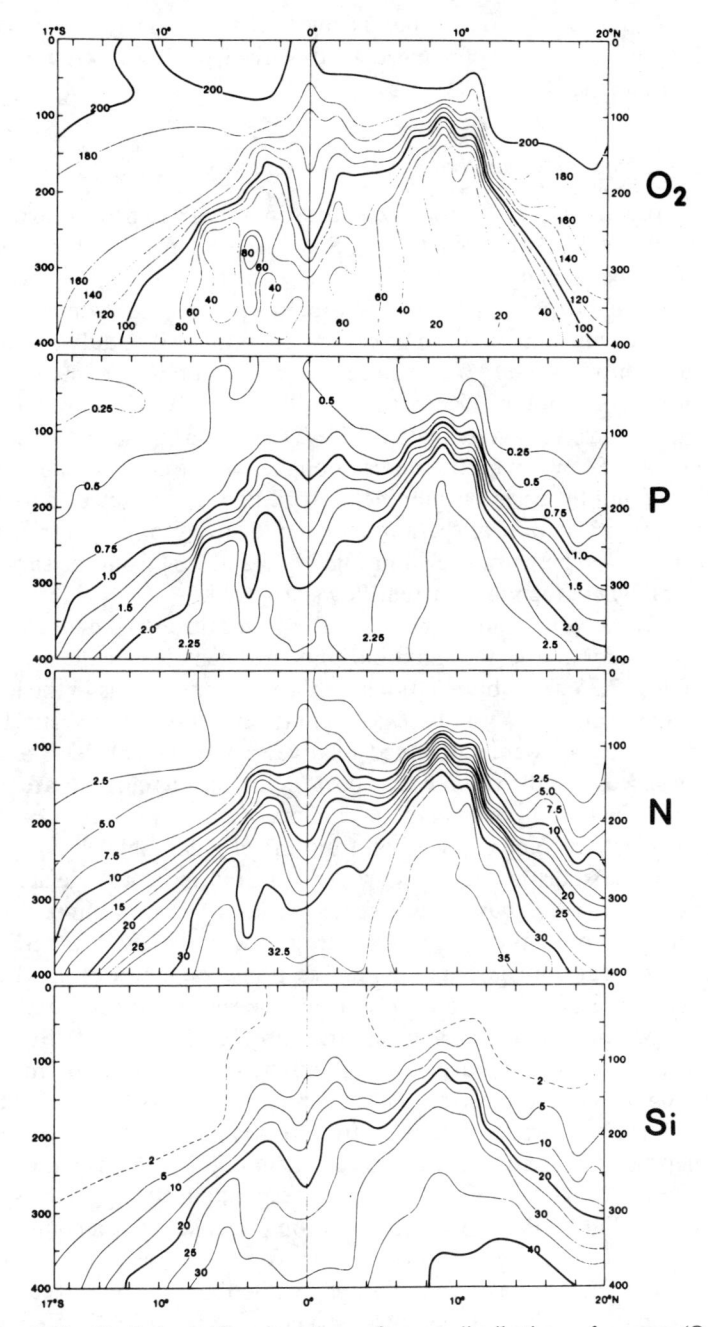

FIG. 7.34. Vertical meridional sections of mean distributions of oxygen (O_2), phosphate (P), nitrate (N) and silicate (Si) in μM/kg between Hawaii and Tahiti for 12 months from April 1979. (Wyrtki and Kilonsky, 1984.)

Section 7.71.) It should be noted that the region between the trade wind zones, the "doldrums", is a region of reduced winds with 10 to 30% calms, not one of continual calm.

7.612 *A meridional section across the upper equatorial Pacific*

The first oceanographic section across the equatorial Pacific was carried out from *Carnegie* in 1929 at about 140°W. Since then a considerable number of studies have been made in the equatorial Pacific, although many only extended over short periods of time. In order to present the main features of this region we will summarize some results of an intensive study during the Hawaii-to-Tahiti Shuttle Experiment in which 43 sections were made during 15 cruises over a period of 17 months from February 1979, and reported by Wyrtki and Kilonsky, 1984 (hereafter WK). The section crossed the equator at 150°, 153° and 158°W but for this description they were treated as one at about 154°W. A 12 months period beginning in April 1979 was selected as the most consistent for calculation of the annual mean distributions of currents and water properties. These are shown in Figs. 7.33 and 7.34 from the surface to 400 m depth where the main current flows occurred.

Fig. 7.33.U shows the annual mean zonal currents through the substantially north-south section, and the main currents are identified in the schematic section of Fig. 7.35 with abbreviations for the main currents as in the legend. A cut-off speed of 2 cm/s was used to separate currents in the vertical. The geostrophic currents were calculated relative to the 1000 db level and the maximum speeds and the volume transports of the various currents are shown in Table 7.4 where positive (negative) values represent flows to the east (west) respectively (values being rounded off from the original table).

The small transport of the SECC was considered to be an artefact of the averaging process as the current was subject to large variations in position so that much of its flow was cancelled by averaging with the SEC1 and SEC2. A better estimate, calculated from individual sections, was considered to be 3 Sv with a maximum of 6 Sv. The EUC was also measured directly during the Shuttle Experiment to have a mean transport of 32.3 ±3.5) Sv and a maximum speed of 102 cm/s at 130 m depth. It was considered to be unusually weak during the 12 month period. The values for transports were estimated to be within 1 Sv for flows more than 3° from the equator.

The meridional slope of the sea surface (ΔD in Fig. 7.33) was characteristic of the equatorial regions with the slopes corresponding closely to the surface currents according to the geostrophic relation. Later we offer a comment on some features of the troughs and ridge of the surface slope.

We will now describe the water property distributions, as typical of the equatorial region. The temperature structure (Fig. 7.33.T) shows the thermocline to be most intense a few degrees north and south of the equator with the isotherms spreading progressively north and south of the 10°

Circulation and Water Masses of the Oceans 241

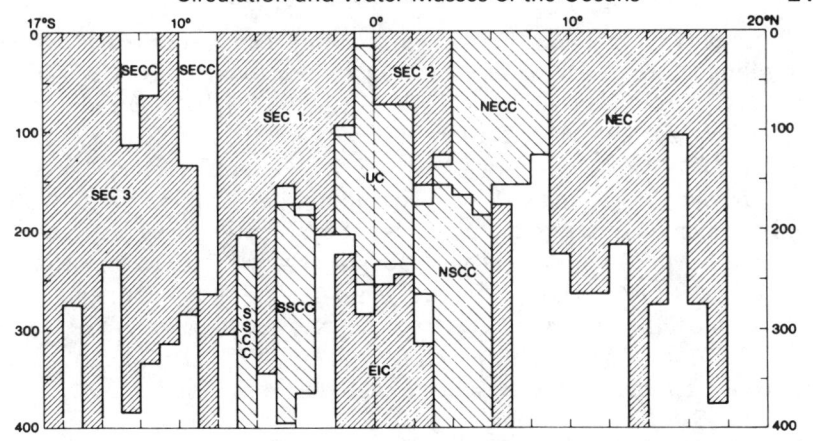

FIG. 7.35. Schematic of mean areas occupied by zonal currents between Hawaii and Tahiti for 12 months from April 1979. Dark (light) shading indicates westward (eastward) flow respectively, blank areas have zonal speeds less than 2 cm/s. NEC = North Equatorial Current, NECC = North Equatorial CounterCurrent, SEC = South Equatorial Current (three sections), SECC = South Equatorial Counter-Current, UC = Equatorial Undercurrent (EUC in our notation), EIC = Equatorial Intermediate Current, NSCC/SSCC = Northern/Southern Subsurface Counter-Currents. (Wyrtki and Kilonsky, 1984.)

TABLE 7.4

Mean zonal transports and current speeds, and cross-sectional areas of currents at about 154°W, Pacific Ocean for 12 months from April 1979. (Current directions are +to the east, −to the west.) (From Wyrtki and Kilonski, 1984.)

Current	Transport	Mean speed	Max. speed	Area
	(Sv)	(cm/s)	(cm/s)	(km^2)
NECC	20	21	>40	92
SEC 1	−27	−14	>−40	193
SEC 2	−15	−30	>−50	50
SEC 3	−13	−5	—	274
SEC Total	−55	−11	—	517
SECC	0.6 (3)	4	—	14
EUC	23	31	>90	73
EIC	−12	−8	—	149
NSCC	9	7	∼10	133
SSCC	4	5	∼10	90

parallels. A feature of the equatorial zone is the spreading of the isotherms at the equator in the location of the EUC. Below the thermocline, between about 5°S and 12°N is a marked *thermostad*, a region of relatively small rate of change of temperature with depth (c.f. the 18°C water in the North Atlantic, Sec. 7.343).

The salinity distribution (Fig. 7.33.S) is quite different. In the *south* is the high-salinity core of the *South Pacific Subtropical Water* with a maximum salinity of 36.2 at 120 m depth and, in the north, the *North Pacific Subtropical Water* of lower salinity, 35.0. The lowest salinity is at the surface in the NECC, while the subsurface low-salinity water entering at about 20°N at 300 m is the *North Pacific Intermediate Water*. The South Pacific Subtropical Water salinity maximum results from an excess of evaporation over precipitation which renders the surface water dense enough in winter to sink to the top of the thermocline with consequent high oxygen content and low nutrients (Fig. 7.34). It enters the EUC and mixes with it, increasing its salinity. The North Pacific Subtropical Water only penetrates to about 12°N. The salinity values and degree of penetration of the subtropical waters varied from cruise to cruise during the Shuttle Experiment.

Dissolved oxygen and nutrient distributions (Fig. 7.34) were similar to each other and to the temperature distribution, all showing spreading of the isopleths in the EUC. Dissolved oxygen decreased and nutrients increased markedly with depth increase. (A useful reference about water properties in the equatorial Pacific is the description by Tsuchiya (1968) frequently quoted in papers about this region.)

7.613 Zonal and temporal variations of the Pacific equatorial current system

The above description based on the Hawaii-Tahiti Shuttle Experiment delineates the currents now recognized between 20°N and 17°S from the surface to 400 m depth and the characteristics of the water masses there. It also gives their mean values during one 12 month period but presents no information on temporal variability (which is being studied from this data set for a future publication). Also it gives no information on possible spatial variations across the width of the Pacific. To fill in these aspects of the equatorial region we will review information from some of the other studies which have been made, starting in the west.

The equatorial system is known to extend from at least 143°E (north of Papua New Guinea) to the Galapagos Islands (90°E) and probably as far as the coast of Ecuador, a distance of some 15,000 km. The EUC is remarkable in that it can be followed over this whole distance (bifurcating round the Galapagos) even though it is so thin, being about 200 m thick and 200 km wide (i.e. thickness to width ratio about 1/1000). (The large vertical to horizontal distortion ($\sim 5000:1$) in sections such as Fig. 7.33 masks the actual thinness of the current.)

A recent investigation in the western Pacific (Lindstrom *et al.*, 1987), using Acoustic Doppler Current Profilers, showed that the probable origin of the EUC is from the newly discovered saline *New Guinea Coastal UnderCurrent* (NGCUC), which is formed by northwest flow along the north coast of Papua

Circulation and Water Masses of the Oceans   243

New Guinea (PNG) from the Coral Sea via the Solomon Sea and the Vitiaz Strait between PNG and New Britain. The NGCUC turns north and then west at about 143°E to feed the EUC. The NGCUC had speeds of 50 cm/s centred at 200 m depth and a transport of 7 Sv at 2°S which were the same values as those of the EUC at the equator. The SEC flowed to the west as long as the zonal component of the wind was westward but reversed to the east within a week of the zonal wind changing to eastward at ~ 3 m/s in November (start of the monsoon season). The EUC was not affected by the wind change. A satellite-tracked drifting buoy drogued at 15 m reversed direction from west to east at the same time at about 7° east of the oceanographic sections. (The reversal of the SEC had been observed frequently before (Hisard and Hénin, 1984) and, on occasion, the EUC split into two cores at about 120 m and 210 m depth west of 170°E during ENSO events between 1928 and 1971.)

Wyrtki (1974) analysed the dynamic topography in the region between 140°E and 140°W to investigate possible seasonal variations and showed that in March–May the SEC and EUC were at their strongest and the NEC and NECC weakest while the opposite was the case in September–December. Delcroix *et al.* (1987) studied the water properties and made direct current measurements at 165°E during six half-yearly cruises between 7–10°N and 20°S from January 1984 to June 1986. The temperature structure observed was basically similar to that described by WK (Fig. 7.33) but the thermocline was generally about 30 m deeper than WK observed and the surface water was warmer by 2K. Delcroix *et al.* showed six salinity sections which again were similar in form to that shown by WK but the South Subtropical Water core was less saline (35.85 compared with 36.2). The measured currents (mean, maximum and minimum) are shown in Table 7.5 with the mean WK values

TABLE 7.5

Transports (means and ranges) and maximum speeds between 7–10°N & 20°S at 165°E, June 1984 to June 1986 (Delcroix, 1987), and WK (1984) values at 154°W, April 1979 to March 1980.

	Delcroix et al. (165°E)			WK (154°W)	
	Transport		Speed	Transport	Speed
	Mean	Range	Max.	Mean	Mean
	(Sv)	(Sv)	(cm/s)	(Sv)	(cm/s)
NECC	27	18–52	60	20	>40
SEC (total)	(−71)	43–90	−50	−50	>−50
SECC	(18)	8–30	30	3	—
NSCC	15	7–26	30	9	~ 10
EUC	20	10–29	50	23	>90
SSCC	9	7–12	20	4	~ 10

Note: Delcroix *et al.* did not give mean values for the SEC and SECC because one January and one June value were missing but we have averaged the remaining four values.

from Table 7.4 for comparison. The EUC was generally weaker in January and stronger in June/July, and deeper (180 m) at 165°E than at 154°W (150 m). Delcroix et al. pointed out that the EUC, NSCC and SSCC can be traced across most of the Pacific, and an NSCC and SSCC have also been identified in the Atlantic but not in the Indian Ocean.

Kessler and Taft (1987) calculated dynamic heights from XBT casts and surface salinity records across the tropical Pacific (from ships of opportunity) from 1979–84, using a mean T–S relation, and estimated transports relative to 450 m to obtain the following values (summarized); NEC: 19 ± 1.7 Sv; NECC: 21.1 ± 5.7 Sv with maxima in December; SEC: 35 ± 5.1 Sv with maxima in May; SECC: 3–9 Sv during the first half of the year and 1 Sv during the second half. In late 1982 (1982–83 El Niño), eastward flow extended from 10°N to 5°S with a total transport of 60–70 Sv of which 42 Sv was attributed to the NECC, while in mid-1983 the NECC transport fell to 2 Sv. The SEC transport fell by 10 Sv in 1982.

Wyrtki (1980) gave a few figures for variability during the Hawaii-Tahiti Shuttle Experiment. The EUC had a maximum core speed of 150 cm/s (compared with its mean of 31 cm/s) and a maximum volume transport of 70 Sv, with 60 Sv for one period of three months (overall mean 23 Sv) so that considerable variations do take place. The maxima occurred in July and the minima in January. The SEC and NECC were out of phase with the EUC.

Halpern (1987) reported temperature and direct current measurements with VACMs on the equator at 110°W and 95°W during 1980–85 which serependitiously included the period of the major El Niño of 1982–83 (see Section 7.642). Excluding the 12 months El Niño period from August 1982, the annual maxima and minima of speeds were as in Table 7.6. (As measurements

TABLE 7.6

Annual ranges of zonal speeds at 95°/110°W (Halpern, 1987) during non-El Niño period, January 1979 to September 1985.

Current	Depth (m)	Minimum/month (cm/s)	Maximum/month (cm/s)
SEC	15	−47 November	+46 April
EUC	25	−25 October	+90 May
EUC	48	+40 October	+130 April
EUC	78	+65 November	+115 April
EUC	155	+40 All year	+40 All year
EIC?	250	−5 March–May	+12 August

were made only at the equator it was not possible to calculate volume transports.)

The temperature at 15 m ranged from 25°C in April to 21°C in October decreasing at 110 m and deeper to an annual range of 1 K or less. During the El Niño year, the temperature at 100 m increased from about 16°C in August

1982 to 27°C in January 1983. At the same time, the upper mixed-layer increased in thickness by 100 m (not associated with any increase in local wind stress). During August–December, 1982, the rate of increase of heat content in the upper 15–200 depth layer was $+500$ W/m^2 and during January–February, 1983, the rate was -460 W/m^2, both values being much larger than the non-El Niño rates (see Figs. 5.7.1 and 5.7.2). At the same time the SEC decreased from -30 cm/s to -3 cm/s at 15 m while in January/February 1983 the zonal current at all depths at 110°W was westward for four weeks and for two weeks at 95°W, i.e. the EUC disappeared for these periods.

This disappearance of the EUC was observed in other locations, e.g. Firing et al. (1983) reported that until July 1982 the EUC at 159°W flowed east at a core speed of 100 cm/s at 150 m depth while westward flow in the SEC was weak. The EUC actually surfaced in April. In September and October, 1982 most of the eastward EUC vanished and instead a westward flow occurred, centred at 150 m, leaving only a small eastward flow at 160–220 m. During November, eastward flow at the surface deepened to over 200 m and by January, 1983 the EUC reappeared with a core speed of 80 cm/s at 130 m and the surface SEC flow was westward again at up to 100 cm/s. It was noted that winds at 173°E changed to blow eastward from June to December, 1982 and at 159°E the normal westward component decreased and the zonal component became eastward in December and fluctuated between westward and eastward during January–March, 1983.

Hayes et al. (1983) made nine oceanographic sections between 10°N and 3°S at 110°W between February 1979 and June 1981 and calculated geostrophic currents. Mean values (and ranges) for transports were: NECC: $+8$ Sv (3–12 Sv), NSCC: $+14$ Sv (11–16 Sv) and the speeds were 38 cm/s (34–51 cm/s). One point which they, and others, made was that, as these two currents flow in the same direction, any division between them must be somewhat arbitrary. Hayes et al. used the $\sigma_\theta = 25.0$ surface as a boundary (at 70–80 m depth). WK chose a sloping division between the two currents at about 170 m where there was generally a speed minimum, and Delcroix et al. used $\sigma_t = 23.5$ at about 150 m. Hayes et al. noted that both the SEC and NECC were maximal in September–November and minimal in March–May.

One feature of the equatorial Pacific is that the depth of the thermocline/pycnocline and the EUC, which are closely associated, rises from 200 m in the west to close to the surface in the east. Figure 7.36 shows the pycnocline from 160°E to 90°W in terms of the thermosteric anomaly, $\Delta_{S,T}$, as observed during a single cruise along the equator.

We started this section with a description of the probable beginning of the EUC in the west; we will finish with one of its termination in the east. Lukas (1989) reviewed the various studies in the region between 5°N–10°S and 100°–80°W and then carried out an analysis of a very large number of oceanographic stations made up to 1975 in that region with the object of tracing the EUC by its characteristic water properties. These were the 13°C

(a) THERMOSTERIC ANOMALY ALONG EQUATOR:

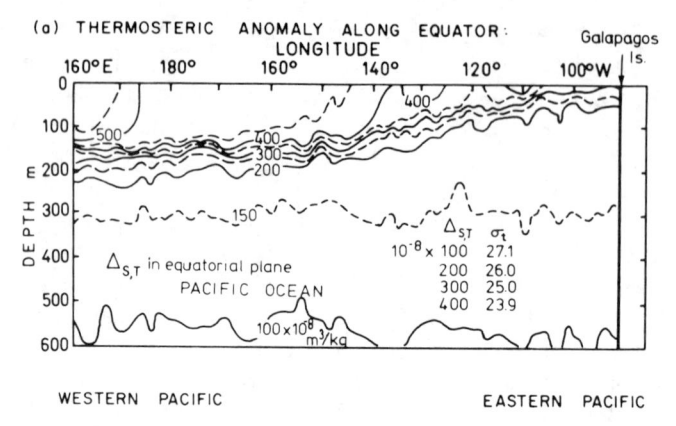

FIG. 7.36. Pacific Ocean—West-east zonal section along equator of thermosteric anomaly observed during a single cruise.

thermostad (between 12° and 14°C) under the thermocline, the high salinity core and the high dissolved oxygen concentration tongue. He showed that the distribution of the high salinity core and oxygen tongue showed that the EUC split at about 92°W and flowed east both north and south of the Galapagos Islands (located at about 0°–1°S, 91°–89°W). The southern part was the stronger, although the high oxygen tongue rapidly disappeared to the east. The two streams reunited east of the islands and some of the EUC waters were recirculated into the SEC. Actually the EUC core was slightly south of the equator from 98°W and its upper part reached the surface there between the two branches of the SEC. This displacement of the EUC to the south of the equator and the Galapagos was attributed to the southerly component of the mean wind stress in the region. By 85°W, the flow from 1.5°N to 3°S was eastward and from 3°N to 5°N was westward. East of the Galapagos, part of the EUC penetrated southeast to the coast of South America at 5°S to form the *Peru CounterCurrent* at the surface and part continued along the equator to the coast and then turned south to form the *Peru-Chile UnderCurrent* (PCUC) near 5°S.

In addition to determining the path of the EUC, there was a sufficient distribution of observations through the year to determine that the EUC east of the Galapagos went through a seasonal cycle, being strongest from January to June when the flow continued along the equator to the coast and then turned south to form the PCUC, and weakest from July to November when the flow was chiefly southeast from the Galapagos to the upwelling area at the coast. This seasonal pulsation progressed from west to east of the study region. Lukas commented that since his study was completed, many other stations had been occupied and current measurements made, including those during the period of the 1982/3 El Niño.

The EUC is "tied" to the equator by the restoring effect of the Coriolis force which, on an eastward flowing current, tends to bring the current back toward the equator if the current drifts away from it either to north or south. However, the current is not always perfectly straight. Direct measurements show significant and varying north or south components of velocity which suggest that the current meanders somewhat about its mean track along the equator.

Once the existence of the EUC had been recognized in the Pacific it was realized that there has been strong indications of its existence for some time previously. These were in the form of strikingly large wire-angles from the vertical when oceanographic gear was being lowered over the side of ships when near the equator. These wire angles were toward the east and were attributed to the strong, shallow surface current (SEC) carrying the ship toward the west relative to weaker or zero currents below the surface layer. It was only after the Undercurrent had been observed directly with the drogued floats that these observations, together with the observations that at times floating fishing gear drifted toward the east, were recognized as being due to a faster eastward flowing Undercurrent.

The discovery of the Pacific EUC naturally started speculation as to the possibility of there being one in the Atlantic. In 1962 Metcalf *et al.* presented observational evidence for such a current in the eastern equatorial Atlantic. About the same time it was pointed out (Montgomery, 1962) that the Atlantic Equatorial Undercurrent had been clearly observed and recognized as such in 1886 by Buchanan, one of the scientists who had earlier taken part in the *Challenger* expedition. Buchanan's accounts were published in well-known geographical journals but had apparently been forgotten until the independent discovery of the Equatorial Undercurrent in the Pacific stimulated a review of the literature. It has been suggested that remarks criticizing the validity of Buchanan's methods, made in Krümmel's *Handbuch der Ozeanographie* which carried much authority early in the century, may have diverted attention from Buchanan's observations.

The equatorial current systems are basically similar in the three oceans but by no means identical since both topography and forcing differ between the oceans. We sketched the Atlantic equatorial circulation in Section 7.33 and in Section 7.71 we will summarize features of the Indian Ocean circulation as influenced by the monsoonal changes of wind systems which are more pronounced there than in the other oceans. We must also remark that a great deal of work has been done on the dynamics of these circulations, including detailed modelling, but these aspects are beyond the scope of this text. We just mention a study by Philander and Pacanowski (1986) on a model of the circulation and heat budget of the tropical Atlantic Ocean and a similar one by Philander *et al.* (1987) on the equatorial Pacific, starting with the basic equations of motion, wind stress and simplified information on heat exchange to derive the main features of the upper-layer zonal circulation, vertical fluxes and heat budgets.

An interesting phenomenon has been observed to occur at the front just north of the equator in the Pacific between the SEC and the NECC. As first reported by Legeckis (1977) the satellite infra-red sea-surface temperature patterns at this location exhibited a cusp-like appearance thought to be an expression of equatorial trapped waves. Such waves would have wavelengths of about 1000 km extending from 95°W to 130°W. This cusp-like pattern was observed to propagate westward at an average phase speed of about 40 cm/s resulting in a period of about 20 to 30 days. Such waves have been observed in all of the satellite imagery (Legeckis et al., 1983) since 1975 except in 1976 and 1982, both of which were years in which El Niño occurred (see Section 7.642) in the eastern tropical Pacific.

Philander (1978) suggested that these wavelike perturbations resulted from a barotropic instability due to the shear between the SEC and the NECC. In his numerical simulations he was able to reproduce the wavelength and frequency characteristics of the waves observed in both the satellite imagery and also in the tracks of a few drifting buoys. A more intensive study of drifting buoy trajectories which was conducted by Hansen and Paul (1984) showed that the cusps in the satellite imagery were associated with a pattern of vigorous mesoscale eddies lying between the equator and 7°N. The spatial structure of this eddy field closely resembles the structure from Philander's (1978) study of barotropic instability. Also consistent with Philander (1978) is the observation that the eddies have a time-scale of about two weeks. The eddies have a braking effect on the mean currents and carry heat towards the equator in an amount about two-thirds of the poleward heat transport effected by the divergent Ekman transport in the near-surface waters.

Incidentally, the search for an equatorial undercurrent in the Atlantic (and Indian) Ocean was done for other reasons than to permit Atlantic oceanographers to keep up with their Pacific colleagues. It was done more to assist in the search for an adequate dynamic explanation for the existence of an undercurrent at the equator. If an undercurrent were found in all the oceans it would suggest that it was a typical feature of the equatorial current system and a common explanation might be adequate. Studies of the difference between the equatorial features as well as their similarities would help in deciding between alternative explanations. On the other hand, if the undercurrent were found only in the Pacific it might be that it was a consequence of some peculiarity of that ocean's circulation and this also might help in deciding on the most satisfactory explanation.

7.614 Convergences and divergences associated with current systems

One other dynamic feature associated with the Coriolis force which affects the distribution of water properties and the plankton, and incidentally of pelagic fishes, in the upper layer is the occurrence of divergences and convergences at the boundary between oppositely directed currents. The reasons are shown schematically in Fig. 7.37. Figure 7.37(a.1) shows two

Circulation and Water Masses of the Oceans

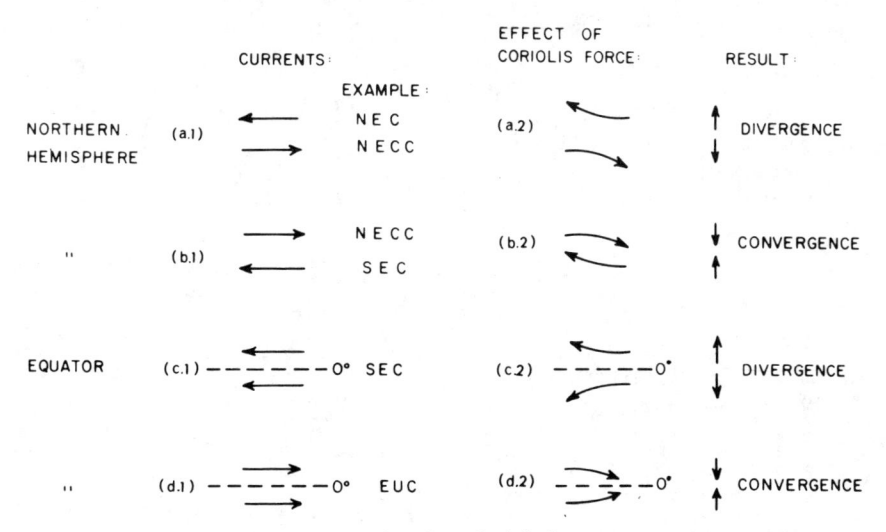

	CURRENTS	EXAMPLE:	EFFECT OF CORIOLIS FORCE:	RESULT:
NORTHERN HEMISPHERE	(a.1) ← →	N E C N E C C	(a.2)	↑ DIVERGENCE ↓
"	(b.1) → ←	N E C C S E C	(b.2)	↓ CONVERGENCE ↑
EQUATOR	(c.1) ← ←	—0° SEC	(c.2) —————0°	↑ DIVERGENCE ↓
"	(d.1) → →	—0° EUC	(d.2) ————0°	↓ CONVERGENCE ↑

FIG. 7.37. Schematic diagram to show how Coriolis force acts on currents to cause convergence or divergence.

oppositely directed currents in the northern hemisphere with the more northerly flowing to the west and the other to the east. Then (a.2) shows that the effect of the Coriolis force, to the right of the direction of motion, is to cause the two currents to *diverge* and consequently subsurface water will rise to replace the diverging surface water. This occurs between the NEC and the NECC. Between two currents flowing as in Fig. 7.37(b.1) a *convergence* occurs as shown in (b.2) and as occurs between the NECC and the SEC (Fig. 7.38) with downward movement of the water. The reader may apply similar reasoning to determine what should happen between currents in the southern hemisphere (Coriolis force acting to the left of the velocity). For currents centred at the equator, flow to the west will cause *divergence* (Fig. 7.37(c.1) and (c.2) as occurs at the surface in the SEC, while flow to the east will result in *convergence* (Fig. 7.37(d.1) and (d.2). This is the main reason why the eastward-flowing EUC tends to remain close to and symmetrical about the equator during its long passage across the Pacific. The convergent tendency for the EUC below the surface gives rise, because of conservation of volume, to upward movement above the current core and downward movement below it, causing the bowing of the isopleths (Figs. 7.33(T) and 7.34).

Another way to show these convergences and divergences is by means of a north-south section across the equator as displayed schematically in Fig. 7.38. This shows schematically the relations between the trade winds and zonal circulation. The annual mean trade winds are shown in direction and speed by the top row of arrows (a) and the surface currents below them (b)—both sets of arrows are understood to be in the horizontal plane. Below this is shown a meridional vertical section (c) with the surface shape (c.f. Fig. 7.33.ΔD) and

250 Descriptive Physical Oceanography

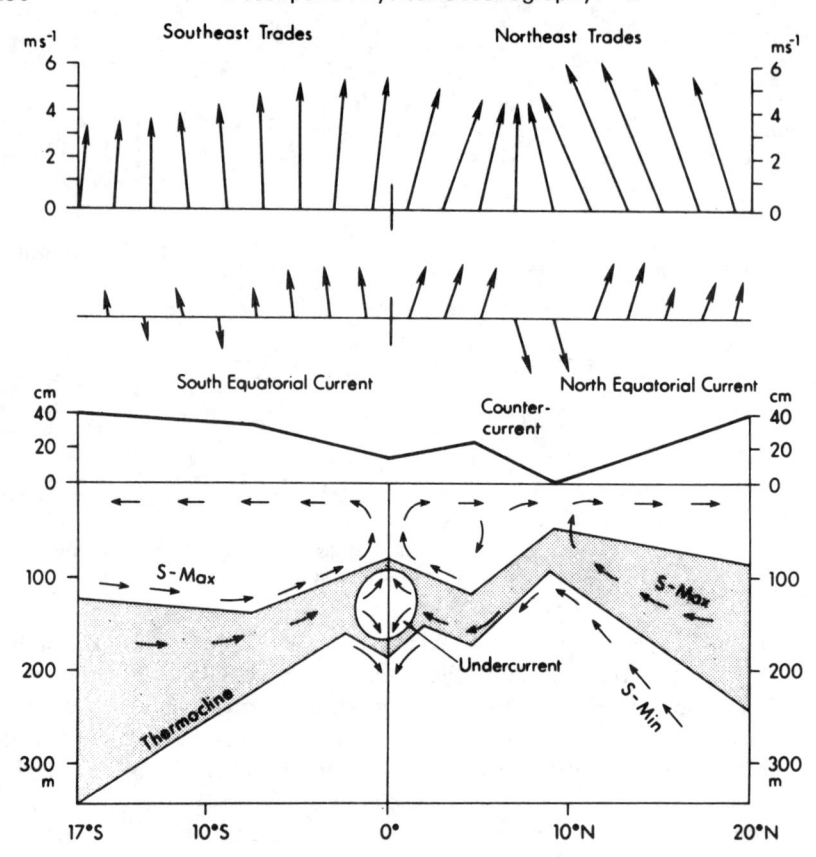

FIG. 7.38. Schematic meridional section across the equator with (top) plan view showing the mean trade winds, surface circulation, and (bottom) vertical section with surface slopes, basic temperature structure and meridional circulation below the surface. (N.B. "Countercurrent" = "NECC" in our notation.) (From Wyrtki and Kilonsky, 1984.)

the basic temperature structure and vertical circulations shown schematically. (Note that the slopes of the sea surface and of the isotherms are grossly exaggerated. The vertical to horizontal scale is exaggerated by about 1 million to 1 for the sea surface section and by about 6000 to 1 for the region below the sea surface.)

The divergence at the equator results from the divergence of the Ekman flow there caused by the south-east Trades which straddle the equator blowing westward. Then there is a surface convergence at about 4°N between the SEC and the NECC and a divergence at 10°N between the NECC and the NEC. WK pointed out that the circulation which they suggested for the upper 100 m or so (Fig. 7.38) differs from that earlier suggested by Defant (in 1939 for the Atlantic, e.g. see Sverdrup et al., 1946, Fig. 172) in that WK found no evidence

Circulation and Water Masses of the Oceans 251

for Defant's anticlockwise vertical circulation cell between 5° and 10°N. They considered that the meridional flow at the surface north of the equator must be everywhere to the north because the trade winds are everywhere directed westward even though they weaken slightly between about 4°N and 10°N (the "doldrums"). They also cited drifting buoy observations which consistently showed a northward component of motion of ~ 5 cm/s between 0° and 20°N. (Convergences or divergences can still occur due to variations of the strength of the Trades with latitude, e.g. the convergence at 4°N where the northward Ekman flow component is greater south of 4°N than north of that latitude.)

Note that Fig. 7.38 is only intended as a schematic diagram to show the main features. Hayes et al. (1983) calculated the surface slope for 16 meridional sections at 110°W and found the position of the equatorial trough at $1.0°S \pm 0.7°$, of the equatorial ridge at $4.9°N \pm 1.3°$ and of the countercurrent trough at $8.7°N \pm 1.8°$, so that the positions of these features do vary somewhat with time and location.

The divergences result in upwelling of water which is often richer in nutrients than the displaced surface water and so biological production is promoted. At convergences, concentration of the upper-layer plankton may occur (because they are brought together horizontally by the flow but resist the downward motion of the water). Both convergence and divergence zones may provide food for fish and hence be profitable fishing areas.

Measurements during the Shuttle Experiment showed that biological productivity was highest at the equatorial upwelling with a secondary maximum at the 10°N upwelling, although the nutrient sections do not show high values at either location. This is possibly because nutrients which are upwelled are consumed too rapidly for their ambient level to increase.

7.615 Deeper currents in the equatorial Pacific

The above descriptions cover the upper 400 m or so but in addition there are significant currents below that depth. In the LIPP (Line Islands Profiling Project) aspect of PEQUOD (Pacific Equatorial Dynamics) Project, direct current measurements were made with a Pegasus acoustical dropsonde profiler from surface to bottom along 3°N to 3°S at 159°W for sixteen months in 1982/3. A complex system of currents was observed between the surface and bottom which was at over 5000 m at 3°S to 3000 m at 3°N. In the first analysis of the results, Firing (1987) described the mean zonal velocity structure. He recognized three main features which appeared to be permanent below 400 m as: the eastward *North* and *South Intermediate CounterCurrents* (NICC, SICC) centred at 1.5°–2° from the equator with main cores at about 700 m depth and maximum mean speeds over 15–20 cm/s. A third current, the westward *South Equatorial Intermediate Current* (SEIC) with its main core at 900 m at about 3°S had a mean maximum speed of 15 cm/s. Mean transports were 3.0 Sv for the NICC, 8.7 Sv for the SICC and 7.1 Sv for the

SEIC, which compared closely in magnitude with the transports of the shallower currents, e.g. the NSCC (8 Sv) and the SSCC (4 Sv). The three currents, identified above on a 16 month mean section were not invariant in time but Firing concluded, from his data and that of earlier measurements in the equatorial region, that they probably are permanent features of the equatorial circulation. Firing also considered doubtful the status of an *Equatorial Intermediate Current* (EIC) below the EUC. Although the sixteen months of measurements were made during the 1982/3 El Niño period, it was felt that that phenomenon probably did not significantly affect these deeper currents.

In a further study of the LIPP/PEQUOD data, Firing (1989) described the zonal equatorial circulation below 1500 m at 159°W. He identified three main currents, referred to by their approximate core depths as the eastward *1700 m current* at 3°S (mean maximum speed 4.5 cm/s, transport >1.5 Sv), the eastward *3000 m current* centred at 1.5°S (3.9 cm/s, 7.4 Sv) and the westward *4000 m current* at 2.5°S (4.3 cm/s, >6.2 Sv). Meridional speeds were not significantly different from zero. Additional evidence for the permanence of the 3000 m current was adduced from previous current measurements at 1°S, 150°W, but there was no other direct evidence for the permanence of the 1700 m and 4000 m currents. Firing (1987) expressed the opinion that "the deep equatorial current system is still poorly described" and that "a 10 year time-series would be ideal for studying annual and interannual variations"— an ambitious suggestion!

7.62 North Pacific circulation

The general pattern of the North Pacific circulation is similar to that in the Atlantic with a concentrated north-eastward-flowing current on the west side while the southward return flow is spread over much of the remainder of the ocean. A difference between these two northern oceans is that the Pacific is essentially closed at the north, with just a small outflow to the Arctic and no return flow therefrom, whereas the Atlantic has a much more open connection with that sea and a considerable volume exchange with it.

7.621 North Pacific gyre and the Kuroshio

The *North Pacific Gyre* may be considered to start with the NEC flowing west and, on approaching the western boundary of the ocean, dividing with some water going south to the NECC and some north (Fig. 7.31). This water continues north-east past Japan as a concentrated current called the *Kuroshio* which is the counterpart of the Florida Current in the Atlantic. After it leaves the Japanese coast to flow east it is called the *Kuroshio Extension* to about 170°E, corresponding to the Gulf Stream and from there it is referred to as the *North Pacific Current*. The volume transport of the Kuroshio itself is about 40 Sv and of the Kuroshio Extension is about 65 Sv, comparable to those of

Circulation and Water Masses of the Oceans 253

the Florida Current and the Gulf Stream. Contributing to the North Pacific Current is the *Oyashio*, coming from the north from the Bering Sea and with some contribution from the Sea of Okhotsk. As the North Pacific Current approaches the North American continent it divides. Part turns south as the *California Current* and eventually feeds into the NEC. The remainder swings north to form the *Alaskan Gyre* in the Gulf of Alaska, and then some of it flows between the Aleutian Islands into the Bering Sea (see Dodimead et al., 1963; Favorite et al., 1976 for a full description). The position of the divergence of the North Pacific Current is at about 45°N in winter and 50°N in summer. The seasonal shift in this position has been clearly demonstrated by the tracks of satellite-tracked drogue buoys released in the North Pacific Current. More of the buoys which reached this divergence in summer turned northward than did those which arrived in winter. Many of the buoys, travelling northward into the Alaskan Gyre, became trapped in closed eddy circulations which occur off the coasts of Alaska and British Columbia. These eddies are related to features in the bottom topography (sea mounts) and may be permanent components of the circulation. Similar eddies are believed to populate the westward-flowing component of the Alaskan gyre which may account for the results of some studies showing return flow in the gyre well to the east of the Aleutians.

The apparent relation of the Kuroshio to fisheries and its probable influence on the climate of Japan and hence on the rice crop, has stimulated Japanese interest in this current, and observations of it started in the early 1900s. Modern scientific studies commenced in the early 1920s and regular monitoring in 1955 with several oceanographic cruises per year, and a number of multiple-ship studies including the international Co-operative Study of the Kuroshio (CSK) commencing in 1965 (e.g. see Stommel and Yoshida, 1972). Water properties were measured and currents observed with the GEK at the surface and with current meters and neutrally buoyant floats below the surface. Analyses of the data obtained from this and earlier studies revealed that the axis of the Kuroshio showed a distinct long-term variation in position. In Fig. 7.39 the full line represents the average position of the current during 1954 to 1959 but this 6-year average conceals many changes. In 1954 and 1955 a great loop to the south developed in area A (Fig. 7.39) and in the region between it and the coast an unusual cold water mass appeared. In 1956 to 1958 the current moved back to the coast but in 1959 the loop to the south developed in area A and with it the inshore cold water mass. (It persisted through 1963, straightened out in 1964 and reappeared as a loop with a cold core in 1969.) In 1955 and 1957 the current swung farther north than the average in area B, while in 1958 and 1959 it swung farther south in area C. Several reasons for these long-period changes and for the appearance of the cold-water mass have been put forward but none is considered entirely satisfactory. The maximum current speeds in the Kuroshio for the period ranged between 75 and 250 cm/s while the width of the current in which the

254 Descriptive Physical Oceanography

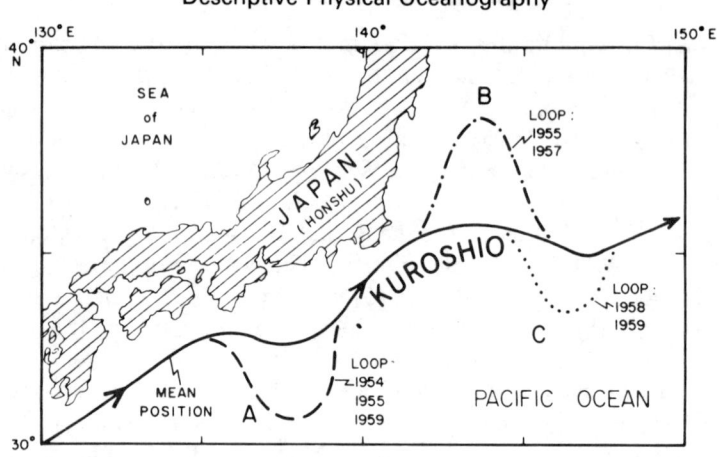

FIG. 7.39. Kuroshio—mean position and fluctuations during 1954 to 1959.

speed was over 100 cm/s was less than 80 km most of the time. There seems to be some indication that the higher current speeds and the greater widths occurred together. If this is also characteristic of the deeper parts of the current it would imply considerable variations in the volume transport. The possible existence of a south-westward countercurrent below the Kuroshio, similar to that below the Gulf Stream, has already been mentioned.

7.622 Influence of the North Pacific Sea-Surface Temperature on the North American climate

Many different mechanisms have been postulated for the longer-term (decadal or longer) changes in weather patterns that have been known as climate changes. Variations in sunspots have long been suspected to be a source of change in the amount of solar radiation and hence the associated climate. Others have suggested that changes in earth-received solar radiation due to atmospheric aerosols from volcanoes cause fluctuations in climate conditions. Focusing on North America, another group has proposed that the presence of large-scale, persistent sea-surface temperature (SST) anomalies (departures from the long-term mean) cause variations in the climate over the North American continent. This group, led by Jerome Namias, has studied in great detail the effects of North Pacific Ocean temperature anomalies on weather patterns over the U.S. and Canada. This consideration for the importance of SST in dictating the climate grew out of Namias' search for long-range weather forecasting techniques while with the U.S. National Weather Service.

Many of the Namias studies tried to establish the atmospheric circulation patterns that occur in response to specific warm and cold SST anomalies located in the eastern North Pacific. This effect of the ocean on the atmosphere is called "feedback" as it is the reverse of the more common driving of the

Circulation and Water Masses of the Oceans   255

ocean by the wind and surface heating. One must remember, however, that this direct driving of the ocean by the atmosphere continues to take place and must be combined with the feedback mechanism to form a very complex system. Thus some of Namias' work examined the way in which the SST anomalies were changed by winds due to the distribution of atmospheric pressure. Climate forecasts, developed by Namias for North America generally gave conditions 3 to 6 months after the occurrence of the strong SST anomaly. Provided in terms of air temperature and precipitation anomalies the Namias climate forecasts often showed some "skill" (i.e. success) in being able to predict climate change. In other cases, however, the predictions were very poor and failed to identify the expected patterns of air temperature and rainfall over North America.

The type of feedback invoked by Namias in his climate change model is typified by a process known as "cyclogenesis" which is the creation of atmospheric cyclones by the presence of anomalously warm SSTs. In those cases where the ocean's surface remains persistently warmer than normal, heat is transferred to the atmosphere from the ocean (assuming that the long-term mean has achieved some level of equilibrium). This added heat in the atmosphere causes the air to rise, lowering the pressure and eventually creating a low-pressure cell or cyclone. This process takes a while to occur and thus cyclogenesis over the ocean requires the sustained presence of anomalously warm water in a particular region. This is essentially the mechanism which both generates and feeds energy into hurricanes (typhoons, cyclones) as they travel over the oceans.

Stimulated by the examples of ocean-atmosphere interaction presented by Namias in his various studies, Davis (1976) made an in-depth study of the statistical relationship between atmospheric pressure and SST. Characterizing a 20-year time-series of monthly values for both fields in terms of empirical orthogonal functions (EOF's, better known as principal components of the data fields) Davis found that there was only very slight skill in forecasting atmospheric pressure changes on the basis of SST anomalies. Instead he statistically confirmed the temporal persistence of the North Pacific SST anomalies in that he found that anomalies could be predicted with some skill from the previous anomaly conditions. He also found that sea-level pressure anomalies could be specified with adequate skill from simultaneous SST conditions but that it was not possible to predict future atmospheric pressure conditions from the SST patterns. He concluded that the statistical connections between sea-level pressure and SST are more likely due to the driving of the ocean by the atmosphere.

7.63 *South Pacific circulation*

The *South Pacific Gyre* is less well documented than that in the North Pacific. The SEC forms the northern part of the South Pacific circulation and carries water into the Coral Sea off north-eastern Australia; in the south, the Antarctic Circumpolar Current forms the easterly flow component of the gyre. The exact character of the connection off Australia is in doubt. Although current atlases show a southward flow (*East Australian Current*) along the east coast of Australia (as in Fig. 7.31) Hamon and Golding (1980) point out that oceanographic studies of the western Tasman Sea between Australia and New Zealand have not shown a continuous current along the coast. When the south-west flows are observed they are generally accompanied by north-west flows offshore. Shipboard XBT surveys along with satellite-tracked buoy records have clearly indicated the presence of large, anticlockwise eddies of 200 to 300 km diameter with speeds of 180 to 200 cm/s and lifetimes of up to a year. The centres of the eddies are well mixed to as much as 300 m. In the austral winter, the surface water in an eddy may be as much as 2K warmer than the surrounding water but in summer no clear temperature pattern related to the eddies may be apparent.

In the south, part of the Antarctic Circumpolar Current turns north up the South American coast as the *Peru Current*. This current has been examined by several expeditions and the coastal side of it, at least, is well described. Its volume transport is only about one-third of that of the Gulf Stream or the Kuroshio. It turns to the west near the equator and contributes to the equatorial current system. The low-temperature tongue extending west along the equator (Figs. 4.1 and 4.2) which looks like a continuation of the Peru Current was attributed by Sverdrup to upwelling of cool, subsurface water associated with the divergence to be expected with a westward current straddling the equator, as explained previously.

7.64 Eastern boundary currents; Peru Current and El Niño

7.641 Peru Current

The *Peru Current* is an example of an eastern boundary current (Wooster and Reid, 1963) which is characteristically broad and slow, contrasting with the narrow and swift western boundary currents. The Peru Current flows equatorward (Fig. 7.31) and as it comes from high latitudes it is cool and of relatively low salinity. Figure 7.40 shows vertical sections of water properties to 1000 m depth and to 1000 km from the coast at 33°S. The isopycnals have the same slopes as the isotherms shown in Fig. 7.40(a) and reveal the equatorward Peru Current above about 500 m depth and the subsurface *Gunther Current* toward the pole near the coast. The whole column is of relatively low salinity (Fig. 7.40(b)) with lowest values near the coast due to the wind-driven upwelling from a few hundred metres depth which occurs there because of prevailing equatorward winds (see Section 8.2). Another feature of the eastern boundary is the low dissolved oxygen content below the surface (Fig. 7.40(c)) near the coast brought from the tropic to the north by the

Circulation and Water Masses of the Oceans 257

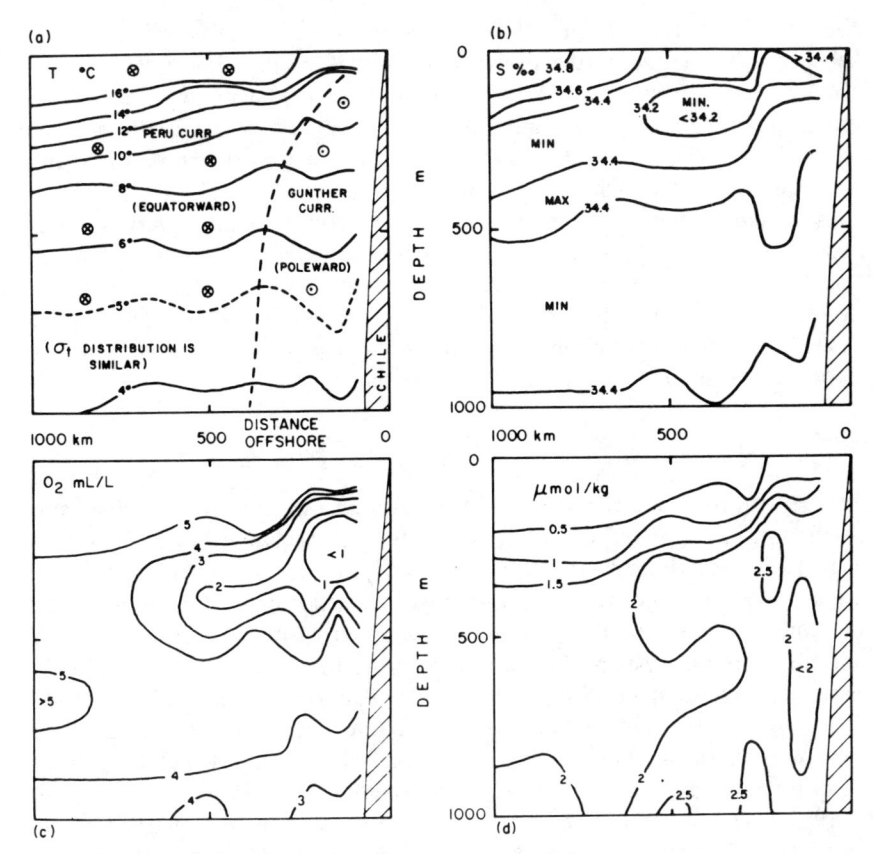

FIG. 7.40. Eastern South Pacific—west–east vertical profiles at 33°S to show (a) temperature with meridional current directions, (b) salinity, (c) dissolved oxygen, (d) phosphorus as phosphate.

coastal undercurrent. High nutrient content, represented here by inorganic phosphorus (Fig. 7.40(d)) and associated with the low oxygen, promotes the biological production which is characteristic of this eastern boundary region. These features described for the ocean off Chile and Peru are found along other eastern boundaries such as in the Benguela Current in the Atlantic off South Africa, in the Canary Current off North Africa and in the California Current in the North Pacific. Upwelling results in cooler water near the coast because temperature always decreases with increasing depth in these regions.

Generally the Peru Current extends to a few degrees south of the equator before turning west and the low temperature of its surface waters is in contrast with higher temperatures north of it. The high temperatures extend farthest south during the southern summer, December to March. At irregular intervals of a few years the high temperatures extend 5 to 10 degrees farther south than usual and the thermocline deepens by 100 m or so, the oceanographic condition known as "*El Niño*". The increase in temperature was thought to kill fish but recent studies have shown that the fish merely descend below the abnormally warm surface layer. The absence of fish in the upper layer leads to the death of many sea birds and a decrease in catch by local fishermen. The temperature increase is also accompanied by increased evaporation and consequent greater precipitation on the neighbouring land, so that floods cause much damage in this region where the normal rainfall is very small. It should be pointed out that every year at this time there is a slight warming of the sea surface along with an increase in precipitation. During El Niño years, however, the temperature increase and the rainfall are far in excess of the norm, producing large anomalies that lead to the consequences such as the floods, collapse of the fishery and bird deaths.

7.642 El Niño and the Southern Oscillation (ENSO) and La Niña

Early ideas about the cause of El Niño centred on local mechanisms, arguing primarily that the alongshore winds off Peru changed to lessen or stop the coastal upwelling. Direct observations, however, demonstrated that the winds did *not* change in direction or magnitude and that upwelling must still be going on in spite of the continued increase in surface temperature and deepening of the thermocline. More intensive studies in the early 1970s, motivated by a major El Niño event in 1972, suggested that the cause for the El Niño might lie in some external factor. The collapse of the Peru/Ecuador anchovy fishery at the time of this 1972 El Niño further motivated scientists to determine the cause of the El Niño. More recently, a very strong El Niño event in 1982 has provided oceanographers and meteorologists with an opportunity to study in depth a very pronounced example. As a result, new models of the phenomenon have been developed and people are now predicting El Niño events six months in advance with some level of confidence.

Along with the faulty argument of a wind shift along the South American coast and a consequent end to the upwelling, early studies suggested a second explanation—that the front between the warm, low-salinity water north of a line from the Galapagos to Ecuador and the cold, more saline Peru coastal water to the south "breaks down" allowing the warm Gulf of Panama water to flow south along the Peru coast causing an increase in coastal SSTs. A related phenomenon is that a reduction in the north-east trade winds in the eastern tropical Pacific will allow the surface waters there to warm up and then be advected eastward by the NECC into the warm pool off Central America.

More recently Wyrtki (1975) proposed a mechanism that relied on the eastward transport of warm water from the western limits of the tropical Pacific back to the eastern boundary of Central and South America. This eastward movement could be accomplished by the EUC, the NECC and its southern counterpart the SECC which may be strongest and shift eastward in El Niño years. As part of his mechanism, Wyrtki postulated that preceding the

El Niño the trade winds blow stronger than normal to build up an excess of warm water in the western Pacific. In El Niño years the winds weaken and the water moves back to the east bringing warmer water that fills the coastal regions along Central and South America, increasing the SST and deepening the thermocline.

In their review of all the El Niños from 1949 to 1980, Rasmussen and Carpenter (1982) developed a description of the canonical El Niño. In this description they confirmed the anomalously stronger easterly trade winds but only to the west of the dateline (180° longitude). During the period from October to November prior to the El Niño, the equatorial easterly wind anomalies are replaced by westerly anomalies indicating that the trade winds have reversed their direction west of the dateline. This change was found to coincide with the occurrence of warm SST anomalies around the equator near the dateline. East of the dateline, the winds follow a different pattern, with the weakening of the easterly trade winds lagging behind rather than leading before the development of warm SST anomalies near the Ecuador-Peru coast. As El Niño proceeds, the positive (warm) SST anomalies, that form along the coasts of Ecuador and Peru, are seen to migrate westward in a narrow band along the equator. This westward spread of the warm SST anomalies coincides with the intensification of westerly wind anomalies along the equator (stronger reversed trade winds blowing from the west). Also at this time there is a shift of the Intertropical Convergence Zone (ITCZ, a meteorological front) that is associated with increased precipitation in the eastern and central tropical Pacific.

The shifts in the wind patterns, during an El Niño event, are the consequence of a shift in the major atmospheric pressure systems in the South Pacific. Known as the *Southern Oscillation*, this shift moves the atmospheric low-pressure cell, that normally lies over northwest Australia in December, to the east bringing higher pressure and warmer temperatures to that part of Australia. At the same time this eastward shift brings more low pressure, and the rainfall associated with it, to the west coast of South America. This shift in atmospheric pressure patterns can be best observed by computing the pressure differences between a station in the west and one in the east. Many people have used the difference between Darwin, Australia and Easter Island but problems with getting timely data from remote Easter Island have prompted a shift to using the pressure difference between Darwin and Tahiti. Called the *Southern Oscillation Index* (SOI), time-series of this pressure difference (Fig. 7.41) indicate that the SOI drops to a minimum during each El Niño event. Many have tried to use the SOI alone to forecast the occurrence of El Niño but often have been led astray since they have not been able to forecast the severity of the event and have therefore often been incorrect about its magnitude and duration. The connection between the El Niño events and marked changes of the Southern Oscillation have led many scientists to refer to strong El Niños as *"ENSO events"*, combining the El Niño with the Southern Oscillation.

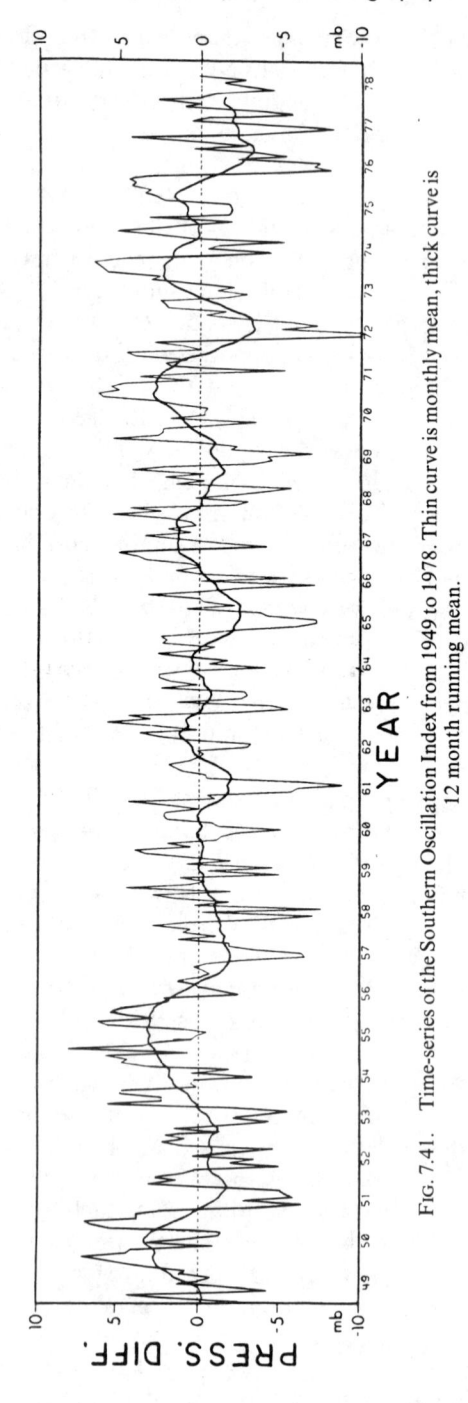

FIG. 7.41. Time-series of the Southern Oscillation Index from 1949 to 1978. Thin curve is monthly mean, thick curve is 12 month running mean.

This business of predicting El Niño is a curious one where there are often many different answers. Using the Wyrtki (1975) hypothesis one would only think it necessary to look for a positive anomaly of the easterly trade winds early in the year in order to predict the subsequent El Niño. In 1982, however, this was not the case and there was no increase in the easterly trade winds preceding the El Niño. Instead, in the early fall the trade winds stopped, reversed and blew strongly from the west all across the Pacific. This had dire consequences for the tropical Pacific east of the dateline where low island atolls were not accustomed to have winds from the west. Warm SSTs associated with these westerly winds led to the death or departure of large numbers of sea birds from islands such as Christmas Island. Increased rains and a sharply anomalous rise in severe weather caused large-scale damage in the central tropical Pacific. Tahiti, which seldom sees severe weather, was hit by nine hurricanes in late 1982 and early 1983.

The El Niño of 1982 also provided a number of other surprises. As described in the canonical El Niño of Rasmussen and Carpenter, the first warm SSTs occurred in the central Pacific but this time they stretched to the east of the dateline and were much larger in amplitude and longer in duration. El Niño "watchers" were curious to see if there would be a subsequent development of warm SST anomalies off Ecuador and Peru later in the year; this finally did occur but a bit later than usual. This warm pool, off western South America, increased in size and amplitude and then finally migrated westward in a strong tongue following the pattern of most previous El Niño events. Similar to other major El Niños, the 1982 event expressed itself all along the coasts of North and South America. The northward extension of warm water along the west coast of North America resulted in the failure of the salmon fishery in Washington State and a consequent boom in the salmon fishery north in Alaska. Fishes normally found off the coast of southern California were being caught by fishermen off Washington and British Columbia.

It is interesting to speculate about the causes of El Niños. One has the problem of whether the SST anomalies occur as a consequence of the wind changes or if they happen first and cause the winds to shift in the type of air-sea interaction discussed in terms of the North Pacific SST anomalies (Section 7.622). Numerical modelling (Gill, unpublished manuscript) has shown that if one introduces warm SST anomalies into the tropics there will be a consequent shift in the wind systems consistent with those observed during El Niño. At the same time, it is true that a shift in the wind will produce the observed changes in SST. In either case, one has the problem of which came first and why. If the wind shift, associated with the Southern Oscillation, has caused the SSTs to increase then what triggers the wind and atmospheric pressure system changes?

Rather than try to dissect these mechanisms, El Niño scientists have taken a variety of model approaches to predicting it. Cane and Zebiak (1986) used a coupled numerical model to provide forecasts of El Niño events while Barnett

(1984) has relied on statistics for prediction. In an earlier study, Barnett (1977) studied the statistical relationships between the atmospheric factors and the SST data. His results suggested that: (a) upwelling effects due to local changes in wind stress were not important, (b) the eastward advection of heat by the NECC was not more important to the eastern tropical Pacific than local heating and (c) about one-half of the SST variation could be predicted from sea-level changes across the Pacific as proposed by Wyrtki (1974). This study then confirmed the greater significance of the eastward advection of warm water over the southward extension of water from the Gulf of Panama. This statistical study simply tested the consistency between the various hypotheses and the observational data but did not decide which of the hypotheses was most correct. In fact it is likely that each El Niño is a unique combination of the various candidate effects and that these effects may vary in importance not only between events but even during a particular event.

One cannot leave the topic of El Niño without some mention of the possible role of Kelvin waves (see Pond and Pickard, 1983, Section 12.10.3). Constrained to propagate eastward along the equator these waves are thought to be linked to the El Niño process. Many have speculated that strong wind effects in the western tropical Pacific might generate such a Kelvin wave which would then propagate eastward along the equator. Some have suggested that whenever a pair of cyclones is located north and south of the equator, providing a sharp increase in westerly winds, that such a Kelvin wave must be created. The examination of sea-level time-series from islands in the tropical Pacific has documented the eastward propagation of such features at a phase speed very consistent with Kelvin waves. Thus it appears that Kelvin waves may be routinely associated with the major wind shifts that occur as part of El Niño but they are not the mechanism to carry all of the warm water back to the coast of South America. This task is left to the strong eastward currents such as the EUC, the NECC and the SECC. As discussed above, the strengthening of the SECC in El Niño years may provide a significant mechanism to transfer the warm water from the tropical western Pacific to the warm pool off Ecuador and Peru.

El Niño is basically an unusual excursion of warm water ($>28°C$) to the east in the equatorial zone, associated with weakened south-east Trade Winds in the east and stronger westerlies in the west. La Niña is the opposite—stronger south-east Trades in the east (and weak westerlies in the far west) with resulting cool water ($<25°C$) extending much further westward along the equator than usual. In essence, El Niño and La Niña may be regarded as the two extremes of the Southern Oscillation as far as the ocean is concerned.

Finally, the question still remains—do the wind changes cause the ocean changes (the oceanographer's point of view) or do the ocean changes cause the wind changes (the meteorologists point of view)?

A comprehensive review of this subject is available in Philander, 1989.

It has been suggested that similar conditions to the *El Niño* probably occur

Circulation and Water Masses of the Oceans   263

annually off other upwelling coasts, such as those of California and Southwest Africa, when the equatorward winds die down seasonally. However, the consequences are small compared with those along the Peru coast where a significant seasonal decrease of wind is unusual.

7.65 Pacific Ocean water masses

The water masses of the Pacific are rather more complicated than those of the Atlantic, probably because the greater size of the Pacific provides more opportunity for different masses to develop and be maintained in the different parts of the ocean (see Fig. 7.42).

7.651 Pacific Ocean upper waters

At the surface, the salinity shows the typical maxima in the tropics with a minimum in the equatorial regions (Fig. 4.9), and with lower values at higher latitudes. The surface salinity in the North Pacific is considerably less than in the North Atlantic, because of the greater runoff and precipitation. In the South Pacific the average salinity is higher than in the North Pacific but slightly less than in the South Atlantic. The surface temperature (Figs. 4.1 and 4.2) shows the usual equatorial maximum, the highest values being in the west at the down-stream end of the westward-flowing equatorial currents. There is also the previously mentioned minimum along the equator in the east attributable to the effect of upwelling. Apart from the perturbation due to the Peru Current and some effect of upwelling along the mid-latitude shores of the American continent, the distribution of the surface isotherms is distinctly zonal.

It is in the upper water masses below the surface layer that the variety of water properties is found. Between depths of about 100 and 800 m there are the *North* and *South Pacific Central* and the *Pacific Equatorial Water Masses* (Fig. 7.19). The Equatorial Water extends almost from shore to shore, but despite its extent of some 12,000 km it has a very uniform T–S relationship across the width of the ocean. This is evident in the mean T–S curves for 5° squares for the mid-Pacific (Emery and Dewar, 1981) shown in Fig. 7.42. These are similar to those for the Atlantic (Fig. 7.22) with a central mean T–S curve and one standard deviation curves on either side of the mean. (The low salinities in the Gulf of Panama referred to earlier are very apparent in this figure.) In the eastern Pacific the Equatorial Water extends from about 20°N to 10°S but it diminishes in north–south extent to the west. Except for the waters in the western South Pacific it is the most saline water in this ocean, but even so is slightly less saline than the Atlantic Central waters (Fig. 7.20). The Equatorial Water is separated from the well-mixed and homogeneous surface layer by a very strong thermocline, and vertical transfer of water properties up or down is inhibited by the stability of this water. For this reason it is often

264   Descriptive Physical Oceanography

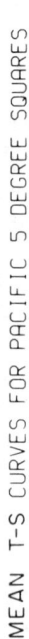

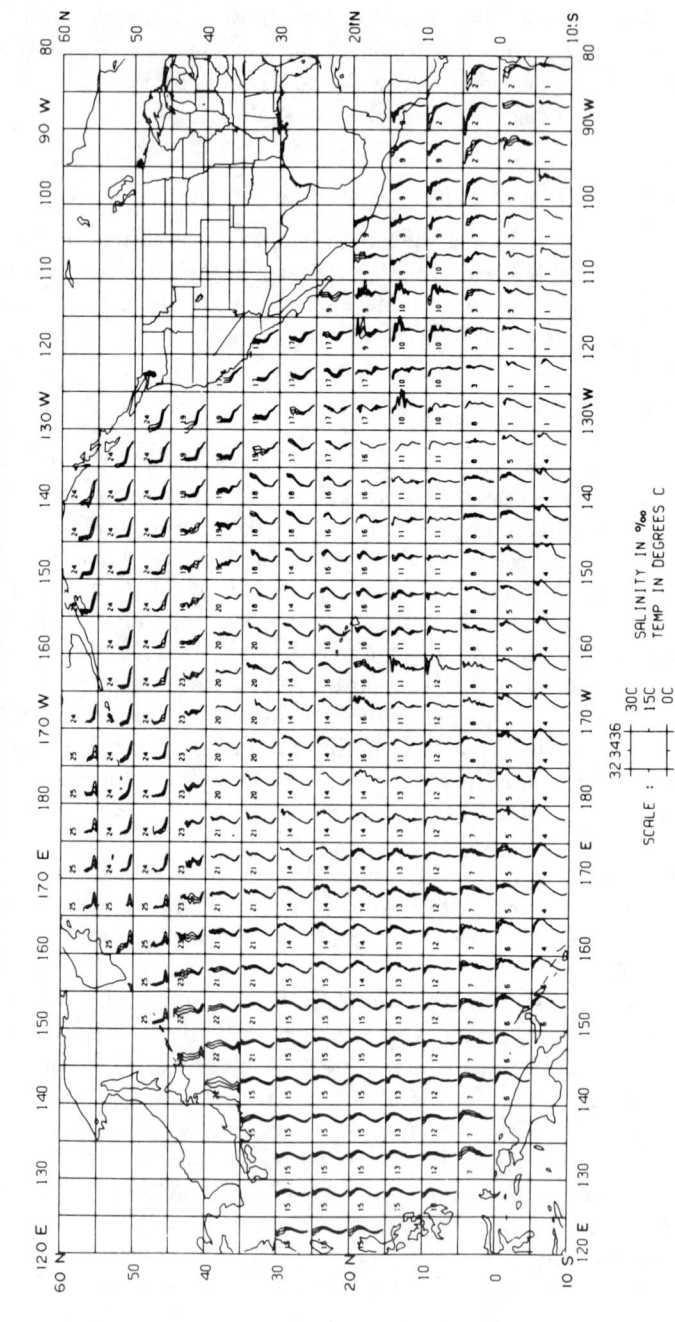

FIG. 7.42. Pacific Ocean: mean T–S curves and one standard deviation curves by 5° squares.

referred to as the "discontinuity layer". The depth of the discontinuity layer decreases from 150 to 200 m in the west to 50 m or less in the east; the layer even seems to reach the surface at times near the American coast. This effect is associated with the *Costa Rica Thermal Dome*, an apt name which describes the three-dimensional shape of the thermocline with the cold water of the lower thermocline occasionally penetrating to the sea surface itself.

At about the level of the thermocline in the Equatorial Waters is the salinity maximum which is characteristic of the equatorial regions (Fig. 4.11). It is of limited areal extent and so does not appear in Fig. 7.19 which shows averages over large areas and it is of too small vertical extent to be visible in the T–S curves of Fig. 7.42. At about 800 m there is a slight salinity minimum which represents the limit of the northward influence of the Antarctic Intermediate Water (Fig. 7.44).

In the Pacific between 30°N and 20°S there are two cores of remarkably low dissolved oxygen values between 200 and 1000 m (Fig. 4.13). These cores are centered at 10°N and 7°S and extend westward right across the Pacific, the values increasing from less than 0.1 mL/L in the east to 3 mL/L in the west. Figure 7.43 shows these oxygen minima plotted on the surface where

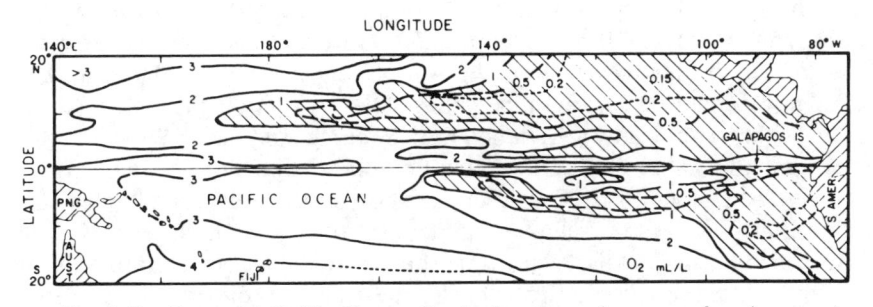

FIG. 7.43. Equatorial Pacific Ocean—dissolved oxygen values on surface $\Delta_{S,T} = 160 \times 10^{-8}$ m^3/kg. (Approximate depths of this surface are: at 1 mL/L about 200 m; 2 mL/L about 250 m; 3 mL/L about 300.)

$\Delta_{S,T} = 160 \times 10^{-8}$ m^3/kg ($\sigma_t = 26.5$). At the centre of the oxygen minimum cores, this surface is at a depth of about 150 m at the extreme east, descending to 250 m in the west. Below these oxygen minimum layers in the upper waters the values rise again in the deep water. Corresponding to the general low oxygen values are high concentrations of nutrients, particularly in the North Pacific. Values of 2 to 3.5 μmol/kg phosphorus as phosphate are typical in the North Pacific compared with values of 0.5 to 2 μmol/kg in the Atlantic. Similar large concentrations of dissolved nitrate and silicate are also found in the North Pacific waters. These materials go into solution during the decomposition of organic matter, e.g. dead plankton. High concentrations may be a result of a copious supply of organic remains from life in the upper waters, or may accumulate as a consequence of the slow movement of the deep water. In the case of the Pacific, the latter is assumed to be the main cause.

The North Pacific Central Water extends from the Equatorial Water to about 40°N and is the least saline of the central water masses of the oceans. Sverdrup distinguishes an extensive western North Pacific Central Water to the west of the Hawaiian Islands from a smaller eastern North Pacific Central Water between these islands and the North American coast, though separated from the coast by another water mass. Examination of the accumulation of data for the North Pacific since 1955 has indicated that the two water masses are less easily distinguished now than they were with the lesser amount of data available to Sverdrup. They are not separated in Fig. 7.19(a).

North of the North Pacific Central Water is a *Pacific Subarctic Water Mass* (Favorite *et al.*, 1976) which extends across the greater part of the ocean. Its characteristic properties are low salinity (33.5 to 34.5) and relatively low temperature (2 to 4°C). It is a much more extensive and important water mass than the Atlantic Subarctic water. It corresponds to the Subantarctic Water in the south but its mode of formation must be somewhat different. The Subantarctic Water is formed between the Subtropical Convergence and the Antarctic Polar Front which are essentially continuous across the South Pacific. In the North Pacific the *Subarctic Convergence* is only clear in the western part of the North Pacific (Fig. 7.31) where the Subarctic Water must be formed by mixing between the warm, saline waters of the Kuroshio Extension and the cold, less saline waters of the Oyashio. In the eastern North Pacific there is certainly a gradation from Subarctic to Central Water but it is less easy to localize than is the Antarctic Polar Front. The Subarctic Water forms a part of the North Pacific Current flowing to the east and it retains its characteristics of low salinity and temperature until it approaches the American coast. Here part of it swings south-east and its temperature starts to rise by heating and the salinity by mixing until it attains the typical Equatorial Water characteristics as it merges into the North Equatorial Current.

South of the Equatorial Water are the western and eastern *South Pacific Central Water Masses* extending to the Subtropical Convergence at 40°S which is usually considered as the oceanographic southern boundary of the Pacific Ocean. In the South Pacific the Central Water Masses are less easily distinguished from the Equatorial Water than is the case in the north Pacific.

Intermediate water masses are found below the Central Waters in both the North and South Pacific (Reid, 1965, 1973). In the latter, the *Antarctic Intermediate Water* is formed by subsurface mixing at and north of the Antarctic Polar Front with fairly well-defined properties, a temperature close to 2.2°C and a salinity of 33.8. This salinity is relatively low and so the Antarctic Intermediate Water gives rise to the minimum in the T–S diagram below the Subantarctic and Central Waters (Fig. 7.19(b)). The Intermediate Water flows north as seen in Fig. 7.44, increasing in salinity by mixing with waters above and below but is limited in northward extent by the Pacific

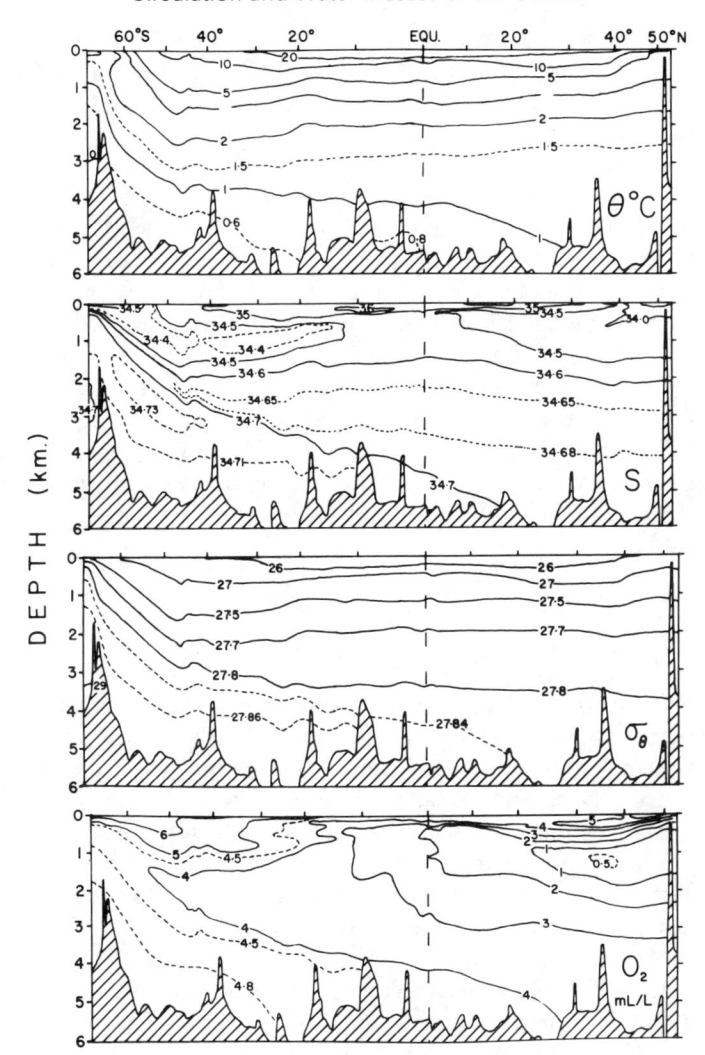

FIG. 7.44. Pacific Ocean—south–north vertical sections at about 160°E of potential temperature, salinity, sigma-theta and dissolved oxygen. (Based on GEOSECS data, Craig et al., 1981.)

Equatorial Water. This contrasts with the situation in the Atlantic where, in the absence of a clearly defined Equatorial Water Mass, the Antarctic Intermediate Water continues north across the equator. In the North Pacific a *North Pacific Intermediate Water*, evident as a salinity minimum, is found below the North Pacific Central Water (Fig. 7.19 and 7.44). In the west it is deepest at about 800 m but rises to about 300 m in the east. It appears to circulate in a clockwise gyre, similar to the surface water, and its relatively high oxygen content indicates frequent replenishment. Sverdrup suggested that it is formed in the west near Japan by sinking at the convergence of the Oyashio and the Kuroshio Extension. Reid (1965) pointed out that the salinity minimum of the North Pacific Intermediate Water occurs on an isopycnal surface of $\sigma_t = 26.8$ and that water of this density very rarely occurs at the surface in the North Pacific. Where this density is found at the surface in the South Pacific (as a possible alternative source) the salinity is higher than in the salinity minimum of the North Pacific Intermediate Water. Reid therefore concluded that this latter water attains its properties below the surface by vertical mixing at the north-west in the Subarctic region where the particular density surface is shallow and the surface waters above are cold, low in salinity and high in oxygen content. It was thought that this was one of the few exceptions to the general rule that subsurface water masses acquire their T–S characteristics at the surface of the sea, but evidence is growing that many recognized water masses are formed below the surface by the mixing of a lower layer with an upper layer of different properties which is itself in contact with the atmosphere.

7.652 Pacific Ocean deep waters

The deep water of the Pacific is characterized by very uniform properties as may be seen in the meridional sections of Fig. 7.44. Between 2000 m and the bottom the temperature range (as far south as the Subtropical Convergence) is only from 1.1 to 2.2°C and the salinity from 34.65 to 34.75. The salinity tends to increase with depth or to remain constant, in contrast to the situation in the Atlantic where there is a marked salinity maximum of the North Atlantic Deep Water at mid-depth with a decrease to lower values in the Bottom Water (Fig. 7.15). The main reason for the uniformity in the Pacific Ocean is that no deep water is formed in it. In the north, the salinity is too low for winter cooling to make the water dense enough to sink to any considerable depth. In the south, although some bottom water probably comes from the Ross Sea, the volume is not large and it appears to be absorbed first into the circumpolar circulation.

The *Pacific Deep* and *Bottom Water*, from 2500 m down, must come from the other oceans, entering from the south-west between New Zealand and Antarctica. It spreads north at 180° to 160°W and, after crossing the equator, branches north-west and north-east along basins into the North Pacific. The *Scorpio* Expedition of 1967 (e.g. Warren, 1973), consisting of two lines of stations right across the Pacific at 28° and 43°S, was designed to investigate the deep waters more thoroughly than before. One particular result was that the distribution of properties indicated a narrow flow between 3000 and 4000 m depth at 28°S along the east side of the Tonga–Kermadec Ridge (about 175°W), northward into the Pacific. The transport was calculated as about 10 Sv compared to some 14 Sv required for a steady-state budget for heat, salt,

Circulation and Water Masses of the Oceans 269

volume and ^{14}C. The evidence for the clockwise flow of the bottom water round the Pacific is the gradual increase of temperature along the route, from about 0.9°C in the Antarctic to 1.5 to 1.6°C in the eastern north Pacific (Mantyla, 1975), and a small decrease in oxygen content (Fig. 7.44). There is a slight indication of a decrease in salinity into the North Pacific but the change is close to the limit of measurement precision. The rise of temperature is attributed to heat flow from the interior of the earth into the sea. The water which penetrates into the North Pacific must rise into the upper layers above 2000 m.

Sverdrup considered, essentially by analogy with the Atlantic, that a slow southward movement of deep water (i.e. between the Bottom Water and the Intermediate Water) must occur in the South Pacific. This southward movement of deep water would be supplied from the northward movement of Bottom Water below it and the Intermediate Water above. The indications are, therefore, that the deep and bottom water movements in the Pacific are very slow and that there is only a small exchange of water between the North and South Pacific in contrast with the large volume exchange at various levels across the equator in the Atlantic.

The water masses of the Pacific and Indian Oceans are very similar and in his analysis of the distribution of temperature and salinity Montgomery (1958) pointed out that this water, with its mean temperature of 1.5°C and salinity of 34.7, forms the largest water mass in the world ocean. For this reason he named it the *(Oceanic) Common Water* from which other water masses are distinguished by differences of temperature and salinity. The Common Water has average properties because it is a mixture of other water masses, chiefly North Atlantic Deep and Antarctic Bottom Waters with some admixture from the Indian and Pacific Oceans.

Although the large body of deep water in the Pacific is usually called the *Pacific Deep Water*, the name in this case indicates the region where the mass is found, rather than where it is formed as is usually the case with water mass names.

Comparison of Fig. 7.44 with Fig. 7.15 will show that in addition to differing from the Atlantic in being more uniform in temperature and salinity distributions, the Pacific water also has lower dissolved oxygen values. In the Atlantic the values range from 3 to 6.5 mL/L in the deep water, whereas they are from 0.5 to 4.8 mL/L in the Pacific. These lower values are consistent with the presumed greater age and slower circulation of the deep water in the Pacific, or perhaps it would be better to say that they are one of the reasons for believing the circulation to be slow.

One feature which is lacking in the Pacific, as compared with the Atlantic and the Indian Oceans, is a source of high salinity, warm water such as comes from the European Mediterranean and from the Red Sea respectively. These water masses are not of major importance in terms of volume but are very useful as tracers of the movement of subsurface waters.

Descriptive Physical Oceanography

7.7 Indian Ocean

The Indian Ocean differs from the Atlantic and Pacific Oceans in its limited northward extent, to only 25°N. The southern boundary of the ocean is usually taken at the Subtropical Convergence at about 40°S.

The Indian Ocean used to be one of the least known oceans but the extensive observations made during the International Indian Ocean Expedition from 1962 to 1965 increased our knowledge immensely, even though investigations emphasized the interesting equatorial zone at the expense of the inhospitable southern part of the Ocean. After this Expedition, Wyrtki brought together the data from that co-operation, and also all available earlier material, in the *Oceanographic Atlas of the International Indian Ocean Expedition* (Wyrtki, 1971) in which the data are displayed in various ways to show seasonal as well as average distributions, and relations between properties in a variety of characteristic diagrams. The following description is largely based on Wyrtki's Atlas and his article (1973).

7.71 Indian Ocean circulation

It is in the wind-driven equatorial current system and in its northern parts that the Indian Ocean circulation differs most from those in the Atlantic and Pacific. Because of the land mass to the north of the ocean there is a seasonal variation in the winds north of the equator (Fig. 7.32). From November to March, these winds blow from the north-east (north-east Trades or *North-east Monsoon*), from May to September they blow from the south-west (*South-west Monsoon*). (The word *monsoon* is derived from an Arabic word meaning winds which change with the seasons.) The south-west monsoon winds are really a continuation across the equator of the south-east Trades which continue throughout the year. The change of wind direction north of the equator then results in a change of currents there. During the north-east monsoon (November to March) there is a westward-flowing *North Equatorial Current* from 8°N to the equator; from the equator to 8°S there is the eastward-flowing *Equatorial Countercurrent* and from 8°S to between 15 and 20°S there is the *South Equatorial Current* to the west (Fig. 7.45(a)). During the south-west monsoon (May to September) the flow north of the equator is reversed and is to the east. This combines with the eastward ECC and the whole eastward flow from 15°N to 7°S is called the *(South-west) Monsoon Current* (Fig. 7.45(b)). The SEC continues to the west south of 7°S but is stronger than during the north-east monsoon.

An *Equatorial Undercurrent* is found in the thermocline east of 60°E during the north-east monsoon period. It is weaker than those in the Pacific and Atlantic. During the South-west Monsoon, with the general flow to the east at the equator, the EUC is not evident. The transition from one monsoon to the other happens relatively rapidly, occurring over a period of 4–6 weeks. This

Circulation and Water Masses of the Oceans

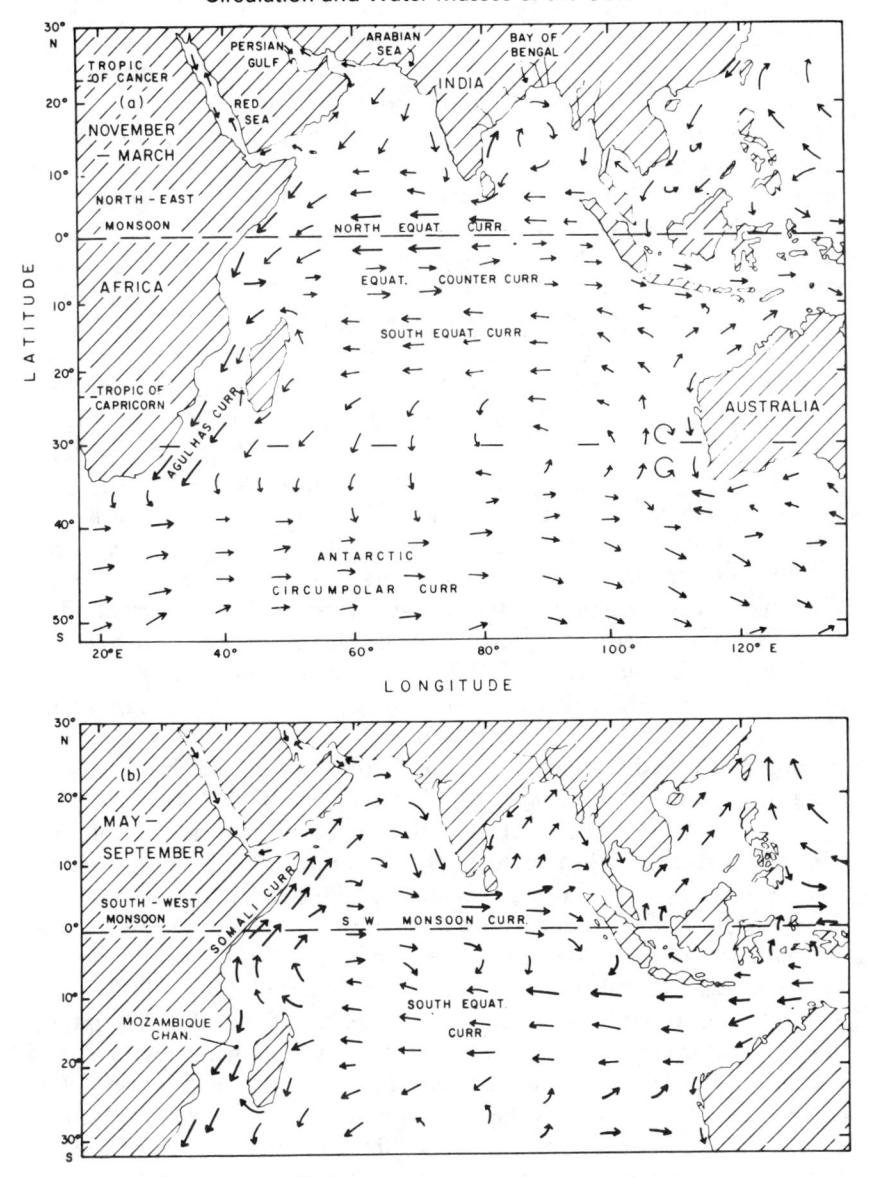

FIG. 7.45. Indian Ocean—surface circulations.

provides oceanographers with an interesting opportunity to observe the temporal development of the driving of ocean currents by the wind during the reversal period. In going from the November–March North-east Monsoon to the South-west Monsoon the change takes place during the month of April. While the wind appears to shift rather abruptly, when the pressure systems shift, the ocean currents change more slowly (Düing, 1970). During this transition period the ocean circulation, particularly in the west-central Indian Ocean, breaks into a series of cold and warm mesoscale eddies which eventually transition into the reversed ocean circulation (Fig. 7.45(b)) in harmony with the reversal of the wind driving. A simple, linear, barotropic model (Düing, 1970) was used to simulate many of these features during the transition and was successful in reproducing the general patterns seen in a collection of oceanographic measurements. The reverse transition takes place in October when the wind systems shift from the South-west to the North-east Monsoon and the pattern of Fig. 7.45(a) returns.

The currents at the African shore are interesting. During the north-east monsoon period (November to March) the SEC, when it reaches the African shore, supplies both the ECC to its north and the *Agulhas Current* flowing south (Fig. 7.45(a)). This current is deep and remarkably narrow, probably only 100 km wide, and flows south close to the African shore with a transport reported to average 50 Sv and to rise to 80 Sv at times. When it reaches the southern tip of Africa, the current turns east into the Circumpolar Current. During the south-west monsoon (May to September), the component of the SEC which turns north supplies the *Somali Current* up the east coast of Africa (Fig. 7.45(b)). This current is notable for its high speeds of up to 200 cm/s, which are comparable to those of the Florida Current, and it has a transport of about 65 Sv, most of it in the upper 200 m. The South Equatorial Current, the Somali Current and the Monsoon Current then comprise a strong wind-driven gyre in the northern Indian Ocean. Strong upwelling occurs at this time along the Somali and Arabian coasts.

Along the eastern boundary of the Indian Ocean, off Western Australia, the *Leeuwin Current*, identified about 1970, differs from other eastern boundary currents. It is a *poleward* flowing current about 100 km wide and 2000 km long along the continental shelf-break from about 22°S, off the Northwest Cape, to the southwestern tip of Australia (Cape Leeuwin) at 35°S. The current then turns eastward toward the Great Australian Bight. Godfrey and Ridgway (1985), Thompson (1987) and Church et al. (1988) have presented descriptions and theoretical explanations. The current reaches its maximum speed of 60 cm/s in May with a poleward flow at 32.5°S of 3–5 Sv in the upper 150 m and there is an equatorward flow of up to 40 cm/s at 300 m depth with a transport of 1–2 Sv below this. At this time, the wind stress is equatorward and therefore the upper flow is against the wind, in contrast to that in other eastern boundary regions. The flow appears to be driven by a steric sea-level slope of about 0.33 m downward from 20°S to 32°S. Another characteristic of the Leeuwin Current is that upwelling does not occur on to the shelf. The isotherms off Western Australia slope strongly downward from about 200 km offshore to the continental slope, in contrast to the situation in the regions off the western United States (30°N), off South Africa (26.5°S) and off South America, where the isotherms slope upward toward the shore and upwelling of

Circulation and Water Masses of the Oceans 273

cool water occurs. In addition, there are both cyclonic and anticyclonic eddies in the Leeuwin Current region, in contrast to that off the east coast of Australia where the eddies are anticyclonic only.

The Leeuwin Current (in the upper 150 m) is warm and of relatively low salinity (35.0), low dissolved oxygen and high phosphate content, while the equatorward cool undercurrent has a core of high salinity (>35.8) at 38°S. The Leeuwin Current therefore transports a significant amount of heat to the south.

It is suggested that the cause of the poleward downslope, to which the Leeuwin Current is attributed, is flow from the Pacific through the Indonesian archipelago to the Indian Ocean north of Western Australia, combined with the seasonal monsoon winds in this area which have peak values to the southeast in January/February and whose stress raises sea-level north-east of Australia and generates a wave of some sort which progresses southward down the Australian coast (since the peak flow occurs progressively later down the coast).

7.72 Indian Ocean water masses

The surface water masses in the open Indian Ocean have typical characteristics, a substantially zonal distribution of the isotherms with a temperature maximum near the equator (Figs. 4.1, 4.4) and a salinity maximum at about 30°S in the eastern ocean (Fig. 4.9). This is somewhat farther south than in the Atlantic and Pacific. In the north, the *Arabian Sea* west of India and the *Bay of Bengal* to the east have very different characteristics. The Arabian Sea has high surface salinity values up to 36.5, due to evaporation, while in the Bay of Bengal the salinity decreases from about 34 at about 5°N to 31 or less in the north. The low values in the Bay of Bengal are due to the very considerable river runoff into it, particularly during the south-west monsoon. Some of this low-salinity water is carried westward while heavy rainfall in the Intertropical Convergence Zone contributes to the low salinity in the northern part of the SEC.

Below the surface layer (Fig. 7.20(b)) and north of 10°S there is the *Indian Equatorial Water* or *North Indian High-salinity Intermediate Water* (Wyrtki, 1973). This water has a relatively uniform salinity of 34.9 to 35.5. Some of it is formed in the Arabian Sea but there are also components from the Red Sea and the Persian Gulf with values to 36.2. It also spreads into the Bay of Bengal. Wyrtki points out that the waters of the northern monsoon gyre are separated from the (southern) subtropical waters by a boundary or front which is inclined from about 100 m depth at 10°S to 1000 m depth at 20°S. North of this front, the water is saline, low in oxygen and high in nutrients while to the south the reverse is the case. South of the equator to about 40°S and to 1000 m is the *Indian Central Water* of the southern gyre of the ocean with salinities from 34.5 to 36; under this is the Antarctic Intermediate Water. Both the *Deep* and *Bottom Waters* are of Atlantic/Antarctic origin, as no waters of these types are formed in the Indian Ocean. As in the Pacific Ocean the circulation of these waters appears to be slow and must be inferred from the property distributions. The Deep Water properties are 2°C and 34.8, essentially those of Montgomery's "Common Water". A flow of Bottom Water to the north in the west and east basins (separated by the mid-ocean ridge) is inferred from a slight increase in potential temperature to the north attributed to geothermal heating from below. A very noticeable feature of the deep waters is the decrease in dissolved oxygen from 5 mL/L in the Antarctic waters to 0.2 mL/L in the Arabian Sea and Bay of Bengal. This is attributed to the recent formation and continued movement of the southern waters contrasted with the isolation and stagnation in the north. This situation is in contrast with that in the Atlantic and Pacific where the most notable regions of low oxygen are on the east side around the equator.

7.8 Red Sea and Persian Gulf

The Red Sea (see Ross in Ketchum, 1983) is a rift valley, resulting from the separation of Africa and the Arabian Peninsular, which is essentially closed at the north (except for the Suez Canal) and opens to the Gulf of Aden, Arabian Sea and the Indian Ocean at the south through the narrow strait of the Bab al Mandab. The depth averages 560 m with maximum values of 2900 m and a sill of about 110 m depth at the Bab al Mandab in the south.

The major feature of the region is the high rate of evaporation of some 200 cm/year while precipitation averages about 7 cm/year and there are no rivers flowing into the Sea. The water structure consists of a shallow upper layer and the main deep water separated by a thermocline/halocline at about 200 m depth. The temperature in summer (June–September) at the surface is 26°–30°C and in winter (October–May) is 24°–28°C while below the thermocline the deep layer is nearly isothermal at 21.6°–21.8°C. The Red Sea is the most saline large body of ocean water with surface layer values of 38–40 summer and winter (although higher values to 42.5 occur in the north) and deep water values of 40.5–40.6. The deep water is formed by winter cooling in the north. The surface layer is saturated with dissolved oxygen but the absolute values are low because of the high temperatures and salinities, i.e. less than 4 mL/L. There is an oxygen minimum of 0.5–1.5 mL/L at 400 m below the thermocline/halocline while the deep water below this has a content of about 2 mL/L.

The circulation has a seasonal variation related to the winds. In summer (South-west Monsoon) the winds are to the south over the whole Sea and the surface flow is southward with outflow through the Bab al Mandab, while there is a subsurface inflow to the north through that strait. In the winter (North-east Monsoon) the winds over the southern half of the Sea change to the north and there is a northward surface flow over the whole of the Red Sea and a subsurface southward flow with outflow through the Bab al Mandab. The outflow here is from an intermediate layer to about 100 m depth and this water can be traced through the Arabian Sea and down the west side of the Indian Ocean to 25°S at about 600 m depth. The residence time for the upper layer has been estimated at 6 years and for the deep water at 200 years.

Notable features of the Red Sea are the hot brine pools found in some of the deepest parts (e.g. see Degens and Ross, Eds., 1969). Temperatures of 58°C and salinities of 320 have been recorded, although this latter figure is not directly comparable to ocean water salinities as the chemical constitution of these brines is very different. They have a much higher content of metal ions. (For comparison, a saturated solution of sodium chloride in water has a salinity in the oceanographic sense of about 270.) Various explanations have been offered for the origin of these brine pools. The one with the fewest arguments against it is that this is interstitial water from sediments, or solutions in water of crystallization from solid materials in the sea bottom, released by heating from below and forced out through cracks into the deep basins of the Red Sea. In this connection it should be noted that the Red Sea is one of the two places where a mid-ocean ridge runs into a continent, and it is a region where the heat flow up through the bottom is much greater than the world average of 4×10^{-2} W/m².

In contrast to the Red Sea, the *Persian Gulf* has a mean depth of only 25 m. Water of temperature between 15° and 35°C and salinity up to 42 is formed here and some is contributed to the Arabian Sea.

CHAPTER 8

Coastal Oceanography

8.1 Introduction

Oceanographic conditions in coastal waters differ in many respects from those in the open sea. Some of the factors causing these differences are the presence of the coast as a boundary to flow, the shallowness of the water over the continental shelf, river runoff and precipitation, and the effects of continental air masses flowing out over the sea. In particular, variations of water properties and motions with position and with time are larger than is generally the case in the open ocean. Many of the world's fisheries are in coastal waters, and other problems such as those concerned with the disposal of sewage or industrial effluent are of immediate importance. Some of the characteristics of coastal oceanography will therefore be described.

The first effect of the shore as a *boundary* is obvious—it limits possible directions of motion so that horizontal flows tend to parallel the coast. It is important to recognize this effect because it represents one of the few situations in which man can exert a significant influence on the ocean. Jetties and breakwaters, designed for the protection of shipping from swell, may also redirect currents. In the past this has often led to unexpected changes such as the deposition of silt and the formation of new shoals, or the removal of beaches when a jetty prevented the longshore movement of sand required to maintain a beach whose material was being eroded away by wave action.

Less immediately obvious are the effects of the coast in causing vertical motions of the water. One of these is the vertical motion which we call the *tide*, i.e. the semidiurnal or diurnal rise and fall of sea level due to the coast blocking the horizontal flows generated by the periodically varying astronomical forces and causing the water to pile up against the coast during the flood or fall away during the ebb. The other effect results when surface water is forced away from a significant length of coast; the presence of the latter prevents replacement by other surface water and, instead replacement takes place from below, called *coastal upwelling* (as distinct from the upwelling which takes place between currents in the open ocean as described in Section 7.614). This process will be discussed in the next section.

In shallow areas, e.g. over the continental shelf, there is a limited reservoir of deep water for mixing and, in consequence, the input of solar energy can cause a greater summer rise of temperature than in the open ocean.

The effects of *tidal currents* are twofold. They may cause large changes twice daily in the volume of water in a harbour or bay, and they may also promote vertical mixing and thus break down the stratification of the water. For instance, where there are strong tidal currents over a rough bottom in shallow water the turbulence causes the heat absorbed in the surface layer to be mixed through a considerable depth. A result will be smaller surface temperature changes compared with those toward the head of a bay nearby where the currents are slight.

The effect of cold, dry, continental *arctic air* flowing out over the ocean and enhancing the heat loss in the north-east Atlantic and Pacific has been discussed in Section 5.381. In other areas, dry air from desert regions may promote evaporation and may also carry terrestrial material as dust to the open ocean.

The direct effect of *river runoff* is to reduce the salinity of the surface layers, and even of the deeper water if there is sufficient vertical mixing. Generally, river runoff has a pronounced seasonal variation and this gives rise to much larger seasonal fluctuations of salinity in coastal waters than in the open ocean. In a coastal region where precipitation occurs chiefly as rain, the seasonal salinity variation will follow closely the local precipitation pattern. In regions where rivers are fed by melt water from snowfields or glaciers, the river runoff increases in the summer to many times the winter rate and causes a corresponding decrease of salinity during this period which lags the snowfall by several months. Since river water frequently carries suspended sediment this causes the coastal waters to have low optical transparency. Sometimes this sediment is carried in the surface low-salinity layer for some distance while the deeper, more saline, water remains clear. The deposition of this sediment causes shoaling and consequent hazards to navigation. Frequently the location of this deposition is influenced by the salinity distribution because increases in salinity can cause flocculation of the sediment and rapid settling. The effect of runoff can often be traced a long way from the coast, both by reduced salinity and by the sediment in the water. Examples of this are found in the Atlantic off the Amazon River, in the north-east Pacific from the many rivers flowing in from the North American continent and in the Bay of Bengal from the large rivers whose runoff is strongly influenced by the monsoon.

Another consequence of reduced salinity in coastal waters is increased stability in the halocline so that summer input of solar heat is retained in the surface layer with resulting high temperatures. In high northern latitudes, ice tends to form first in coastal waters during the cooling season because of the limited reservoir of heat in the shallow waters combined with the low salinity from river runoff.

In low latitude subtropical regions, where precipitation is small, evaporation becomes important and high salinities as well as high temperatures occur in bays and partially surrounded bodies of water.

Along the continental shelf, variations in water conditions may be important to fish populations. Sometimes these variations are seasonal in character, but important consequences of longer-term variations have also been recorded. For instance, cod appear to have a limited tolerance to water temperature and the yield of the cod fishery off West Greenland has been shown to be very sensitive to this factor. A great improvement in the yield occured from 1924 on, when the temperature rose from below 0°C to 1°C. Previously, rich fisheries had been experienced from 1845 to 1850 when there was less ice than usual and presumably higher temperatures (water temperatures are not available for that time), but the yield had declined thereafter.

A particular type of coastal zone which had received little systematic attention from physical oceanographers until recent years was in *coral reef areas*. Some individual studies had been carried out in the lagoons of coral atolls and some decriptive work in coastal coral reef areas, particularly in the Australian Great Barrier Reef (G.B.R.). However, since 1976, Australian oceanographers have carried out extensive studies, some descriptive, many dynamical, in the G.B.R. (Pickard et al., 1986) and French oceanographers have worked systematically in the south-west lagoon off New Caledonia (Rougerie, 1986). Some of these studies will be reviewed in Section 8.5.

When carrying out oceanographic surveys in coastal waters it is necessary to take into account the special characteristics of these regions. Because of the large variations of water properties over small distances it is necessary to arrange oceanographic stations closer together than in the open ocean. For instance, in the open ocean a minimum spacing of 50 to 100 km may be adequate for general surveys but near the coast it may be necessary to reduce this to 5 or 10 km or even less. (Of course, when studying small eddies or features like fronts or the edge of the Gulf Stream, close spacing will be needed even in the open ocean.) In addition, seasonal variations are much greater near shore and in order to obtain a complete picture of oceanographic conditions it is necessary to extend observations over one or more full years. In doing this a number of oceanographic cruises are necessary because single winter and summer cruises will usually be insufficient to describe the full range of water characteristics and the times and extent of changeover periods between winter and summer extremes. Because tidal currents may play an important part, it is desirable to plan time-series observations over one or more tidal days (25 hours) at key locations. It is not uncommon to find as large variations in 25 hours at a single location in coastal waters as occur over distances of 50 to 100 km simultaneously in the same region. When commencing a study of a previously undescribed region it is advisable to plan a number of such time-series studies ("anchor stations") early in the project.

8.2 Coastal Upwelling

In Section 7.12 we described the wind-driven Ekman circulation as it applies to the open ocean. Ekman's initial study assumed an infinite ocean so that the upper-layer transport to the right of the wind direction (in the northern hemisphere) could be continually supplied by horizontal flow from the left (for continuity) without any divergence developing as long as the wind stress was uniform over the area. However, a special case occurs when the wind blows toward the equator parallel to a coast at the eastern boundary of an ocean. In this case, the Ekman transport in the upper layer will be away from the coast and the only place from which water can be supplied is from below the Ekman layer, i.e. *upwelling* must occur into that layer. The width of the upwelling zone is generally of the order of 100 km and the upwelling speed some 5–10 m/d. Since the water temperature decreases as depth increases in these regions, the upwelled water is cooler than the original surface water and a characteristic band of low temperatures develops close to the coast. This may be seen off the Pacific coast of North America in Fig. 4.2 and off the Atlantic coast of South Africa in Fig. 4.1. Often the upwelled water also has greater concentrations of nutrients (phosphates, nitrates, silicates, etc.) than the original surface water which had been depleted by biological demands. The result is that upwelling is important in replenishing the surface layers with these components which are needed for biological production. This process is particularly important for the fishery off the South American coast.

It is important to note that the water which upwells is not deep water. Comparison of the properties of the upwelled water with those in the water column before the start of upwelling has shown that the water comes to the surface from depths which are usually between 50 and 300 m. It was for this reason that the earlier reference was to the upwelling of "subsurface" water only. Off the North American coast from British Columbia to California (i.e. 55° to 30°N) relatively cool water is usually found from April to August within a region 80 to 300 km wide from the shore (Fig. 4.2). The cold water is often patchy, rather than being in a uniform band, and comes from depths of no more than 300 m. Off the South American coast, upwelling has been observed between 5° and 35°S and comes from an average depth of 130 m with a maximum of 350 m.

In addition to cooling the upper layer, the upwelling normally causes a decrease in salinity because in most areas salinity decreases with increased depth. The exception is off the North American coast where salinity increases with depth (the north-east Pacific being a region of low surface salinity because of river runoff and precipitation) and so upwelling increases surface salinity.

The earlier studies of upwelling areas were made off South-west Africa and in the Pacific off South America. Because of the importance of the upwelling process to the fisheries, a new study was started in 1971. The overall goal is to understand the Coastal Upwelling Ecosystem (CUE, e.g. see Barber, 1977) well enough to be able to predict the future response of the system by monitoring a few key ocean parameters. The physical studies are concerned with the quantitative response of the eastern boundary waters to the wind stress. The main areas for intensive study have been off the west coasts of North and South America, particularly off Oregon and Peru respectively, in the eastern North Atlantic and a major operation was carried out off the north-west coast of Africa in 1974. Earlier studies were reviewed by Smith (1968) and a more recent study by Bryden (1978). Many of the CUE studies (physical, chemical and biological) were reported in "*Coastal Upwelling*" edited by Richards (1981).

The time scale for changes in the upper water is the same as that for the development of the Ekman circulation. In one example, off the coast of Oregon, the surface temperature dropped by 6K in two days after a longshore wind started.

The description of coastal upwelling which we have given above presents only the bare essentials and several additional features should be recognized. The upwelling results in the isopycnal and isothermal surfaces rising toward the sea surface as the coast is approached, giving rise to the water property changes near the coast as we have mentioned. Then, as the water is forced away from the coast, a slope down toward the coast develops and geostrophy requires a longshore current to flow equatorward in the upper layer. In addition, poleward currents can develop in the deeper water along the coast (e.g. see Bowden, 1983, Chapter 5).

8.3 Ekman Circulation in Shallow Water

So far we have discussed the Ekman circulation in an upper layer driven by the frictional stress of the wind on the surface of the sea, but there may also be an Ekman layer at the bottom. If there is a current flowing in the lower part of the water column over the sea bottom, which we will assume for simplicity to be flat, then there will be a bottom frictional stress on the water which will tend to retard its flow. The current might be due to a slope on the water surface or to a distribution of density giving rise to a geostrophic flow. The effect of the frictional stress at the bottom will be to reduce the speed of the current near the bottom and hence reduce the magnitude of the Coriolis force (to the right in the northern hemisphere) so that the current will rotate to the left to renew the balance between friction, pressure and Coriolis force. The result will be an Ekman current spiral in the bottom layer with the current rotating to the left as the bottom is approached, rather than to the right, as depth increases from the surface in the upper layer (e.g. Pond and Pickard, Sec. 9.44).

Hitherto we have tacitly assumed that the total water depth was much larger than the Ekman depth, D, but in shallow water it is possible for the top and bottom Ekman layers to overlap, so that the right-turning tendency in the top layer (northern hemisphere) will overlap the left-turning tendency in the

Coastal Oceanography 281

bottom layer. The opposing right- and left-turning effects will tend to cancel more and more as the water depth decreases. If there is a wind stress at the top surface which would give rise to an Ekman layer depth D in deep water, then in water of depth h, the angle α between the wind and the surface flow is approximately:

h/D	α	Net flow direction in the water column:
1 or more	45°	At 90° to right of wind,
0.5	45°	About 60° to right of wind,
0.25	22°	About 25° to right of wind,
0.1	3°	About 6° to right of wind.

That is, when the only driving force is the wind stress at the surface then as long as the water depth h is greater than the Ekman depth D an Ekman spiral current system will occur in the upper layer with the net flow to the right of the wind direction (northern hemisphere), but as water depth decreases the net flow is more and more in the direction of the wind and in very shallow water the entire wind-driven flow is substantially in the direction of the wind.

8.4 Estuaries

Oceanographically the term *estuary* has a wider meaning than the conventional one of the tidal region at the mouth of a large river. Cameron and Pritchard (1963) define an estuary as "a semi-enclosed coastal body of water having a free connection to the open sea and within which the sea-water is measurably diluted with fresh water deriving from land drainage". They restrict the definition to coastal features and exclude large bodies of water such as the Baltic Sea. The river water which enters the estuary mixes to some extent with the salt water therein and eventually flows out to the open sea in the upper layer. A corresponding inflow of sea-water takes place below the upper layer. The inflow and outflow are dynamically associated so that while an increase in river flow tends to reduce the salinity of the estuary water it also causes an increased inflow of sea-water which tends to increase it. Thus an approximate steady state prevails.

8.41 Types of estuaries

In terms of shape, Pritchard (1952, also Dyer, 1973) distinguishes three types, the coastal plain, the deep basin and the bar-built estuary. The first of these is the result of land subsidence or a rise of sea-level flooding a river valley; examples are the St. Lawrence River valley and Chesapeake Bay in North America. Typical examples of the second type are the fjords of Norway, Greenland, Canada, South America and New Zealand. Most of these have a sill or region toward the seaward end which is shallower than both the main basin of the fjord and the sea outside and so restricts the exchange of deep water. The third type is the narrow channel between the shore and a bar which has built up close to it by sedimentation or wave action. The inland end of an estuary is called the *head* and the seaward end the *mouth*. Positive estuaries have a river or rivers emptying into them, usually at the head, and it is this which gives rise to the characteristic features of water property distributions in such estuaries.

Estuaries have been classified by Stommel (see Dyer, 1973) in terms of the distribution of water properties as (a) vertically mixed, (b) slightly stratified, (c) highly stratified and (d) salt wedge estuaries. The stratification referred to is of salinity, because it is typical of estuaries that the density of the water is determined mainly by salinity rather than by temperature. The salinity distributions in these four types are shown schematically in Fig. 8.1 in two ways. In the left-hand column of graphs the salinity distributions are shown as vertical profiles at each of four stations between the head and the mouth of the estuary as shown in the schematic plan view at the top of the figure. The right-hand column shows simplified longitudinal sections of salinity from head to mouth for the full depth of the estuary.

The *vertically mixed estuary* (type A, Fig. 8.1) is generally shallow and the water is mixed vertically so that it is homogeneous from surface to bottom at any particular place along the estuary. The salinity increases with distance along the estuary from head to mouth. The river water in such an estuary flows toward the mouth while the salt may be considered to progress from the sea toward the head by eddy diffusion at all depths. In the right-hand figure, the vertical isohalines indicate the homogeneity of the water at each location while the straight arrows indicate that the direction of net flow of the water is seaward at all depths. (The circular arrows symbolize the mixing taking place at all depths.) The Severn River in England is an example of such a vertically mixed estuary.

In the *slightly stratified estuary* (type B), which is usually also shallow, the salinity increases from head to mouth at all depths. The water is essentially in two layers with the upper layer a little less saline than the deeper one at each position along the estuary, with a mixing layer between them (symbolized by the circular arrows in Fig. 8.1B). In this type of estuary there is a net seaward (outward) flow in the upper layer and a net inward flow in the deeper layer as shown by the straight arrows in the vertical salinity section. In addition to these flows at both levels there is the vertical mixing of both fresh and salt water giving rise to the longitudinal variation of salinity in both layers. The James River in Chesapeake Bay is an example of this type.

In the *highly stratified estuary* (type C), of which the fjords are typical, the upper layer increases in salinity from near zero in the river at the head to a value close to that of the outside sea at the mouth. The deep water, however, is of almost uniform salinity from head to mouth. Again there is a net outflow in the upper layer and inflow in the deeper water as shown by the straight arrows in the salinity section. In these estuaries there is a very strong halocline

Coastal Oceanography

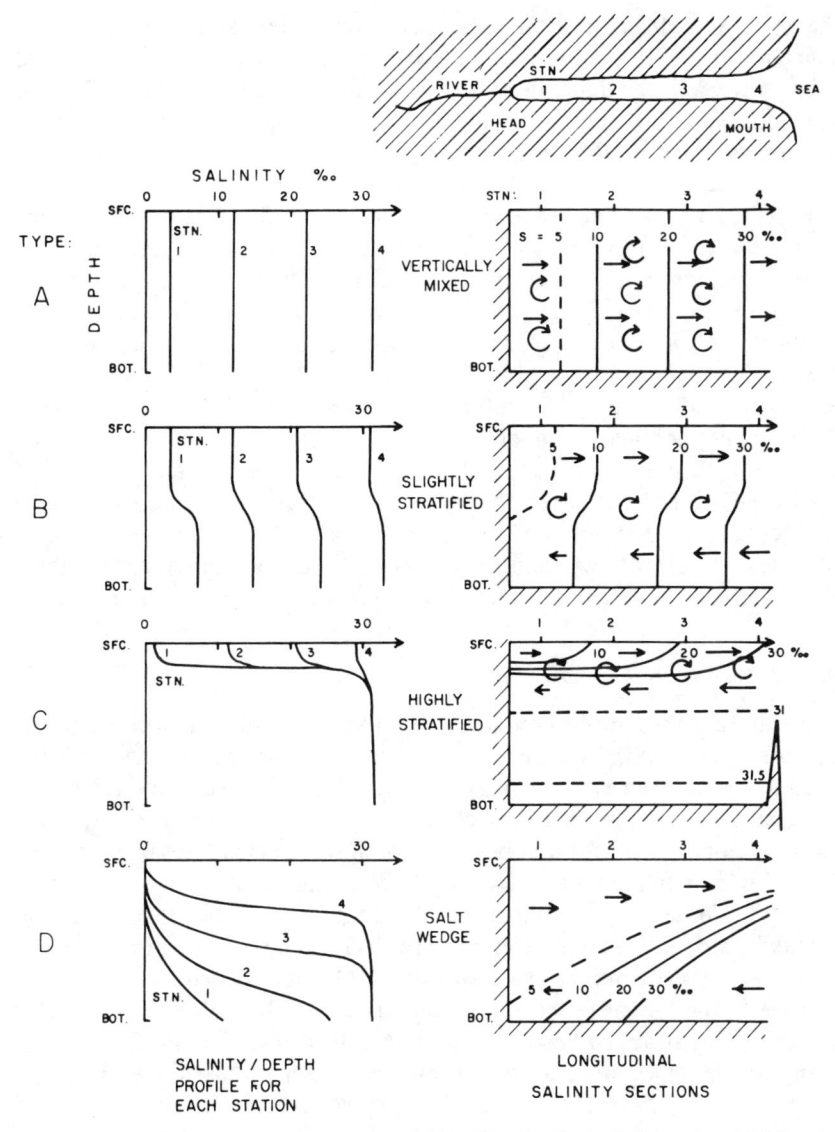

FIG. 8.1. Typical salinity/depth profiles (left) and longitudinal salinity sections (right) in estuaries (schematic).

between the upper water and the deep water, particularly at the head where vertical salinity gradients of 10 to 20 per metre may occur in summer during the period of greatest river runoff. There is vertical mixing but this results predominantly in an upward movement of salt water from below into the upper layer, with little downward movement of fresh water. One explanation for this almost unidirectional mixing is that internal waves are generated by the velocity shear between the upper low salinity layer and the deeper more saline water, and that the tops of these waves break and throw off a "spray" of saline water into the upper layer into which it mixes. There is no breaking at the bottom of the internal waves and therefore no spray of fresh water downward into the saline water.

For the *salt wedge estuary* (type D) the longitudinal section indicates the reason for its name. The saline water intrudes from the sea as a wedge below the river water. This situation is typical of rivers of large volume transport such as the Mississippi or the Fraser Rivers. It should be noted that as usual the section in Fig. 8.1 is exaggerated in the vertical direction; the salt wedge is really of a much smaller angle than shown in the figure so that the isohalines are, in fact, almost horizontal.

The salt wedge estuary has features in common with the stratified estuaries. There is a horizontal gradient of salinity at the bottom as in a slightly stratified estuary and a pronounced vertical salinity gradient as in a highly stratified estuary. The distinction is in the lack of saline water at the surface until it reaches the sea at the mouth of the estuary, because of the large river flow. One other feature of this type of estuary is that the salt wedge migrates up and down the estuary as the tide floods and ebbs, sometimes by several kilometres.

8.42 Estuarine circulation

A feature of the stratified estuaries is that the depth of the halocline, i.e. the thickness of the upper, low salinity layer, remains substantially constant from head to mouth of an estuary for a given river runoff. If the estuary width does not change much, the constancy of depth of the upper layer means that the cross-sectional area of the upper layer outflow remains the same while its volume transport increases because of the entrainment of salt water from below. In consequence the speed of the outflowing surface layer increases markedly along the estuary from head to mouth. The increase in volume and speed can be very considerable, the outflow at the mouth being as much as 10 to 30 times the volume flow of the river. In his classical study of Alberni Inlet, a typical highly-stratified fjord-type estuary in British Columbia, Tully (1949) demonstrated the above features. He also showed that the depth of the upper layer decreased as the river runoff increased up to a critical value and thereafter increased as runoff increased.

The circulation described, with outflow of the upper layer and inflow below it is referred to as an *estuarine circulation* and has to be considered when practical problems of disposing of industrial effluent are under consideration. The circulation is dependent on several factors, among them the sill depth, the river runoff and the character of the outside water density distribution. If the sill is so shallow that it penetrates into the low-salinity outflowing upper layer the full estuarine circulation cannot develop and the subsurface inflow of saline water does not occur regularly. In consequence the deep water is not exchanged regularly and tends to become stagnant. This situation occurs in some of the smaller Norwegian fjords but is by no means typical of deep basin estuaries. Most of the fjords in that country and almost all of those on the west coasts of North and South America and New Zealand have sills which are deeper than the upper layer. In consequence the estuarine circulation is developed sufficiently to effect continual renewal of the deep water so that stagnation does not occur (Pickard, 1961; Pickard and Stanton, 1980). The rate of renewal is proportional to the circulation which is itself proportional to the river runoff. Fjord estuaries with small river runoff show more evidence of limited circulation in the form of low oxygen values than do those with large runoff. The depth of the sill has little effect as long as it is greater than the depth of the low-salinity outflowing upper layer.

The other major factor influencing the exchange of the deep basin water is a seasonal variation in density structure of the outside sea-water. Although the downward mixing of fresh water in an estuary is small it does occur to some extent. In consequence the salinity, and therefore the density of the basin water, tends to decrease slowly. If a change then occurs in the outside water such that the density outside becomes greater than that inside at similar levels above the sill depth, there will be an inflow of water from the sea. The inflowing water is likely to sink, although not necessarily to the bottom, in the estuary basin and so displace upward and outward some of the previously resident water. In this way the basin water becomes refreshed. In deep-sill estuaries this refreshment may occur annually, but in shallow-sill estuaries it may occur only at intervals of many years and the disturbance to the biological regime may be cataclysmic on these occasions (by displacing upward into the biotic zone the low-oxygen water from the bottom). It is this type of basin-water replacement which has been well documented for some Norwegian fjords (with very shallow sills), but it should not be considered characteristic of all fjord estuaries.

The above remarks should be regarded only as a brief description of some of the salient characteristics of estuaries, and it must be realized that the property distributions shown in Fig. 8.1 are smoothed and schematic. Real distributions show fine and meso-scale structure and detailed features, some general and some local. In particular, because the density structure is determined largely by the salinity distribution, temperature maxima and minima are quite common in the water column. The discussion in detail of the circulation is a matter for the dynamical oceanographer and will not be considered here. The mechanics of the process of mixing between fresh and salt water, in which tidal movements appear to play a large part as well as the effects of internal waves, are in the same category and much research remains to be done in this field. Reviews of both the descriptions and the dynamics of estuaries will be found in the texts by Dyer and by Officer listed in the Suggestions for Further Reading. A recent summary of the development of estuarine circulation ideas and of investigations of circulation in a coastal region (the Middle Atlantic Bight and

Gulf of Maine, from about 36° to 43° N) has been given by Beardsley and Boicourt (1981). A comprehensive review of the physical oceanography of fjord estuaries was presented by Farmer and Freeland (1983).

It should be pointed out that estuarine characteristics and processes are observed in ocean areas as well as by the coast. In the north-east Pacific and in the Bay of Bengal where there is considerable river runoff, the density of the upper layer is controlled by the salinity rather than by temperature as is usually the case in the open ocean. The upper, low-salinity layer of perhaps 100 m depth in the north-east Pacific is less dense than the deeper, more saline water and the stability in the halocline between them inhibits mixing. In consequence the summer input of heat is trapped in the surface layer and a marked seasonal thermocline develops as shown in Fig. 4.6. In the Arctic Sea, the formation of the subsurface Arctic Water has been explained (Section 7.521) as the result of circulation and mixing processes similar to those described above for a coastal estuary.

8.5 Coral Reefs

8.51 Introduction

Prior to about 1970, most physical oceanography in coral reef areas had been carried out as ancilliary to biological or chemical studies; only about 15 papers whose main theme was physical had been published and these were basically descriptive. Studies of the dynamics only started in the late 1970s and most of such studies have been carried out in the Great Barrier Reef of Australia (G.B.R.). To the present time, of some 200 publications on the physical oceanography of coral reef regions, 70% refer to the G.B.R. and only about 30% to other reef areas.

The early work (descriptive) for the G.B.R. was reviewed by Pickard (1977), and work done during 1976–86 (mostly dynamical) by Pickard et al. (1990). An account of two years of study of the physics, chemistry and biology of the South-west Lagoon of New Caledonia was presented by Rougerie (1986). A review of the physical oceanography of coral reef regions on a world-wide scale has been presented by Andrews and Pickard (1989).

8.52 Topography

Coral reefs are features of many coastal regions between the tropics, along the continental shelves, around islands and also on the tops of shoals and seamounts in the open ocean with little or no emergent land in their vicinity, e.g. atolls. Reefs act as complex barriers to flow in their neighbourhood. Living coral cannot withstand exposure above water and so it only occurs below low-tide level; it also requires light for growth and so generally cannot survive below about 50 m depth.

Along a land boundary, two types of reef are recognized, the *fringing reef* which extends out from the shore and the *barrier reef* which is located away from the shore with a relatively reef-free region (*lagoon*) between it and the shore. Despite its name, a barrier reef is rarely continuous for very long distances but consists of a series of reefs with gaps between them, and water exchange with the ocean can take place through the gaps as well as over the reefs. In some cases the outer reefs may be long (parallel to the coast) and narrow; in others, the "barrier" may consist of a number of individual reefs dotted over a relatively wide band (50 km or more) parallel to the shore, with irregular passages between them connecting the lagoon with the open ocean outside. Open ocean reefs, based on shoals or sea-mounts, usually extend round the shoals and the "lagoon" is the body of shallow water within the reef perimeter which usually has gaps which permit some direct exchange with the ocean. *Atolls* are such reefs where sufficient material has collected on parts of the reef to raise the level a few metres above sea level and on which shrubs and trees may grow.

8.53 Water properties

The most extensive and long-term measurements of water properties have been made in the G.B.R. and in the south-west lagoon of New Caledonia.

The annual variation of water temperature is generally approximately sinusoidal with the maximum in the local summer and closely correlated with the air temperature. The annual range of temperature variation decreases toward the equator.

Salinity variations are less regular than those of temperature. Near land, decreases occur due to local precipitation and to river runoff associated with monsoons, while for atolls only precipitation is effective. Increases of salinity are due to evaporation. An increase from an oceanic value of 35.7 near the pass into Canton Island lagoon to 39.5 at the back of the lagoon some 15 km away was recorded but this is probably an extreme example. Figure 6.13 shows seasonal variations of both temperature and salinity in the form of T-S-t diagrams for the G.B.R.

Water depths in lagoons and around reefs are generally small, e.g. less than 50–100 m, and the water is usually unstratified, i.e. well-mixed vertically due to turbulence from wind-wave effects and the rough character of the bottom over the reefs. Some stratification in the upper 10–20 m occurs near river mouths during periods of heavy runoff but even then the variations in the vertical are generally less than 1K in temperature and 1 in salinity.

Along the G.B.R., intrusions of cooler, more saline water can be evident near the bottom of passes through the outer reefs with $\Delta t = -5K$ and $\Delta S = +1$ relative to the upper-layer reef-area waters. In the very shallow water over fringing reefs, diel (day-night) variations of as much as 10–12K have been observed and attributed to solar heating during the day and radiant cooling at night.

It is probable that the relatively reef-free lagoon between the shore and offlying barrier reefs is due in part to the fact that coral is intolerant both of the fresh water and of the silt which is carried in by rivers.

8.54 Tides

The surface tides along mainland shore reefs are typical of continental shelf areas, with semi-diurnal and longer-period constituents. Across wide barrier reefs, the phases of the tide are generally later at the shore than at the ocean side of the reefs (by 20 min to 2 hours for the G.B.R.) but the amplitude is little changed. An exception occurs near 22°S in the G.B.R. where the barrier reef is very wide and dense, and the tide at the shore approaches the region from north and south along the lagoon rather than directly across the barrier reef. These two flows combine to give a maximum range of about 9 m at the shore (Broad Sound) compared with only about 3 m at the ocean side of the reef, and a lag of about 2 hours from ocean to shore. (In contrast, the tide at Tahiti, which is close to an amphidromic point, has a mean range of only 0.24 m and is solar in character with high waters at noon and midnight and low waters at 0600 and 1800 h.) In many atolls, the tidal range in the lagoon is significantly less and the times of high and low water later than in the ocean because of flow limitations over the reefs and through the narrow passes between the reefs.

Internal tides have been observed near the shelf break. These are vertical oscillations, often of many tens of metres range, of the water below the surface and are most easily observed in the vertical motion of isotherms, particularly in the thermocline zone. Such internal tide motions can raise the nutrient-rich sub-pycnocline ocean water on to the shelf and through passes into the reef area, providing nutrients for the corals.

8.55 Currents

To describe the currents we will divide them into three classes: drift or long-period (periods of weeks or more), weather band (periods of days) and tidal.

Drift currents are those generated by steady wind stress (e.g. the trade winds) or long-shore pressure gradients. The oceanic equatorial currents generated by the trade winds cause flow over the mid-ocean reefs, e.g. at Bikini. In the central G.B.R., a 2.5 year time-series of current measurements showed equatorward currents of 20 cm/s during the south-east Trade Wind season while at other times there was a poleward drift of about 30 cm/s attributable to the downward slope to the south associated with the southward flow of the *East Australian Current* outside the Reef.

Weather band currents associated with continental shelf waves have also been documented for the central and southern G.B.R. They are a consequence of fluctuations of wind stress as weather systems move eastward with their centres over the southern part of Australia. They have periods of 10–20 days and speeds of some 500 km/d (equatorward). As their vertical range is only 10–30 cm (near the shore and diminishing to zero outside the reef) they are not evident to the eye and can only be identified by analysis of tide or current records. Water particle speeds are some 20–40 cm/s or 17–35 km/d. This is very much less than the phase speed of the waves but, on account of their relatively long period, it can result in long-shore displacements of water of 100–200 km which can be very significant in transporting pollutants or plankton over such distances in the reef area.

Tidal flows through reef passes of some 200 cm/s are common and values as high as 370 cm/s (13 km/h) at Aldabra Atoll have been recorded. It should be noted that although such tidal speeds through passes can be large, the inflowing water then spreads out in the lagoon and the distance of penetration of ocean water during the flood (which only lasts about 6 h) may be only a few kilometres. This is small compared with the diameter of many atoll lagoons and therefore tidal flows may only have a limited effect on water exchange and flushing. Also, it has been observed that there is often little mixing between the intruding ocean water and the resident lagoon water.

The term *reef flat* refers to extensive areas of coral of relatively uniform height, and the water depths over them are generally only a few metres. Flow over them may be due to drift currents, shelf waves, and tidal currents, and also to the local wind stress. Note that as the water in these areas is shallow, bottom friction will be more important than Coriolis force and the wind-driven flow will be downwind, not to the left or right of the wind direction as is the case in deep water (see Section 8.3). Tidal currents over the reef flats may have speeds of 100 cm/s or more. A feature to note about such flows (over the surrounding reef) is that they will not necessarily be into a lagoon during the flood or our during the ebb. For instance, in the G.B.R. the tide wave approaches from the north-east so that the tidal flow in the area during the flood will be to the south-west and can be over a surrounding reef *into* its lagoon on the north-east side but *out* of the lagoon on the south-west side at the same time.

Another contribution to the water in a reef lagoon is by *wave-overtopping* when ocean waves or swells break on the outside of a reef and so generate a slope across the reef which causes flow across it. This component can contribute as much transport across a reef as the other mechanisms combined.

8.56 Circulation in lagoons

The circulation in individual reef lagoons may be forced by some or all of the current mechanisms described above. In the extensive open lagoon between the shore and the barrier reef in the G.B.R., steady drift currents due to wind stress and pressure gradients, and periodic currents due to tides and continental shelf waves, all contribute to the water circulation. In the New

Caledonia lagoon, about 20 km wide by 80 km long inside a narrow barrier reef, the tide appears to be the main contributor to water circulation during light wind conditions but during strong south-east Trade Winds the wind stress superimposes a general north-west motion over the barrier reef and across the lagoon.

Within the G.B.R., individual reefs form partial obstacles to the general flow, "partial" because, except at very low water, some flow occurs continually over the coral reefs. Eddies often form downstream of reefs, particularly during flows associated with tidal currents. Such eddies can both increase mixing on the small scale (10's to 100's of metres) but can also form closed volumes in which plankton can be held for hours or longer. Shedding of eddies behind reefs probably does not occur very often because flow speeds are not great enough.

For the roughly circular lagoons within individual reefs in the G.B.R., again the drift and periodic currents contribute to the circulation together with inflow due to wave-overtopping. Studies within atoll lagoons have demonstrated that in addition to inflow due to ocean currents, tides and wave-overtopping, wind stress causes downwind flow in the upper layer at speeds of about 3% of the wind speed while a compensating upwind flow develops in the deeper water.

Residence times for water within lagoons cover a wide range. For lagoons of 2–10 km diameter in the G.B.R., times of 0.5 to 4 d have been estimated, for Bikini Atoll 40 to 80 d and for very shallow lagoons such as at Fanning Atoll (18 km long but only a few metres deep) periods of up to 11 months were estimated. Rougerie (1986) estimated residence times in the New Caledonia lagoon as 2 to 28 d, depending on the particular area and the runoff, wind and tide characteristics.

CHAPTER 9

Some Directions for Future Work

IN THIS BOOK, an endeavour has been made to describe the kind of physical information available about the oceans at the present time. It has been illustrated with descriptions of some of the features of the ocean waters and of their circulation, but the reader will realize that these are only samples. For more complete information it would be necessary to consult more extensive texts, reviews and journal articles, such as those listed in the Bibliography. To conclude this presentation it will be pertinent to assess the state of our present knowledge and to indicate some of the information which we still need.

For the upper layer of the ocean, it is safe to say that we are acquainted with the main features of the circulation and of the distribution of water properties, but there are few oceanographers who are satisfied with the extent of their knowledge of even limited areas.

One of the first features of the upper-water structure described in this book was the thermocline. This occurs over most of the ocean but the reason for its continued existence is by no means clear. In low latitudes there is a net annual input of heat through the surface and together with the mixing due to the wind this might be expected to produce a steady deepening of the thermocline. The evidence available over the past 100 years or so is that this is not occurring. As an explanation for this Stommel (1958) suggested that the downward mixing of heat must be balanced by a net upward flow of cool deep water displaced by winter cooled water sinking in the North and South Atlantic. This is a very reasonable suggestion for the basic mechanism but it still leaves to be explained the differences in depth and in temperature gradient in the thermocline zone between different regions. Unfortunately the rate of upward flow is estimated to be so small as to be unmeasurable with present instruments (it is of the order of 1 cm/day) and therefore we can only infer it from heat budget calculations. These are only very approximate, since our techniques for measuring the component terms in the heat budget leave much to be desired in fundamental soundness, accuracy and convenience in use. In fact, one of the major problems in physical oceanography is to accurately resolve the terms in the heat budget. It is well known that the ocean stores heat energy which it exchanges with the atmosphere; these exchanges must have an impact on the earth's climate but our imperfect knowledge of how the ocean stores, transports and exchanges this heat prevents us from developing reliable prediction methods for climatic change. It should be noted that Iselin's suggestion (Section 7.353) for the mode of formation of Central Water Masses provides an alternative mechanism for the formation of the thermocline.

Many of the problems of the upper layers are to understand how oceanic motions are driven by the atmosphere; these are mainly problems in dynamical oceanography which we will not discuss here explicitly. To solve these problems requires quantitative observation of the behaviour of the surface and subsurface characteristics and currents. For example, the main character of the circulation has been well established and we have many series of measurements of water characteristics and volume transports from transequatorial sections such as the Hawaii-Tahiti Shuttle Experiment, but there is still much to be learned about their long-term variations, particularly for the deeper currents (vide Firing's suggestion at the end of Section 7.615), and their possible relation to the ENSO phenomenon. Many of the deeper currents have only been identified during the past decade and the South Pacific is still very imperfectly known, particularly the eastern half including the details of the very important *El Niño* phenomenon, and the currents on the east and west of Australia remain to be fully described.

Upwelling of subsurface water is important as it often brings to the surface the nutrients which promote the growth of plankton and hence of fish, while the strip of cool water along the coast affects the climate there and over the neighbouring land. Many of the regions of persistent upwelling are known, but there may be others which are at present unrecognized. In the exchange of water within the upper layer, downwelling in regions of convergence is also important. Such regions are less easy to locate than upwelling regions but are still important in the circulation of the upper layer. Our great lack of information here is of the rate of vertical motion in both processes, and there does not seem to be in view at present any practical direct method to measure it in the upper layers.

In general our picture of the upper ocean has evolved to the point where we understand the basic features and can describe the large-scale long-term distribution of these features. At the same time we have come to realize that the ocean varies over a wide range of space and time scales making our large-scale description useful in heuristic terms but not adequate to describe the ocean at any particular state or even to properly estimate the level of variation. It is the task of the modern descriptive oceanographer to improve his observations of the ocean to the point where these scales and modes of variability can be defined as motivation for the development of new dynamical theories to explain them.

This observational task requires new skills and approaches which emphasize the need for greater spatial and temporal resolution. While we have methods for sampling intensively in time at a particular location (moored instruments) it is much more difficult to observe changes in both space and time with adequate resolution. Recent improvements in underway sampling, such as the XBT, have made it easier to cut down on the ship time needed to survey an area. In addition, developments in the interpretation of infra-red satellite images (such as in Plates 11 and 12) have provided more useful descriptions of oceanic processes associated with surface temperature patterns. The availability of regular satellite altimeter measurements has offered new hope for being able to monitor space-time changes of the sea surface to infer geostrophic ocean currents much as the meteorologist studies atmospheric pressure maps to determine geostrophic winds.

For the deep water we are even more poorly informed than for the upper water. For the Atlantic Ocean we do have a good idea of the general direction of movement of the main water masses, but have a much poorer knowledge of even the direction of flow of the deep waters of the Pacific and Indian Oceans. Stommel (1958, and with Arons, 1960a, 1960b) put forward a hypothesis, which has since been elaborated by others, that in the deep water there is a strong flow along the west sides of the oceans, with branches eastward and poleward to supply the upward flow needed to maintain the thermocline. This west-side flow is mainly southward from the Labrador Sea in the Atlantic and northward from the Southern Ocean in the Pacific and Indian Oceans. With regard to the speed of motion of the deep water we have little information for any ocean. Estimates of the age of the Atlantic subsurface water masses, from ^{14}C information have decreased from 700 years in 1960 to only 250 to 500 years by 1983, as our knowledge of the ocean structure and motion has improved. Even the age gives no more than an indication of the average speed. The measurements with Swallow floats, limited as they are, demonstrate that the instantaneous speeds are very variable. As stated before, the classical concept of a sluggish flow of deep water is certainly unrealistic and it is more likely that jets or filaments of fast-moving water are a basic feature of the deep circulation.

Another way of saying that we do not know much of even the mean speed of the deep water is to say that we have little idea of the rate of formation of water masses. For instance, is the North Atlantic Deep Water produced each year in small quantities (from the Greenland Sea) or is it formed cataclysmically at irregular intervals as Worthington has suggested? Or do both processes play their part and, if so, to what extent?

The discovery in the mid-1960s that the ocean is populated with mesoscale eddies has added new insight into the nature of the space-time variability discussed above. The exact definition of mesoscale eddies is not very clear and the term has come to be applied to any deviation from a mean having mesoscale time and space structure. Some of these features are clearly closed eddies or "rings" and have been studied as to their contributions to the horizontal advection of heat, momentum and water properties, including the biota. Of particular interest have been rings that have been spawned by meanders of major currents such as the Gulf Stream, the Kuroshio and the Antarctic Circumpolar Current. Smaller eddies are generated as shear instabilities from currents and fronts while others form as a consequence of atmospheric energy exchange. Studies of the distribution of variability of temperature and altimetric sea height have provided us with regional and global maps of "eddy energy" distributions. Maximum eddy energies appear to be located adjacent to the strongest currents and along zonal bands in the tropical oceans. These representations are biased to the larger portions of the mesoscale and there are likely to be a great many smaller features that populate the open ocean. Studies such as POLYMODE have clearly revealed that some regions may not be generally populated by strong mesoscale eddies.

The T–S diagram has been described with its use for identifying subsurface water masses. These are believed to be formed at the surface, often in winter, as a result of heat and water exchange with the atmosphere. Over the period of 100 years for which oceanographic data suitable for comparison are available there have been marked variations in climatic conditions. It is remarkable that despite this the various water masses in the ocean have retained their oceanographic characteristics virtually unchanged. The "18° water" in the Atlantic is an example of what is almost a water type which has maintained its characteristics since the time of the *Challenger* expedition in 1873. It is clear that our knowledge of the heat budget requires considerable refinement in terms of input/output through the sea surface, of advection by ocean currents, and of the balance between advection and diffusion in the vertical direction. It is hoped that the data from SEASAT, TOPEX and other satellites and other ocean monitoring efforts, such as WOCE (described later), will provide us with continuing large-scale coverage of the oceans to help us to understand both the overall picture and the seasonal variations for the ocean as well as providing information to help to understand the effect of ocean temperature anomalies on the atmosphere and our climate.

The physical properties of sea-ice and the reasons for its movement and distribution are neither fully described nor understood. To nations in the higher northern latitudes, Canada, Scandinavia and Russia, the ability to move shipping in the North Atlantic, Arctic and adjacent waters is very important. Although significant advances have been made in forecasting ice conditions, this cannot be done with certainty or very far ahead. The TIROS satellites from 1961 improved our observing capabilities immensely, even though cloud cover caused observing limitations. Passive microwave radiometers, such as the Scanning Microwave Radiometer (SMR) flown on Nimbus-7 (1978–86) and the Special Sensor Microwave Imager (SSM/I) flying on the Defence Meteorological Satellite Program (DMSP) satellites, are capable of providing all-weather images of sea-ice parameters on a daily basis. The only drawback to this system is the lower spatial resolution (about

Some Directions for Future Work 295

20–30 km) inherent with a passive microwave sensor. Synthetic Aperture Radar (SAR) overcomes this spatial resolution problem and can provide, for example, 25 m resolution images of sea-ice. Sequential SAR images provide detailed descriptions of ice motion and automated techniques have been developed for objectively computing the motion experienced between successive images.

In connection with the propagation of sound in the ocean, both for military purposes and for fish detection, there are still a vast number of data required on temperature distribution both on the macroscopic scale and also on the smaller scales associated with turbulence. Acoustic signals have also been used with the technique of acoustic tomography to monitor changes in the ocean temperature between moored sound sources and receivers.

Today's descriptive oceanographer needs data in two forms: simultaneous coverage over large areas of the ocean (together with meteorological data over the same area) and time-series observations over periods up to a year in selected regions. It is clear that conventional ship-based measurements will never be able to satisfy the need for simultaneity and that there are not now, nor ever will be, enough oceanographers to carry out all of the data collection needed. In response, new systems such as the drifting buoys, autonomous deep floats, ships-of-opportunity, and satellites are becoming increasingly important in collecting the data needed to monitor the space-time variability that is so important in developing our understanding of the ocean.

CTDs are in wide use today and are very helpful both in providing an immediate readout of water properties and in enabling data to be fed directly into a computer for correction and data assembly for reports and analysis. However, it is still considered good practice to check the CTD frequently against bottle-cast samples of both temperature and salinity to maintain a satisfactory level of absolute accuracy. Dissolved oxygen sensors are coming into use but have not yet reached the same level of acceptance as the temperature and salinity instruments. Immediate indication and recording as a function of depth permits the oceanographer to see on the spot what information he is getting, and to adjust his survey as necessary to make the best use of his ship time, and to follow particular features in space and time. Unfortunately, however, differences in instrument and calibration procedures have kept many of these CTD data from being submitted to the large data archives. As relatively few bottle casts are collected any more, our general ocean coverage is not increasing significantly. Combined with a general decrease in large-scale oceanographic surveys this lack of data submission has limited our historical data file on the ocean's density structure to measurements collected predominantly before 1970. Although the use of air dropped XBTs (AXBTs) has made rapid upper ocean surveys possible, the potential of aircraft and satellites for use as platforms for oceanographic studies has still to be fully realized. The successful development of new instruments requires the combination of an assessment of the kind and accuracy of data required, the services of skilled instrument designers, and a full appreciation of the conditions under which the instruments have to be used at sea.

Finally, we draw attention to the dynamical oceanographer's technique for investigating the circulation by solving the equations of motion by numerical methods (*numerical modelling*) which is being increasingly used both for large- and small-scale motion studies because the non-linear character of the equations makes it very difficult to obtain analytic solutions to them (i.e. solutions in the form of algebraic/trigonometric expressions for velocity and density distributions as a function of space and time). To use the modelling techniques the dynamical oceanographer needs from his descriptive colleagues the best avilable information on the density distribution (from temperature and salinity distributions), on air/sea exchange processes and on circulation, both to initiate his modelling studies and to test his results. Such procedures may be used to determine if the known ocean develops when the modelling procedure is started from an ocean initially at rest (*prognostic models*), or may be used to determine the effects on an existing circulation of changing the conditions or acting processes (*diagnostic models*). It is primarily for the latter that more complete descriptions of the ocean's characteristics are needed, to provide the most realistic and complete starting conditions for a model. Results of such studies may suggest ways of deploying our limited observational facilities to acquire information for more advanced models or theories, or may suggest means for solving practical problems associated with the sea. A short account of the techniques used in numerical modelling and of some of the results obtained is given in Pond and Pickard (1983).

A major programme planned to obtain a global picture of the oceans is the World Ocean Circulation Experiment (WOCE) which has been under consideration and planning for some 20 years and is part of the World Climate Research Programme. Its first goal is (1) to develop models useful for predicting climate change and to collect the data necessary to test them. Specific objectives within this goal are to determine (1a) the large-scale fluxes of heat and fresh water, their divergences over five years, and their annual and interannual variability, (1b) the dynamical balance of the World Ocean circulation and its response to changing surface fluxes, (1c) components of ocean variability on scales of months to years, mega-metres to global scale, and the statistics on smaller scales, (1d) the rates and nature of formation, ventilation and circulation of water masses that influence the climate system on time scales from ten to one hundred years. Its second goal is to determine the representativeness of the WOCE data for the long-term behaviour of the ocean, and to find methods for determining long-term changes in the ocean circulation. Specific objectives are (2a) to determine the representativeness of the specific WOCE data sets, (2b) to identify those oceanographic parameters, indices and fields that are essential for continuing measurements in a climate observing system on decadal time-scales and (2c) to develop cost-effective techniques suitable for deployment in an on-going climate observing system.

Some Directions for Future Work 297

Some aspects, such as numerical modelling and instrument development have been under way for many years already, while the main field observation programme was planned to start about 1990 and to last for five years or more in the three major oceans. More details of the plans may be found in "World Ocean Circulation Experiment Implementation Plan', Vol. 1. (World Climate Research Programme, 1988) and in the article by Mayes (1989) which describes the current state of the programme.

It is clear that there are many intriguing problems yet to be solved in physical oceanography. There are extensive programmes of observation to be planned and instruments to be developed so that they may be carried out. The study and interpretation of the data and comparison with theoretical studies will yield a fuller understanding of the ocean circulations and, we can be sure, will reveal further problems to be investigated. Much of the research will be carried out by oceanographers who have no other motive than to improve their understanding of the behaviour of the oceans. The individual who likes to see some practical application for the new knowledge can be assured that it will at once be seized upon by those concerned with urgent practical questions in the develoment of fisheries, the solution of waste disposal problems and in coastal engineering.

APPENDIX

Units used in Descriptive Physical Oceanography

Introduction

Formerly, physical oceanographers used a mixed system of (mostly) metric units but as the International System of Units (SI) is coming into general use it is used in this text. In SI there are *base units* and their multiples and *derived units*; a number of temporary units are accepted but will be phased out eventually. In the following sections, the base and derived units appropriate to our use will be listed and then these will be related to the older mixed units and to some specialized oceanographic units. The basic references used for SI practice are *The International System of Units (SI)*, reference CAN3-Z234.2-76, and *Canadian Metric Practice Guide*, reference CAN3-Z234.1-76, prepared by the Canadian Standards Association.

Relevant SI Units

Base Units

Quantity	Base unit	Abbreviation
Length	metre	m
Mass	kilogram	kg
Time	second	s
Thermodynamic temperature	kelvin	K

Multiples and Derived Units

Note that only units relevant to physical oceanography are given:

Quantity	Unit	Abbreviation or equivalent
Length	1 micrometre	$\mu m = 10^{-6}$ m
	1 centimetre	$cm = 10^{-2}$ m
	1 kilometre	$km = 10^{3}$ m

Units used in Descriptive Physical Oceanography

Volume	$1\ m^3$	
	1 litre (L) = 1000 millilitres (mL) = $10^{-3}\ m^3$	
	(the litre is not used for high-precision measurements)	
Mass	1 gram	$g = 10^{-3}\ kg$
	1 tonne	$t = 10^3\ kg$
Time	1 minute	$min = 60\ s$
	1 hour	$h = 60\ min = 3600\ s$
	1 day (mean solar)	$d = 24\ h = 86,400\ s$
	1 year	yr (is not defined in SI, here taken as 365 d)
Frequency	1 hertz	$Hz = 1$ vibration/s
Speed	$1\ m/s = 100\ cm/s$	
	(Note that the commonly used unit of 1 cm/s for current speeds is equivalent to 0.86 km/d or roughly 1 km/d or 300 km/yr.)	
	1 knot (kn) = 51 cm/s = 1.85 km/h	
Density (ρ)	$1\ kg/m^3 = 10^3\ g/m^3$	
Relative density (d)	(formerly "specific gravity")—the value for liquids is given relative to pure water.	
Force	1 newton	N = force required to give 1 kg mass an acceleration of $1\ m/s^2$.
Pressure (p)	1 pascal	$Pa = 1\ N/m^2$
Energy	1 joule	$J = 1\ N.m$
Power	1 watt	$W = 1\ J/s$
Temperature (t)	The Celsius temperature (t°C) is the difference between the thermodynamic temperature TK and the temperature 273.15 K (unit = 1 K). (Temperature differences are stated in this text as K.)	

Units used in Physical Oceanography and Some Numerical Examples

Volume transport 1 Sv = 1 sverdrup = $10^6\ m^3/s$ (Not approved by IAPSO but used in this text for conciseness.)

Salinity (S_A, S) The definitions of absolute salinity (S_A) and of practical salinity (S) (PSS78) are given in Chapter 3. Practical salinity is used in this text and expressed as a pure number, e.g. $S = 35.05$, but the form $S = 35.05$ psu is sometimes used nowadays.

Density (sea water (ρ) is a function of salinity, temperature and pressure, i.e. $\rho = \rho_{S.T.P.}$. For sea-water of $S = 35$, $t = 10.00°C$ at standard atmospheric pressure $p = 101.325$ kPa (i.e. at zero hydrostatic pressure), the value of $\rho_{35,10,0} = 1026.97\ kg/m^3$. Strictly speaking the measured oceanographic data for sea-water are for relative density (to pure water) which is dimensionless, but in physical equations it must be treated as density (dimensions = kg/m^3).

Sigma-t (σ_t) This is introduced for convenience (Chapter 3). For the above sample of sea-water $\sigma_t = 26.97$ (it is usual to omit the units (kg/m^3) when stating the values for σ_t, which is usually used for descriptive purposes.

Specific volume ($\alpha = 1/\rho$) Unit = 1 m^3/kg = 10^3 cm^3/g. For the above seawater sample, $\alpha_{35,10,0} = 0.973\ 74 \times 10^{-3}$ m^3/kg. Also, $\alpha_{35,0,0} = 0.972\ 64 \times 10^{-3}$ m^3/kg.

Specific volume anomaly (δ) The SI unit is m^3/kg, the mixed unit is cm^3/g.

Thermosteric anomaly ($\Delta_{S,T}$) $\Delta_{S,T} = [(1000/\rho) - 0.972\ 64] \times 10^{-3}$ m^3/kg so that the value of $\Delta_{S,T}$ for the above sea-water sample is 109.83×10^{-8} m^3/kg. In the mixed units system, the corresponding value is 109.83×10^{-5} cm^3/g which is often stated as 109.83 centilitres/tonne (cL/t) to avoid writing the power of ten (1 cL/t = 10^{-8} m^3/kg). In some texts, δ_t or Δ_T is used for the thermosteric anomaly.

Pressure The appropriate SI unit for oceanography is 1 kPa = 10^3 Pa. The previous basic unit was 1 bar, approximately equal to standard atmospheric pressure, and a unit used in oceanography was 1 decibar (dbar), equal to 10 kPa, which is about the pressure due to 1 m depth of seawater. (The bar, dbar are not part of the SI.) The pressure in the open ocean at a depth of 1000 m is about 10,100 kPa (= 1010 dbar). Standard atmospheric pressure = 101.325 kPa   (= 1.013 25 bar = 1013.25 mb = 760 mm Hg).

Heat energy The appropriate SI unit is 1 joule (J) = 0.239 calorie using the conversion factor 1 calorie = 4.185 J.

Heat flow The SI unit is 1 J/s = 1 watt (W).

Heat-flow density The SI unit is 1 W/m^2[= 2.39×10^{-5} cal/(s cm^2)]. A related mixed system unit was 1 langley (Ly) = 1 cal/cm^2 and radiation heat flow was expressed in units of 1 Ly/min = 698 W/m^2 or 1 Ly/d = 0.484 W/m^2.

Latent heat The SI unit is J/kg or kJ/kg. 1 cal/g = 4.185 kJ/kg so that for pure water, the latent heat of melting of 80 cal/g = 335 kJ/kg and the latent heat of evaporation at 10°C of 591 cal/g = 2.47×10^3 kJ/kg.

Electrical power The SI unit = 1 kW = 1.34 h.p.

Electrical conductance   The SI unit is 1 siemens/m (S/m) = 1 mho/m = 0.01 mho/cm.

Dissolved oxygen Usually quoted in units of 1 mL/L (= 1.43 mg/L = 0.089 mg at/L).

The appropriate SI unit for dissolved oxygen is micromole/kg (μmol/kg) and 1 mL/L = 43.3 μmol/kg (for mean sea-water of $S = 34.7$, $t = 3.5°$C, $\sigma_t = 27.96$).

Bibliography

Two lists are given. The first is a general list of textbooks and sources of further information while the second is a list of specific journal articles or more specialized texts referred to in the body of our text.

Suggestions for Further Reading

The following texts may be consulted for more detailed information on some of the topics discussed in the present book:

APEL, J. R.; *Principles of Ocean Physics*, Academic Press, 1987, p. 634. A detailed treatise at an advanced level.

BOWDEN, K. F.; *Physical Oceanography of Coastal Waters*, Ellis Horwood Ltd., Chichester, U.K. 1983, p. 302. (Distrib. J. Wiley & Sons, Ltd.). An introduction to coastal oceanography, both descriptive and dynamical.

BUDYKO, M. I.; *Climate and Life*, Academic Press, 1974, p. 508. A broad review of the climate of the earth with discussion of the relations between it and life. (English translation of the 1971 Russian text.)

DEFANT, A.; *Physical Oceanography*, Pergamon Press, 1960, p. 1319. An advanced level text. Volume 1, Pt. 1 is descriptive while Pt. 2 and all of Vol. 11 are dynamical.

DIETRICH, G. and K. KALLE; *General Oceanography—an Introduction*, Wiley, 1963, p. 588. A fairly comprehensive text on descriptive and dynamic oceanography with a little geology and biochemistry.

DYER, K. R.; *Estuaries—a Physical Introduction*, Wiley. 1973, p. 140. A summary of the descriptive and dynamic oceanography of estuaries.

GILL, A. E.; *Atmosphere-Ocean Dynamics*, Academic Press, 1982, p. 662. A unified comprehensive approach to the study of atmospheric and oceanic circulations, basically mathematical but with excellent verbal descriptions.

GROSS, M. GRANT; *Oceanography—a View of the Earth*, Prentice-Hall, 2nd edn., 1977, p. 497. A survey of most aspects of oceanography, physical, geological, chemical and biological.

HARVEY, J. G.; *Atmosphere and Ocean; our Fluid Environments*, Artemis Press, 1976, p. 143. A coherent description of the characteristics and dynamics of the atmosphere and ocean and of their interaction, with simple mathematical developments.

HILL, M. N. (Ed.); *The Sea: Ideas and Observations*, Interscience, Vol. 1, 1962, p. 864. Physical oceanography. A collection of advanced papers on dynamical oceanography and energy transmission. Vol. 11, 1963, p. 554. Composition of sea-water. Comparative and descriptive oceanography. Further advanced papers on chemical synoptic and biological oceanography.

JERLOV, N. G.; *Marine Optics*, Elsevier, 1976, p. 231. An up-date of his earlier *Optical Oceanography* (1968) being a review of what is known about light in the sea.

KENNETT, J. P.; *Marine Geology*, Prentice-Hall, 1981, p. 813. Updates Shepard's texts, advanced.

KRAUS, E. G.; *Atmosphere—Ocean Interaction*, Clarendon Press, 1972, p. 271. A fairly advanced discussion of radiation exchange at the surface of the ocean and of the exchange of energy and momentum between the atmosphere and ocean.

LAUFF, G. H. (Ed.); *Estuaries*, Amer. Assn. Adv. Sci., Publ. No. 83, p. 757. A collection of papers on a variety of aspects of the oceanography of estuaries.
MCLELLAN, H. J.; *Elements of Physical Oceanography*, Pergamon, 1965, p. 150. An introduction to descriptive and dynamic oceanography.
NEUMANN, G. and W. J. PIERSON; *Principles of Physical Oceanography*, Prentice-Hall, 1966, p. 545. Moderately advanced text on descriptive and dynamic oceanography.
OFFICER, C. B.; *Physical Oceanography of Estuaries and Associated Coastal Waters*. Wiley, 1976, p. 465. A moderately advanced account of the description and dynamic theory with applications to typical areas around the world.
PEDLOSKY, J.; *Geophysical Fluid Dynamics*, Springer Verlag, 1979, p. 624. An advanced text of interest to graduate students in physical oceanography.
PERRY, A. H. and J. M. WALKER; *The Ocean–Atmosphere System*, Longman, 1977, p. 160. A good account of the ocean–atmosphere system with many illustrations and references to original literature.
POND, S. and G. L. PICKARD; *Introduction to Dynamical Oceanography*, Pergamon, 1983, p. 329. An introduction to the basic principles and equations of dynamic oceanography, the role of the non-linear terms, currents without and with friction, numerical modelling, waves and tides.
SHEPARD, F. P.; *Submarine Geology*, Harper & Row, 1973, p. 517. A description of the sediments and structure of the ocean basins and of their origins.
SHEPARD, F. P.; *Geological Oceanography*, Crane, Russak, 1977, p. 214. A well-illustrated account of many developments in this field.
STEWART, R. H.; *Methods of Satellite Oceanography*, University of California Press, 1985, p. 370. A moderately advanced but readable text.
STOMMEL, H.; *The Gulf Stream*, University of California Press, 2nd edn., 1965, p. 248. Both a description of this ocean feature and an excellent introduction to physical oceanography for upper-year undergraduates and graduate students in physics.
SVERDRUP, H. U., M. W. JOHNSON and R. H. FLEMING; *The Oceans, their Physics, Chemistry and General Biology*, Prentice-Hall, New York, 1946, p. 1087. A comprehensive reference book on all aspects of oceanography to about 1942.
TCHERNIA, P.; *Descriptive Regional Oceanography*, Pergamon, 1978, p. 250. Describes the physical oceanography of the oceans and seas in more detail than the present text with numerous illustrations and fold-out charts.
URICH, R. J.; *Principles of Underwater Sound*, McGraw Hill, 1982, p. 384. A review of this subject, moderately advanced.
VAN DORN, W. G.; *Oceanography and Seamanship*, Dodd Mead, 1974, p. 481. An interesting and simply written account of how knowledge of the ocean is applied to practical seamanship.
VON ARX, W. S.; *An Introduction to Physical Oceanography*, Addison Wesley, 1962, p. 422. A stimulating introduction to many aspects of physical oceanography, with special emphasis on current measurement and the use of laboratory scale-models. Also includes study questions and a chronological list of significant events in the marine sciences.
WARREN, B. A. and C. WUNSCH (Eds.); *Evolution of Physical Oceanography*, MIT Press, 1981, p. 623. An excellent series of survey articles on the general circulation, physical processes, techniques and ocean/atmosphere interaction tracing the development of these fields and laying out the state of our knowledge now. Prepared in honour of Henry Stommel. Advanced but essential to the library of any serious physical oceanographer.
WEYL, P. K.; *Oceanography—an Introduction to the Marine Environment*, Wiley, 1970, p. 535. Mostly physical, chemical and geological with some biology and ecology.
Oceanography—Readings from Scientific American; Freeman, 1971, p. 417. A collection of stimulating articles on many aspects of oceanography.
Ocean Science—Readings from Scientific American; Freeman, 1977, p. 307. Contains articles additional to those in *Oceanography*, 1971.

Four accounts of the development of oceanography are:
DEACON, M. B.; *Scientists and the Sea 1650–1900; a Study of Marine Science*, Academic Press, 1971, p. 398.
RAITT, H. and B. MOULTON; *Scripps Institution of Oceanography; the First Fifty Years*, Ward Ritchie, 1967, p. 217.
SCHLEE, S.; *The Edge of an Unfamiliar World: a History of Oceanography*, Dutton, 1973, p. 397.

SHOR, E. N.; *Scripps Institution of Oceanography; Probing the Oceans 1936–1976*, p. 502.

Some information on the properties of sea-water and on methods will be found in the following publications:
IVANOFF, A.; *Introduction à l'Océanographie, Propriétés Physiques et Chimiques des Eaux de Mer*, Vol. 1, Vuibert, Paris, 1972, p. 206. A compact summary of these properties and of distributions in the oceans. Volume 2 on the heat budget and on electrical and acoustical properties is planned.
RILEY, J. P. and G. SKIRROW (Eds); *Chemical Oceanography*, Academic Press, 1965, Vol. 1, p. 712. Includes a chapter with references on the physical properties of sea-water.
Handbook of Oceanographic Tables; U.S. Naval Oceanographic Office, Special Publication 68, Washington, DC, 1966, p. 427. A collection of tables of use to oceanographers.
International Oceanographic Tables, Vol. 1; National Institute of Oceanography of Great Britain, and Unesco, 1966, p. 128. For converting conductivity ratio to salinity of sea-water. Nowadays, such conversions are usually done by computing from formulae such as those in UNESCO, 1983, listed in the Journal References.
LAFOND, E. C.; *Processing Oceanographic Data*, U.S. Naval Oceanographic Office Publication 614, Washington, DC, 1951, p. 114. A compilation of tables needed for correcting thermometers, calculating density and specific volume, etc.
Instruction Manual for obtaining Oceanographic Data; U.S. Naval Oceanographic Office, Washington, DC, Publ. 607. 3rd edn. 1968. reprint 1970, p. 210. A description of routine oceanographic procedures at sea (bottle sampling, mechanical BT, coring, etc.) and of the older standard instruments.

The classical papers in oceanography are scattered through many scientific journals and reports of expeditions. Many of the recent papers may be found in:
Deep-Sea Research, Pergamon Press, Oxford (since 1953).
Estuarine and Coastal Marine Science, Academic Press, London (since 1973).
Journal du Conseil (from 1926) and *Annales Biologiques* (from 1939), Conseil Perm. Intern, pour l'Explor. de la Mer, Copenhagen.
Journal of Fisheries and Aquatic Science, formerly *Journal of the Fisheries Research Board of Canada*, Ottawa (since 1934).
Journal of Geophysical Research, Amer. Geophys. Union, Washington, DC (since 1959).
Journal of Marine Research, Sears Foundation for Marine Research, New Haven, Connecticut (since 1939).
Journal of Physical Oceanography, Amer. Met. Society, Lancaster, Pa (since 1971).
Limnology and Oceanography, Amer. Soc. of Limnol. and Oceanogr., Lawrence, Kansas (since 1956).

Two annual reviews of aspects of oceanography are:
Oceanography and Marine Biology, M. BARNES (Ed.), Aberdeen University Press (from 1963).
Progress in Oceanography, M. V. ANGEL and J. O'BRIEN (Eds.), Pergamon Press (from 1964).

An interesting quarterly with simply written but authoritative articles on many aspects of oceanography is:
Oceanus, Woods Hole Oceanographic Institution, Woods Hole, Mass. (since 1952).

References to Journal and Review Articles

The location for the references (by author and date) made in the text are presented here for the reader who wishes to pursue a topic further. A few of the references are to articles of historic interest, e.g. first announcements of newly recognized ocean features or developed techniques, others are to review articles or to those having more detail than can be included in an introductory textbook. We make no claim that this is a list of the "260 most important articles' in descriptive oceanography. The reader should note that two useful sources for references are *The Oceans* by Sverdrup, Johnson and Fleming (1946) for material published before about 1942 and *Evolution of Physical Oceanography* edited by Warren and Wunsch (1981) for later material.

AAGAARD, K. and P. GRIESMAN (1975) Toward new mass and heat budgets for the Arctic Ocean. *Journal of Geophysical Research*, **80**, 3821–3827.

AAGAARD, K., J. H. SWIFT and E. C. CARMACK (1985) Thermohaline circulation in the Arctic Mediterranean Seas. *Journal of Geophysical Research*, **90**, 4833–4846.
ANDREWS, J. C. and G. L. PICKARD (1989) The physical oceanography of coral-reef systems. Chapter 1 in *Coral Reefs* (Ecosystems of the World, No. 25), Elsevier (in press).
ANGSTRÖM, A. (1920) Application of heat radiation measurements to the problems of the evaporation from lakes and the heat convection at their surfaces. *Geografische Annalen*, **3**, p. 16.
BAILEY, W. B. (1957) Oceanographic features of the Canadian Archipelago. *Journal, Fisheries Research Board of Canada*, **14**, 731–769.
BAINBRIDGE, A. E. (1976) *GEOSECS Atlantic Expedition*, Vol. 2 *Sections and Profiles*, National Science Foundation, Wash., DC, p. 198.
BAKER, D. J. (1981) Ocean instruments and experiment design. Ch. 14, pp. 396–433, in *Evolution of Physical Oceanography*, B. A. WARREN and C. WUNSCH (Eds.), MIT Press.
BARBER, R. T. (1877) The JOINT-1 expedition of the Coastal Upwelling Ecosystems Analysis programme. *Deep-Sea Research*, **24**, 1–6.
BARNETT, T. P. (1977) An attempt to verify some theories of El Niño. *Journal of Physical Oceanography*, **7**, 633–647.
BARNETT, T. P. (1984) Prediction of the El Niño of 1982–83. *Monthly Weather Review*, **112**, 1403–1407.
BEARDSLEY, R. C. and W. C. BOICOURT (1981) On estuarine and continental-shelf circulation in the Middle Atlantic Bight. Ch. 7, pp. 198–233 in *Evolution of Physical Oceanography*, B. A. WARREN and C. WUNSCH (Eds.), MIT Press.
BERLIAND, T. G. (1960) Methods of climatological computation of total incoming solar radiation. (In Russian) *Meteorologica Gidrologica*, **6**, 9–12.
BJERKNES, V., J. BJERKNES, H. SOLBERG and T. BERGERON (1933) *Physikalische Hydrodynamik*. Springer-Verlag, p. 797.
BJERKNES, V. and J. W. SANDSTRÖM (1910) *Dynamic Meteorology and Hydrography, Part I, Statics*. Carnegie Institute, Washington, Publ. No. 88, p. 146.
BOWEN, I. S. (1926) The ratio of heat losses by conduction and by evaporation from any water surface. *Physical Review*, **27**, 779–787.
BROECKER, W. S. and T.-H. PENG (1982) p. 690. *Tracers in the Sea*, Lamont-Doherty Geological Observatory, Columbia University, NY.
BROWN, N. L. and B. V. HAMON (1961) An inductive salinometer. *Deep-Sea Research*, **8**, 65–75.
BRYDEN, H. L. (1978) Mean upwelling velocities on the Oregon continental shelf during summer 1973. *Estuarine and Coastal Marine Science*, **7**, 311–327.
BRYDEN, H. L. and R. D. PILLSBURY (1977) Variability of deep flow in the Drake Passage from year-long current measurements. *Journal of Physical Oceanography*, **7**, 803–810.
BUDYKO, M. I. (1963) *Atlas of the heat balance of the earth* (in Russian), Academy of Sciences, Moscow, p. 69. (Also *Guide to the Atlas of the heat balance of the earth*, Trans. by I. E. Donehoo, U.S. Weather Bureau, WB/T-106, Washington, DC, p. 25.)
BUNKER, A. F. (1976a) Energy exchange charts for the North Atlantic Ocean. *Bulletin of the American Meteorological Society*, **57**, 670–678.
BUNKER, A. F. (1976b) Computation of surface energy flux and annual air-sea interaction cycles of the North Atlantic Ocean. *Monthly Weather Review*, **104**, 1112–1140.
BUNKER, A. F. (1988) Surface energy fluxes of the South Atlantic Ocean. *Monthly Weather Review*, **116**, 809–823.
CAMERON, W. M. and D. W. PRITCHARD (1963) Estuaries. Ch. 15, pp. 306–324, in *The Sea: Ideas and Observations*, M. N. HILL (Ed.), Vol. 2, Wiley-Interscience.
CANE, M. and S. E. ZEBIAK (1986) A theory for El Niño and the Southern Oscillation. *Science*, **220**, 1085–1087.
CARMACK, E. C. and K. AAGAARD (1973) On the deep water of the Greenland Sea. *Deep-Sea Research*, **20**, 687–715.
CARRITT, D. E. and J. H. CARPENTER (1958) The composition of sea-water and the salinity–chlorinity–density problem, pp. 67–86 in *Physical and Chemical Properties of Sea-water*, National Academy of Science-National Research Council, Publ. 600.
CHENEY, R. C. and J. G. MARSH (1981) Seasat altimeter observations of dynamic topography in the Gulf Stream. *Journal of Geophysical Research*, **86**, 473–484.

CHENEY, R. E., J. G. MARSH and B. D. BECKLEY (1983) Global mesoscale variability from collinear tracks of Seasat altimeter data. *Journal of Geophysical Research*, **88**, 4343–4354.

CHURCH, J. A., G. R. CRESSWELL and T. J. GOLDING (1988) The Leewin Current. *Poleward Flow along Eastern Boundaries*, S. NESHYBA, C. N. K. MOOERS and R. L. SMITH, Eds. Springer-Verlag Lecture Notes (in press).

CLARKE, R. A., H. HILL, R. F. REINIGER and B. A. WARREN (1980) Current system south and east of the Grand Banks of Newfoundland. *Journal of Physical Oceanography*, **10**, 25–65.

COACHMAN, L. K. and K. AAGAARD (1974) Physical oceanography of Arctic and Subarctic seas. Ch. 1, pp. 1–72 in *Marine Geology and Oceanography of the Arctic Seas*, Y. HERMANN (Ed.), Springer-Verlag.

COCHRANE, J. D. (1958) The frequency distribution of water characteristics in the Pacific Ocean. *Deep-Sea Research*, **5**, 111–127.

COLLINS, M. J. and W. J. EMERY (1988) A computational method for estimating sea-ice motion in sequential Seasat synthetic aperture radar imagery by matched filtering. *Journal of Geophysical Research*, **92**, 9241–9251.

CORNILLON, P., C. GILMAN, L. STRAMMA, O. BROWN, R. EVANS and J. BROWN (1987) Processing and analysis of large volumes of satellite-derived thermal infra-red data. *Journal of Geophysical Research*, **92**, 12,993–13,002.

CORNUELLE, B. *et al.* (1985) Tomographic maps of the ocean mesoscale. Part 1: Pure acoustics. *Journal of Physical Oceanography*, **15**, 133–152.

COX, R. A., M. J. MCCARTNEY and F.CULKIN (1970) The specific gravity/salinity/temperature relationship in natural sea-water. *Deep-Sea Research*, **17**, 679–689.

COX, R. A. and N. D. SMITH (1959) The specific heat of sea-water. *Proceedings, Royal Society of London*, Series A, **252**, 51–62.

CRAIG, H., W. S. BROECKER and D. SPENCER (1981) GEOSECS *Pcific Expedition*, Vol. 4, Sections and Profiles, National Science Foundation, Washington, DC, p. 251.

DAUPHINEE, T. M. and H. P. KLEIN (1977) The effect of temperature on the electrical conductivity of sea-water. *Deep-Sea Research*, **24**, 891–902.

DAVIS, R. E. (1976) Predictability of sea-surface temperature and sea-level pressure anomalies over the North Pacific Ocean. *Journal of Physical Oceanography*, **6**, 249–266.

DEACON, G. E. R. (1937) The hydrology of the Southern Ocean. *Discovery Reports*, **15**, 1–124.

DEGENS, E. T. and D. A. ROSS (Eds.) (1969) *Hot Brines and Recent Heavy Metal Deposits in the Red Sea*. Springer-Verlag, p. 600.

DELCROIX, T., G. ELDIN and C. HÉNIN (1987) Upper ocean water masses and transports in the western tropical Pacific (165°E). *Journal of Physical Oceanography*, **17**, 2248–2262.

DIETRICH, G. (1969) Atlas of the hydrography of the northern North Atlantic Ocean based on the Polar Front Survey of the International Geophysical year, winter and summer 1958. *Conseil International pour l'Exploration de la Mer*, p. 140.

DODIMEAD, A. J., F. FAVORITE and T. HIRANO (1963) Review of oceanography of the Subarctic Pacific region. Part 2, pp. 1–195, of *Salmon of the Pacific Ocean*, Bulletin 13, International North Pacific Fisheries Commission.

DONN, W. L. and D. SHAW (1966) The heat budgets of an ice-free and an ice-covered Arctic Ocean. *Journal of Geophysical Research*, **71**, 1087–1093.

DORONIN, Y. P. and D. E. KHESIN (1975) *Sea Ice*. Amerind Publishing Company (Trans., 1977), p. 323.

DÜING, W. (1970) The monsoon regime of the currents in the Indian Ocean. East-West Centre Press, University of Hawaii, Honolulu, p. 68.

DÜING, W. and D. JOHNSON (1972) High resolution current profiling in the Straits of Florida. *Deep-Sea Research*, **19**, 259–274.

EKMAN, V. W. (1905) On the influence of the earth's rotation on ocean currents. *Royal Swedish Academy of Science, Arkiv för matematik, astronomi och fysik*, **2**, No. 11, 1–53.

EKMAN, V. W. (1908) Die Zusammendrückbarkeit des Meereswassers. *Conseil Permanent International pour l'Exploration de la Mer, Publications de Circonstance*, No. 43, p. 47.

EKMAN, V. W. (1953) Studies on ocean currents. Results of a cruise on board the *Armauer Hansen* in 1930, Parts I and II. *Geofysiske Publicasjoner*, **19**, p. 106 and 122.

EMERY, W. J. (1977) Antarctic Polar Frontal Zone from Australia to the Drake Passage. *Journal of Physical Oceanography*, **7**, 811–822.

EMERY, W. J. (1980) The *Meteor* Expedition, an ocean survey, pp. 690–702 in *Oceanography, the Past*, M. SEARS and D. MERRIMAN (Eds.), Springer-Verlag.

EMERY, W. J. and J. S. DEWAR (1982) Mean temperature–salinity, salinity–depth and temperature–depth curves for the North Atlantic and the North Pacific. *Progress in Oceanography*, Pergamon, **11**, 219–305.

EMERY, W. J. and J. MEINCKE (1986) Global water masses: summary and review. *Oceanologia Acta*, **9**, 383–391.

EMERY, W. J., T. C. ROYER and R. W. REYNOLDS (1985) The anomalous tracks of North Pacific drifting buoys, 1981 to 1983. *Deep-Sea Research*, **32**, 315–347.

EMERY, W. J., A. C. THOMAS, M. J. COLLINS, W. R. CRAWFORD and D. L. MACKAS (1986) An objective method for computing advective surface velocities from sequential infrared satellite images. *Journal of Geophysical Research*, **91**, 12,865–12,878.

ESTERSON, G. L. (1957) Induction conductivity indicator. A new method for conductivity measurement at sea. *Chesapeake Bay Institute, Technical Report* No. 14, p. 183.

EVANS, R. H., K. S. BAKER, O. B. BROWN and R. C. SMITH (1985) Chronology of warm-core ring 82B. *Journal of Geophysical Research*, **90**, 8803–8811.

FARMER, D. M. and L. ARMI (1989) The flow of Atlantic Water through the Strait of Gibraltar *and* The flow of Mediterranean Water through the Strait of Gibraltar. *Progress in Oceanography*, **21**, 1–105.

FARMER, D. M. and H. J. FREELAND (1983) The physical oceanography of fjords. *Progress in Oceanography*, **12**, 147–219.

FARADAY, M. (1832) Bakerian Lecture—Experimental researches in electricity. *Philosophical Transactions*, Royal Society of London, Part 1, 163–177.

FAVORITE, F., A. J. DODIMEAD and K. NASU (1976) *Oceanography of the Subarctic Pacific Region, 1960–71*. Bulletin 33, International North Pacific Fisheries Commission, p. 187.

FEDEROV, K. N. (1976) *The Thermohaline Finestructure of the Ocean* (Trans. D. A. BROWN), Pergamon, p. 170.

FIRING, E. (1981) Current profiling in the NORPAX Tahiti Shuttle. *Tropical Ocean-atmosphere Newsletter*, No. 5, January 1981, pp. 1, 8, 9. (Unpublished manuscript.)

FIRING, E. (1987) Deep zonal currents in the central equatorial Pacific. *Journal of Marine Research*, **45**, 791–812.

FIRING, E. (1989) Mean zonal currents below 1500 m near the equator, 159°W. *Journal of Geophysical Research*, **94**, 2023–2028.

FIRING, E., R. LUKAS, J. SADLER and K. WYRTKI (1983) Equatorial Undercurrent disappears during 1982/83 El Niño. *Science*, **212**, 1121–1123.

FLEMING, R. H. (1939) Tables for Sigma-T. *Journal of Marine Research*, **2**, 9–11.

FOFONOFF, N. P. (1956) Some properties of sea water influencing the formation of Antarctic bottom water. *Deep-Sea Research*, **4**, 32–35.

FOFONOFF, N. P. (1977) Computation of potential temperature of seawater for an arbitrary reference pressure. *Deep-Sea Research*, **24**, 489–491.

FOFONOFF, N. P. (1981) The Gulf Stream System. Ch. 4, pp. 112–139, in *Evolution of Physical Oceanography*, B. A. WARREN and C. WUNSCH (Eds.), MIT Press.

FOFONOFF, N. P. (1985) Physical properties of seawater: a new salinity scale and equation of state for seawater. *Journal of Geophysical Research*, **90**, 3332–3342.

FORCHHAMMER, G. (1865) On the composition of sea-water in the different parts of the ocean. *Philosophical Transactions*, Royal Society of London, **155**, 203–262.

FUGILISTER, F. C. (1955) Alternative analyses of current surveys. *Deep-Sea Research*, **2**, 213–229.

FUGILISTER, F. C. (1960) *Atlantic Ocean Atlas of Temperature and Salinity Profiles and Data from the I.G.Y. of 1957–58*. Woods Hole Oceanographic Institution Atlas Series No. 1, p. 209.

FUGILTER, F. C. and L. V. WORTHINGTON (1951) Some results of a multiple-ship survey of the Gulf Stream. *Tellus*, **3**, 1–14.

GARRETT, C. J. R. (1976) Generation of Langmuir circulations by surface waves—a feedback mechanism. *Journal of Marine Research*, **34**, 117–130.

GERSTNER, F. (1802) Theorie der Wellen u.s.w. Abh. Kgl. Böhm. Ges. Wiss. Prague.

GILL, A. E. (1973) Circulation and bottom water production in the Weddell Sea. *Deep-Sea Research*, **20**, 111–140.

GLOERSEN, P., H. J. ZWALLY, A. T. C. CHANG, D. K. HALL, W. J. CAMPBELL and R. O. RANSEIER (1978) Time dependence of sea-ice concentration and multi-year ice fraction in the Arctic Basin. *Boundary Layer Meteorology*, **13**, 339–359.

GODFREY, J. S. and K. R. RIDGWAY (1985) The large-scale environment of the poleward-flowing Leeuwin Current, western Australia: Longshore steric height gradients, wind stresses and geostrophic flow. *Journal of Physical Oceanography*, **15**, 481–495.
GORDON, A. L. and J. C. COMISO (1987) Recurring polynas over the Cosmonaut Sea and the Maud Rise. *Journal of Geophysical Research*, **92**, 2819–2833.
GORDON, A. L., D. T. GEORGI and H. W. TAYLOR (1972) Antarctic polar front zone in the western Scotia Sea—Summer 1975. *Journal of Physical Oceanography*, **7**, 309–328.
GUELKE, R. W. and C. A. SCHONTE-VANNECK (1947) The measurement of sea-water velocities by electromagnetic induction. *Journal of the Institute of Electrical Engineering*, **94**, 71–74.
HALPERN, D. (1987) Observations of annual and El Niño thermal and flow variations at 0°, 110°W and 0°, 95°W during 1980–1985. *Journal of Geophysical Research*, **92**, 8197–8212.
HAMON, B. V. (1955) A temperature–salinity–depth recorder. *Conseil Permanent International pour l'Exploration de la Mer, Journal du Conseil*, **21**, 22–73.
HAMON, B. V. and N. L. BROWN (1958) A temperature–chlorinity–depth recorder for use at sea. *Journal of Scientific Instruments*, **35**, 452–458.
HAMON, B. V. and T. J. GOLDING (1980) Physical oceanography of the Australian region. Commonwealth Scientific and Industrial Research Organisation, Fisheries and Oceanography Report 1977–79, 1–8.
HANSEN, D. V. and C. A. PAUL (1984) Genesis and effects of long waves in the equatorial Pacific. *Journal of Geophysical Research*, **89**, 10,431–10,440.
HAYES, S. P., J. M. TOOLE and L. J. MANGUM (1983) Water-mass and transport variability in the equatorial Pacific. *Journal of Physical Oceanography*, **13**, 153–168.
HELLAND-HANSEN, B. (1916) Nogen hydrografiscke metodor. *Forh. Skandinaviske Naturforske möte*, **16**, 357–359.
HELLAND-HANSEN, B. (1934) The Sognefjord section, pp. 257–274 in the *James Johnstone Memorial Volume*, Liverpool University Press.
HELLERMAN, S. and M. ROSENSTEIN (1983) Normal monthly wind stress over the world ocean with error estimates. *Journal of Physical Oceanography*, **13**, 1093–1104.
HISARD, P. and C. HÉNIN (1984) On the weakening of the Equatorial Undercurrent during the 1982–83 ENSO event. *Tropical Ocean-Atmosphere Newsletter*, No. 26, 1–2. (Unpublished manuscript.)
HOWE, B. M. and P. F. WORCESTER (1987) Ocean acoustic tomography: mesoscale velocity. *Journal of Geophysical Research*, **92**, 3785–3805.
ISELIN, C.O'D. (1936) A study of the circulation of the western North Atlantic. *Papers in Physical Oceanography and Meteorology*, **4**, p. 101.
ISELIN, C.O'D. (1939) The influence of vertical and lateral turbulence on the characteristics of the waters at mid-depths. *Transactions, American Geophysical Union*, **20**, 414–417.
ISEMER, H. J. and L. HASSE (1987) *The Bunker Climate Atlas of the North Atlantic Ocean*, Vol. 2, Springer-Verlag, p. 252.
JARRIGE, F. (1973) Temperature inversions in the equatorial Pacific, pp. 47–53 in *Oceanography of the South Pacific 1972*, R. FRASER (Ed.), New Zealand National Commission for Unesco.
JENKINS, G. M. and D. G. WATTS (1969) *Spectral Analysis and its Applications*. Holden-Day, p. 525.
KESSLER, W. S. and B. A. TAFT (1987) Dynamic heights and zonal geostrophic transport in the central tropical Pacific during 1979–84. *Journal of Physical Oceanography*, **17**, 97–122.
KETCHUM, B. H. (Ed.) (1983) Estuarine and enclosed seas. *Ecosystems of the World 26*, Elsevier, p. 500.
KINDER, T. H. and H. L. BRYDEN (1987) The 1985–86 Gibraltar Experiment. Data collection and preliminary results. *Eos, Transactions, American Geophysical Union*, **68**, 786–795.
KIRWAN, A. D., G. MCNALLY and S. PAZAN (1978) Wind drag and relative separations of undrogued drifters. *Journal of Physical Oceanography*, **8**, 1146–1150.
KNUDSEN, M. (1900) Ein hydrographische Lehrsatz. *Annalen der Hydrographie und Marinen Meteorologie*, **28**, 316–320.
KNUDSEN, M. (Ed.) (1901) *Hydrographical Tables*. G.E.C. Gad, Copenhagen, p. 63.
KOSSINNA, E. (1921) Die Tiefen des Weltmeeres. *Institute für Meereskunde, Veroff. Geogr. naturwiss.*, **9**, 70.
KREMLING, K. (1972) Comparison of specific gravity in natural sea-water from hydrographical tables and measurements by a new density instrument. *Deep-Sea Research*, **19**, 377–383.
KULP, J. L., L. E. TRYON, W. R. ECKELMANN and W. A. SNELL (1952) Lamont natural radiocarbon measurements. *Science*, **116**, 409–414.

LAEVASTU, T. (1963) Energy exchange in the North Pacific; its relations to weather and its oceanographic consequences. Part 1: Formulas and nomograms for computation of heat exchange components over the sea. Hawaii Institute of Geophysics Report No. 29, 15 pp., 11 nomograms.
LANGMUIR, I. (1938) Surface motion of water induced by wind. *Science*, **87**, 119–123.
LAPLACE, P. S. (1775) Recherches sur plusiers points du système du monde. *Mémoires de l'Academie Royale des Sciences, Paris*, **88**, 75–182.
LAZIER, J. R. N. (1973) The renewal of Labrador Sea water. *Deep-Sea Research*, **20**, 341–353.
LEGECKIS, R. (1977) Long waves in the eastern equatorial Pacific Ocean: a view from a geostationary satellite. *Science*, **197**, 1181–1197.
LEGECKIS, R., W. PICHELL and G. NESTERCZUK (1983) Equatorial long waves in geostationary satellite observations and in a multi-channel sea surface temperature analysis. *Bulletin of the American Meteorological Society*, **64**, 133–139.
LEVITUS, S. (1982) *Climatological Atlas of the World Ocean*, NOAA Prof. Paper No. 13, U.S. Government Printing Office, Washington, DC, 173 pp.
LEVITUS, S. (1986) Annual cycle of salinity and salt storage in the World Ocean. *Journal of Physical Oceanography*, **16**, 322–343.
LEVITUS, S. (1987) A comparison of the annual cycle of two sea surface temperature climatologies of the world ocean. *Journal of Physical Oceanography*, **17**, 197–214.
LEVITUS, S. (1988) Ekman volume fluxes for the world ocean and individual ocean basins. *Journal of Physical Oceanography*, **18**, 271–279.
LEWIS, E. L. (1980) The Practical Salinity Scale 1978 and its antecedents. *IEEE Journal of Oceanic Engineering*, **OE-5**, 3–8.
LEWIS, E. L. and N. P. FOFONOFF (1979) A practical salinity scale. *Journal of Physical Oceanography*, **9**, 446.
LEWIS, E. L. and R. G. PERKIN (1978) Salinity: its definition and calculation. *Journal of Geophysical Research*, **83**, 466–478.
LEWIS, M. R., N. KURING and C. YENTSCH (1988) Global patterns of ocean transparency:implications for the new production of the open ocean. *Journal of Geophysical Research*, **93**, 6847–6856.
LINDSTROM, E., R. LUKAS, R. FINE, E. FIRING, S. GODFREY, G. MEYERS and M. TSUCHIYA (1987) The western equatorial Pacific Ocean circulation study. *Nature*, **330**, 533–537.
LONGUET-HIGGINS, M. S., M. E. STERN and H. STOMMEL (1954) The electric field induced by ocean currents and waves with applications to the method of towed electrodes. *Papers in Physical Oceanography and Meteorology*, **13**, No. 1, p. 37.
LUKAS, R. (1986) The termination of the Equatorial Undercurrent in the eastern Pacific. *Progress in Oceanography*, **16**, 63–90.
LYNN, R. J. and J. L. REID (1968) Characteristics and circulation of deep and abyssal waters. *Deep-Sea Research*, **15**, 577–598.
MALMGREN, F. (1927) On the properties of sea-ice. *Norwegian North Polar Expedition with the Maud, 1918–1925, Scientific Research*, **1**, No. 5, p. 67.
MAMEYEV, O. I. (1975) *Temperature–Salinity Analysis of World Ocean Waters*. Elsevier, p. 374.
MANN, C. R. (1967) The termination of the Gulf Stream and the beginning of the North Atlantic Current. *Deep-Sea Research*, **14**, 337–359.
MANTYLA, A. W. (1975) On the potential temperature in the abyssal Pacific Ocean. *Journal of Marine Research*, **33**, 341–354.
MAURY, M. F. (1874) *The Physical Geography of the Sea*. Nelson, p. 493.
MAYES, D. (1989) U.S. WOCE program ready to begin. *Eos, Transactions, American Geophysical Union*, **70**, 874–885.
MENARD, H. W. and S. M. SMITH (1966) Hypsometry of ocean basin provinces. *Journal of Geophysical Research*, **71**, 4305–4325.
MERZ, A. and G. WÜST (1922) Die Atlantische Vertikalzirkulation. *Zeitschrift der Gesellschaft für Erdkunde zu Berlin*, Jahrgang 1922, pp. 1–35.
METCALF, W. G. (1960) A note on water movement in the Greenland–Norwegian Sea. *Deep-Sea Research*, **7**, 190–200.
METCALF, W. G., A. D. VOORHIS and N. C. STALCUP (1962) The Atlantic Equatorial Undercurrent. *Journal of Geophysical Research*, **67**, 2499–2508.
MILLERO, F. J. (1967) High precision magnetic float densimeter. *Review of Scientific Instruments*, **38**, 1441–1444.

MILLERO, F. J., C.-T. CHEN, A. BRADSHAW and K. SCHLEICHER (1980) A new high pressure equation of state for sea-water. *Deep-Sea Research*, **27A**, 255–264.
MILLERO, F. J., A. GONZALEZ and G. K. WARD (1976) The density of sea-water solutions as a function of temperature and salinity. *Journal of Marine Research*, **34**, 61–93.
MILLERO, F. J., G. PERRON and J. E. DESNOYERS (1973) The heat capacity of sea-water solutions from 5 to 35°C and 0.5 to 22‰ chlorinity. *Journal of Geophysical Research*, **78**, 4499–4507.
MIZEX GROUP (1989) MIZEX EAST 83/84: The summer marginal ice zone program in the Fram Strait/Greenland Sea. *Eos, Transactions, American Geophysical Union*, **67**, 513–517.
MODE GROUP (1976) Ocean eddies. *Oceanus*, **19**, 1–86.
MODE GROUP (1978) The Mid-Ocean Dynamics Experiment. *Deep-Sea Research*, **25**, 859–910.
MONTGOMERY, R. B. (1938) Circulation in the upper layer of the southern North Atlantic deduced with the aid of isentropic analysis. *Papers in Physical Oceanography and Meteorology*, **6**, No. 2, p. 55.
MONTGOMERY, R. B. (1958) Water characteristic of Atlantic Ocean and of World Ocean. *Deep-Sea Research*, **5**, 134–148.
MONTGOMERY, R. B. and WARREN S. WOOSTER (1954) Thermosteric anomaly and the analysis of serial oceanographic data. *Deep-Sea Research*, **2**, 63–70.
MUNK, W. H., W. C. O'REILLY and J. L. REID (1988) Australia–Bermuda sound transmission experiment (1960) revisited. *Journal of Physical Oceanography*, **18**, 1876–1898.
MURRAY, J. W., H. W. JANNASCH, S. HONSO, R. F. ANDERSON, W. S. REEBURGH, Z. TOP, G. E. FRIEDRICH, L. A. COLDISPOTI and E. IZDAR (1989) Unexpected changes in the oxic/anoxic interface in the Black Sea. *Nature*, **338**, 411–413.
NAMIAS, J. (1972) Large-scale and long-term fluctuations in some atmospheric and oceanic variables, pp. 27–48 in Nobel Symposium 20: *The Changing Chemistry of the Oceans*, D. DRYSSEN and D. JAEGER (Eds.), Wiley-Interscience.
NAMIAS, J. (1975) Stabilization of atmospheric circulation patterns by sea-surface temperature. *Journal of Marine Research*, **33** (Supplement), 53–60.
NANSEN, F. (1912) Die Bodenwasser und die Abkülung des Meeres. *Internationale Revue der gesamten Hydrobiologie und Hydrographie*, **5**, No. 1, p. 42.
NEIMANN, A. C. and D. A. GILL (1961) Circulation of the Red Sea in early summer. *Deep-Sea Research*, **8**, 223–235.
NEWTON, I. (1687) *Philosophia naturalis principia mathematica*, London.
NINNIS, R. M., W. J. EMERY and M. J. COLLINS (1986) Automated extraction of pack-ice motion from Advanced Very High Resolution Radiometer imagery. *Journal of Geophysical Research*, **91**, 10,725–10,734.
NOWLIN, W. D., T. WHITWORTH and R. D. PILLSBURY (1977) Structure and transport of the Antarctic Circumpolar Current from short-term measurements. *Journal of Physical Oceanography*, **7**, 787–802.
OLSON, D. B., R. W. SCHMITT, M. KENNELLY and T. M. JOYCE (1985) A two-layer diagnostic model of the long-term physical evolution of warm-core ring 82B. *Journal of Geophysical Research*, **90**, 8813–8822.
OORT, A. H. and T. H. VANDER HAAR (1976) On the observed annual cycle in the ocean–atmosphere heat balance over the northern hemisphere. *Journal of Physical Oceanography*, **6**, 781–800.
PARKER, C. E. (1971) Gulf Stream rings in the Sargasso Sea. *Deep-Sea Research*, **18**, 981–993.
PETERSON, R. G. (1988) On the transport of the Antarctic Circumpolar Current through Drake Passage and its relation to wind. *Journal of Geophysical Research*, **93**, 13,993–14,004.
PHILANDER, S. G. H. (1978) Instability of zonal equatorial currents. *Journal of Geophysical Research*, **83**, 3679–3682.
PHILANDER, S. G. H. (1989) *El Niño, La Niña and the Southern Oscillation*. Academic Press, San Diego, p. 283.
PHILANDER, S. G. H. and PACANOWSKI, R. C. (1986) The mass and heat budget of the tropical Atlantic Ocean. *Journal of Geophysical Research*, **91**, 14,212–14,220.
PHILANDER, S. G. H., W. J. HURLIN and A. D. SIEGEL (1987) Simulation of the seasonal cycle of the tropical Pacific Ocean. *Journal of Physical Oceanography*, **17**, 1986–2002.
PICKARD, G. L. (1961) Oceanographic features of inlets in the British Columbia mainland coast. *Journal, Fisheries Research Board of Canada*, **18**, 907–999.

PICKARD, G. L. (1977) A review of the physical oceanography of the Great Barrier Reef and Western Coral Sea. *Australian Institute of Marine Science*, Monograph No. 2, p. 134.

PICKARD, G. L. and B. R. STANTON (1980) Pacific fjords—a review of their water characteristics, pp. 1–51 in *Fjord Oceanography*, H. J. FREELAND, D. M. FARMER and C. D. LEVINGS (Eds.), Plenum Press.

PICKARD, G. L., J. C. ANDREWS and E. WOLANSKI (1989) A review of the physical oceanography of the Great Barrier Reef 1976–86. *Australian Institute of Marine Science and Great Barrier Reef Marine Park Authority*. (In press.)

POLLACK, M. J. (1958) Frequency distribution of potential temperature and salinities in the Indian Ocean. *Deep-Sea Research*, **5**, 128–133.

POLLARD, R. T. (1977) Observations and theories of Langmuir circulations and their role in near-surface mixing, pp. 235–251 in *A Voyage of Discovery: George Deacon 70th Anniversary Volume*, Supplement to *Deep-Sea Research*.

PREISENDORFER, R. W. (1986) Secchi disk science: visual optics of natural waters. *Limnology and Oceanography*, **31**, 909–926.

PRITCHARD, D. W. (1952) Estuarine hydrography. *Advances in Geophysics*, **1**, 243–280.

RASMUSSEN, E. M. and T. H. CARPENTER (1982) Variations in tropical sea-surface temperature and surface wind fields associated with the Southern Oscillation/El Niño. *Monthly Weather Review*, **11**, 354–384.

REED, R. K. (1985) An estimate of the climatological heat fluxes over the tropical Pacific Ocean. *Journal of Climate and Applied Meteorology*, **24**, 833–840.

REID, J. L. (1959) Evidence of a South Equatorial Countercurrent in the Pacific Ocean. *Nature*, **184**, 209–210.

REID, J. L. (1965) Intermediate Waters of the Pacific Ocean. *Johns Hopkins Oceanographic Studies*, No. 2, p. 85.

REID, J. L. (1973) Northwest Pacific Ocean waters in winter. *Johns Hopkins Oceanographic Studies*, No. 5, p. 96.

REID, J. L. and R. J. LYNN (1971) On the influence of the Norwegian–Greenland and Weddell Seas upon the bottom waters of the Indian and Pacific Oceans. *Deep-Sea Research*, **18**, 1063–1088.

RICHARDS, F. A. (1981) *Coastal Upwelling*, American Geophysical Union, Washington, DC, p. 529.

RICHARDSON, P. L. (1980) Gulf Stream ring trajectories. *Journal of Physical Oceanography*, **10**, 90–104.

RICHARDSON, P. L., R. E. CHENEY and L. V. WORTHINGTON (1978) A census of Gulf Stream rings, Spring 1975. *Journal of Geophysical Research*, **83**, 6136–6144.

RING GROUP (1981) Gulf Stream cold-core rings: their physics, chemistry and biology. *Science*, **212**, 1091–1100.

ROBINSON, A. R. and STOMMEL, H. (1959) The oceanic thermocline and associated thermohaline circulation. *Tellus*, **3**, 295–308.

ROSSBY, T. and D. WEBB (1970) Observing abyssal motion by tracking Swallow floats in the SOFAR channel. *Deep-Sea Research*, **17**, 359–365.

ROSSBY, T., D. DORSON and J. FONTAINE (1986) The RAFOS system. *Journal of Atmospheric and Oceanic Technology*, **3**, 672–679.

ROTSCHI, H. (1970) Variation of equatorial currents, pp. 75–83 in *Scientific Exploration of the South Pacific*, W. S. WOOSTER (Ed.), U.S. National Academy of Science.

ROTSCHI, H. (1973) Hydrology at 170°E in the South Pacific, pp. 113–128 in *Oceanography of the South Pacific 1972*, R. FRASER (Ed.), New Zealand National Commission for Unesco.

ROUGERIE, F. (1986) Le lagon sud-ouest de Nouvelle-Calédonie: spécificité, hydrologique, dynamique et productivité. *Éditions de l'ORSTOM, Institut Français de Recherche Scientifique pour de Développement en Cooperation, Collection ÉTUDES et THÈSES*, p. 234.

SANKEY, T. (1973) The formation of deep water in the Northwestern Mediterranean. *Progress in Oceanography*, B. A. WARREN (Ed.), **6**, 159–179.

SAVCHENKO, V. G., W. G. EMERY and O. A. VLADIMIROV (1978) A cyclonic eddy in the Antarctic Circumpolar Current south of Australia: results of Soviet-American observations aboard the R/V *Professor Zubov*. *Journal of Physical Oceanography*, **8**, 825–837.

SCHMITT, R. W. and D. B. OLSON (1985) Wintertime convection in warm-core rings: thermocline ventilation and the formation of mesoscale lenses. *Journal of Geophysical Research*, **90**, 8825–8837.

SCHMITT, R. W., P. S. BOGDEN and C. E. DORMAN (1989) Evaporation minus precipitation and density fluxes for the North Atlantic. *Journal of Physical Oceanography*, **19**, 1208–1221.

Bibliography

SMITH, E. H., F. M. SOULE and O. MOSBY (1937) *Marion and General Greene* expeditions to Davis Strait and Labrador Sea 1928–1935. Scientific results, part 2, physical oceanography. *U.S. Coast Guard Bulletin* No. 19, p. 259.
SMITH, R. L. (1968) Upwelling. *Oceanography and Marine Biology, Annual Reviews*, **6**, 11–46.
SMITH, S. D. (1988) Coefficients for sea surface wind stress, heat flux and wind profiles as a function of wind speed and temperature. *Journal of Geophysical Research*, **93**, 15,467–15,472.
STOKES, G. (1847) On the theory of oscillatory waves. *Cambridge Transactions*, **8**, pp. 212.
STOMMEL, H. (1948) The westward intensification of wind-driven currents. *Transactions, American Geophysical Union*, **29**, 202–206.
STOMMEL, H. (1958) The abyssal circulation. *Deep-Sea Research*, **5**, 80–82.
STOMMEL, H. and A. B. ARONS (1960a) On the abyssal circulation of the World Ocean, I. Stationary planetary flow patterns on a sphere. *Deep-Sea Research*, **6**, 140–154.
STOMMEL, H. and A. B. ARONS (1960b) On the abyssal circulation of the World Ocean, 11. An idealized model of the circulation pattern and amplitude in oceanic basins. *Deep-Sea Research*, **6**, 217–233.
STOMMEL, H., P. NIILER and D. ANATI (1978) Dynamic topography and recirculation of the North Atlantic. *Journal of Marine Research*, **36**, 449–468.
STOMMEL, H. and K. YOSHIDA (Eds.) (1972) *Kuroshio: Physical Aspects of the Japan Current*. University of Washington Press, p. 517
STRAMMA, L., P. CORNILLON, R. A. WOLLER, J. F. PRICE and M. G. BRISCOE (1986) Large diurnal sea surface temperature variability: satellite and in situ measurements. *Journal of Physical Oceanography*, **16**, 827–837.
STRICKLAND, J. D. H. and T. R. PARSONS (1972) *A Practical Handbook of Sea-water Analysis*. Fisheries Research Board of Canada, Bulletin 167 (Second Edition), p. 310.
STUIVER, M., P. D. QUAY and H. G. OSTLUND (1983) Abyssal ^{14}C and the age of the world oceans. *Science*, **219**, 849–851.
SWALLOW, J. C. (1955) A neutral-buoyancy float for measuring deep currents. *Deep-Sea Research*, **3**, 74–81.
SWALLOW, J. C. and L. V. WORTHINGTON (1957) Measurements of deep currents in the western North Atlantic. *Nature*, **179**, 1183–1184.
SWALLOW, J. C. and L. V. WORTHINGTON (1961) An observation of a deep counter-current in the western North Atlantic. *Deep-Sea Research*, **8**, 1–19.
TAIT, R. I. and M. R. HOWE (1968) Some observations of thermohaline stratification in the deep ocean. *Deep-Sea Research*, **15**, 275–280.
TAIT, R. I. and M. R. HOWE (1971) Thermohaline staircase. *Nature*, **231**, 178–179.
THOMPSON, R. O. R. Y. (1987) Continental-shelf-scale model of the Leewin Current. *Journal of Marine Research*, **45**, 813–827.
THOULET, J. and A. CHEVALLIER (1889) Sur la chaleur spécifique de l'eau de mer à divers degrés de dilution et de concentration. *Comptes Rendues de l'Académie des Sciences, Paris*, **108**, 794–796.
TSUCHIYA, M. (1968) Upper waters of the intertropical Pacific Ocean. *The Johns Hopkins Oceanographic Studies*, No. 4, The Johns Hopkins Press, 50 pp.
TULLY, J. P. (1949) Oceanography and prediction of pulp-mill pollution in Alberni Inlet. *Fisheries Research Board of Canada*, Bulletin 83, pp. 169.
TURNER, J. S. (1974) Double-diffusive phenomena. *Annual Review of Fluid Mechanics*, **6**, 37–56.
TURNER, J. S. (1981) Small-scale mixing processes. Ch. 8, pp. 236–262, in *Evolution of Physical Oceanography*, B. A. WARREN and C. WUNSCH (Eds.), MIT Press.
UNESCO (1983) Algorithms for computation of fundamental properties of seawater. *Unesco technical papers in marine science*, No. 44, p. 53.
VAN RIEL, P. M. (1934) The bottom configuration in relation to the flow of bottom water. *Snellius Expedition*, Vol 2, Ch. 2, p. 63.
VASTANO, A. C. and S. E. BORDERS (1984) Sea surface motion over an anticyclonic eddy on the Oyashio Front. *Remote Sensing of the Environment*, **16**, 87–90.
WALLACE, W. J. (1974) *The Development of the Chlorinity/salinity Concept in Oceanography*. Elsevier, p. 227.
WARREN, B. A. (1973) Transpacific sections at latitudes 43°S and 28°S: the *Scorpio* Expedition, II. Deep water. *Deep-Sea Research*, **20**, 9–38.

WARREN, B. A. (1981) Deep circulation of the world ocean. Ch. 1, pp. 6–41, in *Evolution of Physical Oceanography*, B. A. WARREN and C. WUNSCH (Eds.), MIT Press.
WERTHEIM, G. K. (1954) Studies of the electrical potential between Key West, Florida and Havana, Cuba. *Transactions, American Geophysical Union*, **35**, 872–882.
WILLIAMS, A. J. (1975) Images of ocean microstructure. *Deep-Sea Research*, **22**, 811–829.
WILSON, B. W. (1960) Speed of sound in sea-water as a function of temperature, pressure and salinity. *Journal of the Acoustical Society of America*, **32**, 641–644.
WILSON, T. R. S. (1975) Salinity and the major elements in sea-water. Ch. 6, pp. 365–413 in *Chemical Oceanography*, Vol. 1, 2nd edn., J. P. RILEY and G. SKIRROW (Eds.), Academic Press.
WOOSTER, W. S. and J. L. REID (1963) Eastern boundary currents. Ch. 11, pp. 253–280, in *The Sea: Ideas and Observations*, Vol. 11, M. N. HILL (Ed.), Wiley-Interscience.
World Ocean Circulation Experiment Implementation Plan, Vols. 1 and 2, WCRP–11, July, 1988. WOCE International Planning Office, Wormley, England.
WORTHINGTON, L. V. (1954) Preliminary note on the time-scale in the North Atlantic circulation. *Deep-Sea Research*, **1**, 244–251.
WORTHINGTON, L. V. (1955) A new theory of Caribbean bottom water formation. *Deep-Sea Research*, **3**, 82–87.
WORTHINGTON, L. V. (1976) On the North Atlantic circulation. *Johns Hopkins Oceanographic Studies*, No. 6, p. 110.
WORTHINGTON, L. V. (1981) The water masses of the world ocean: some results of a fine-scale census. Ch. 2, pp. 42–69, in *Evolution of Physical Oceanography*, B. A. WARREN and C. WUNSCH (Eds.), MIT Press.
WUNSCH, C. (1981) Low frequency variability in the sea. Ch. 11, pp. 342–375, in *Evolution of Physical Oceanography*, B. A. WARREN and C. WUNSCH (Eds.), MIT Press.
WÜST, G. (1935) Die Stratosphäre. *Wissenshaftliche Ergebnisse der Deutschen Atlantischen Expedition* Meteor *1925–1927*, **6**, 1 Teil, 2 Lief., p. 180. (*The Stratosphere of the Atlantic Ocean*, W. J. EMERY (Ed.), 1978, Amerind, p. 112.)
WÜST, G. (1957) Stromgeschwindigkeiten und Strommengen in den Tiefen des Atlantischen Ozeans. *Wissenschaftliche Ergebnisse der Deutschen Atlantischen Expedition* Meteor *1925–1927*, **6**, 261–420.
WÜST, G. (1961) On the vertical circulation of the Mediterranean Sea. *Journal of Geophysical Research*, **66**, 3261–3271.
WÜST, G. and A. DEFANT (1936) Atlas zur Schichtung und Zirkulation des Atlantischen Ozeans. *Wissenschaftliche Ergebnisse der Deutschen Atlantischen Expedition* Meteor *1925–1927*, **6**, 103 plates.
WYRTKI, K. (1961) The thermohaline circulation in relation to the general circulation of the oceans. *Deep-Sea Research*, **8**, 39–64.
WYRTKI, K. (1965) The average annual heat balance of the North Pacific Ocean and its relation to the ocean circulation. *Journal of Geophysical Research*, **70**, 4547–4559.
WYRTKI, K. (1971) *Oceanographic Atlas of the International Indian Ocean Expedition*. National Science Foundation, Wash., DC, p. 531.
WYRTKI, K. (1973) Physical oceanography of the Indian Ocean, pp. 18–36 in *The Biology of the Indian Ocean*, Springer-Verlag.
WYRTKI, K. (1974) Sea level and the seasonal fluctuations of the equatorial currents in the western Pacific Ocean. *Journal of Physical Oceanography*, **4**, 91–103.
WYRTKI, K. (1975) *El Niño*—the dynamic response of the Pacific Ocean to atmospheric forcing. *Journal of Physical Oceanography*, **5**, 572–584.
WYRTKI, K. (1980) The Hawaii–Tahiti Shuttle Experiment. *Tropical Ocean–atmosphere Newsletter*, No. 4, Oct. 1980, pp. 7, 8. (Unpublished manuscript.)
WYRTKI, K., E. FIRING, D. HALPERN, R. KNOX, G. J. MCNALLY, W. C. PATZERT, E. D. STROUP, B. A. TAFT and R. WILLIAMS (1981) The Hawaii to Tahiti Shuttle Experiment. *Science*, **211**, 22–28.
WYRTKI, K. and B. KILONSKY (1984) Mean water and current structure during the Hawaii–Tahiti Shuttle Experiment. *Journal of Physical Oceanography*, **14**, 242–254.
YOUNG, F. B., H. GERARD and W. JEVONS (1920) On electrical disturbances due to tides and waves. *Philosophical Magazine*, Series 6, **40**, 149–159.

Index

Absorption
  atmosphere 72, 73
  water 30–32
Accuracy, definition 11
Acoustic Doppler Current Profiler 111
Acoustic Tomography 123, 124, 295
Adiabatic change 19
Adjacent seas, circulation 218, 219
Advection 41, 69, 187
Age of water 137–139, 269, 293
Aircraft 94, 128, 188
Albedo 231
Anchoring, deep 112
Anemometers 110
Anoxic conditions 211, 214
Antarctic
  Convergence 173
  Polar Front 173, 174
  Polar Frontal Zone 174, 175, 180, 198
  Zone 173, 177, 178, 180
Antarctic Ocean, currents, water masses see
  Southern Ocean
Anticyclonic, definition 172
Arabian Sea 157, 273–275
Arctic Sea 5, 78, 215, 219–226, 228, 286
  budgets 225, 226
  Canadian Basin 220, 221, 223–225
  circulation 215
  Eurasian Basin 220–222, 224, 225
  water masses
    Arctic 220–223
    Atlantic 221–224
    Deep or Bottom 150, 221, 222, 224, 225
    T-S diagram 222, 224
ARGOS system 105, 106, 136
Atlantic Ocean 5, 6, 162, 163, 180–208
  currents, water masses see under
    Equatorial, North Atlantic, South Atlantic
Atolls see Reefs
Attenuation coefficient 30, 31, 125–127

AVHRR 130, 131, 132
AXBT 102, 129, 295

Baffin Bay 215–218, 220
Baffin Deep Water 217, 218
Baffin Land Current 216, 232
Baltic Sea 208, 213, 214, 218, 219
Baroclinic, barotropic, definitions 117
Basin 3, 9
  Canadian 220, 223–225
  Eurasian 220–225
Bathythermograph 98
Bay of Bengal 157, 273, 274, 277, 286
Beach 7
Benguela Current see South Atlantic
Bering Sea 5, 223, 253
Bering Strait 183, 220, 223
Biological processes 57, 251, 257, 258, 261, 279
Black Sea 58, 67, 68, 150, 208, 211–213, 218
  Deep water 211
Bottle cast 128, 167, 295
Bottom, deep-sea 9, 176
Bottom material 10, 11
Bottom water, definition 172
Bowen's Ratio 81–83
Box model 186
Brazil Current see South Atlantic
Break-in-slope 8
Brunt-Väisälä frequency 17, 56, 167
Budget
  heat 68–91
  salt 65, 66
  volume 64–68
  see also Arctic, Black, Mediterranean Seas
Bulk modulus 22
Buoy
  drifting 36, 94, 103, 105, 106, 248, 251
  satellite tracked 103, 105, 253

Index

Buoyancy 165
  flux 165, 167, 170
  frequency 56

Cabbeling 169, 170
California Current *see* North Pacific
Carbon-14 138
Caribbean Sea 5, 185, 218
Challenger Expedition 3, 191, 247, 294
Characteristic diagrams 144–153
Chlorinity 14, 15, 98
Circulation 1, 56, 170
  abyssal 41
  deep water *see* Atlantic, Pacific, Indian Oceans
  estuarine 212, 213
  gyral 164
  lagoon 289, 290
  mechanisms 154–173
  thermohaline 154–158, 164, 190
  upper-layer 171, 172
  wind-driven 158–164
Climate 292
  and SST 254, 255
Cloud factor 73, 74, 77
Coastal waters 3, 276–290
Coastal Zone Colour Scanner (CZCS) 132
Cold wall 190
Colour of sea-water 32, 33
Common Water 269, 274
Compressibility 19, 22, 25, 49, 64, 199
Condensation 68
Conduction
  eddy 81
  heat 68, 81, 83
Conductivity, electrical 13–16
Conductivity Ratio 99
Conservation (budget)
  of heat 68–91
  of salt 66
  of volume 64–68
Conservative properties 23
Continental
  rise 7, 9
  shelf 7, 8, 278, 289
  slope 7, 8
Continuity, equation of 64
Convection 81, 154, 165
Convergence 42–44, 61, 160, 248–251
  Subtropical 203, 266, 270
Coral reef 278, 286–290
  atolls 286, 287, 290
  currents 288
  lagoon circulation 289
  residence times 290
  tides 288
  topography 286, 287

types 287
  water properties 287, 288
Core method 140, 147, 148, 224
Coriolis force 93, 116, 159, 160, 161, 164, 175, 230, 246, 248, 280
Costa Rica Thermal Dome 265
CTD 16, 96, 98, 100, 128, 168, 295
Current
  data plots 142, 143
  eastern boundary 256–263
  Eulerian description, measurement 102, 107–112
  geostrophic 41, 164
  Lagrangian description, measurement 102–107
  measurement 102–112, 124, 135
  meters 92, 102–115
  western boundary 41, 102, 164
  see also under ocean regions
Cyclogenesis 255
Cyclonic, definition 157

Data presentation 139–153
Davis Strait 217
Deep scattering layer 25
Deep-sea bottom 7, 9
Deep Water
  flows 172
  formation, volumes 293
  zone, definition 172
Density, sea-water 13, 15–23, 299
  distribution 49, 51, 53, 54, 56, 57
  measurement 22, 101
  potential 20, 49, 56, 143, 196, 197
  relative 299
Depth
  ocean measurement 95
  of frictional influence 160
  of no motion 117, 183
Determination, definition 12
Diagrams, characteristic 144–153
Dicothermal layer 45
Diel migration 25
Diffusion
  double 168, 169, 216
  eddy 41, 60, 79, 80, 187, 191
  molecular 79, 165, 168, 169
Dimensions, ocean 5–10
Discontinuity layer 265
Divergence 43, 61, 248–251, 253, 279
Doldrums 240
Doppler Current Meter, Acoustic 110, 111
Downwelling 42, 43, 61, 160
Drag, Chesapeake Bay 108
Drag coefficient 81
Drake Passage 174–178
Drifters, drift pole 103

Index

Drogues 103, 104
Dynamic height 93, 115, 117, 119, 244
Dynamic metre 120
Dynamic topography 93, 117, 119–122, 243
Dynamical oceanography 3

East Australian Current *see* South Pacific
Eastern boundary currents 256–263
Echo sounding 10, 24, 25, 29
Eddies 94, 175, 187, 188, 195, 248, 256, 272, 273, 278, 290, 293, 294
Eddy
 coefficients 83, 84
 conduction 81
 transfer 82
 viscosity, friction 160, 161
18° Water 190, 193, 294
Ekman
 circulation 159–163, 279, 280
  shallow water 280, 281
 current meter 108
 depth 160, 280, 281
 flow, transport 160, 248, 250, 279
 layer 160, 161, 279, 280, 281
 pumping 161
 spiral 159–161, 280, 281
 volume flux 161–163
Electromagnetic current meter 111, 112
El Niño 244, 258–262, 292
Energy
 long-wave 75–78
 short-wave 30, 71–75
ENSO phenomenon 4, 258, 259, 262
Equation of state (EOS 80) 17, 101
Equatorial
 currents
  Atlantic
   Equatorial Undercurrent 172, 184, 247
   North Equatorial 185
   North Equatorial Counter 172, 184
   North and South Counter 184
   South Equatorial 172, 185
  Indian
   Equatorial Counter 270
   Equatorial Undercurrent 172, 270
   North Equatorial 270
   South Equatorial 270, 272
   South-west Monsoon 272
  Pacific
   Equatorial Intermediate 244
   Equatorial Undercurrent 172, 235, 240–247, 249, 258, 262
   Intermediate and Deep 251, 252
   North Equatorial 86, 172, 235, 243, 244, 252, 266

   North Equatorial Counter 235, 243, 244, 248, 258, 262
   North Subtropical Counter 241, 243, 245
   South Equatorial 172, 235, 243–246, 248, 256
   South Equatorial Counter 240, 244, 258, 262
   South Subtropical Counter 241, 243
  speeds 238, 241, 243, 245, 251, 252
  variations, spatial, temporal 242–245
  volume transports 241, 243, 244, 245, 251, 252
 water masses
  Pacific Equatorial 263, 266
  Indian Equatorial 273
  meridional sections, currents, water properties, Pacific 240–242, 249, 250
 region, definition 37
 water
  Indian 205, 273
  Pacific 204, 263, 265
Errors, systematic, random 12
Estimation, definition 12
Estuary 57, 281–286
 circulation 65, 218, 284–286
 types 281–284
Evaporation 49, 51, 65–68, 78, 83, 84, 156, 158, 191, 211, 258, 277
 heat loss by 78–81
 minus precipitation (E-P) 49, 51, 208, 213, 218, 274

Falkland Current *see* South Atlantic
FGGE 90
Fjord 10, 57, 65, 218, 282, 284–286
Float
 neutrally buoyant 105, 106, 253
 'pop-up' 106
 RAFOS 106
 SOFAR 106
 Swallow 24, 105, 106, 190
Floor, ocean 7–11
Florida Current *see* North Atlantic
Flushing time 66, 209

GARP 3, 90
GATE 3, 90, 93
GEK 111
Geopotential 115, 117
GEOSAT 133
GEOSECS 138, 139, 196
Geostrophic
 equation 116, 117
 method, currents 22, 93, 115–122, 148, 157, 181, 186, 187, 190, 220, 238, 245, 280

315

Index

Gibraltar, Strait of 158, 167, 207, 209
Graphical presentation, data 139–153
Great Barrier Reef 278, 286, 288, 290
Greenland Sea 156, 198, 214–216, 224
 Deep Water 215, 216, 225
 Gyre 214, 216
Gulf of Mexico 185
Gulf Stream
 eddies and rings 131, 192–195
 system 185–195
 temperature and salinity 190–192, 201, 206
 volume transport 190
 see also North Atlantic Gyre 172

Halocline 52, 222, 274, 277, 282, 284
Hawaii-Tahiti Shuttle 238–242, 244, 251
Heat budget 4, 31
 Arctic 225, 226
 Atlantic 87, 88, 247
 Pacific 84–87, 245, 247
 terms 68–89
 variations 75–77
 World Ocean 88–91
Heat
 latent 78
 units 300
Hudson Bay 215, 217, 219
Hurricanes 255, 261
Hydrogen sulphide 56, 212, 213
Hydrography 2

Ice see Sea-ice
Icebergs 233, 234
IIOE 93
Image navigation 130
Indian Ocean 3, 5, 6, 162, 163, 270–274
 currents
  Agulhas 272
  Leeuwin 272, 273
  Somali 272
  Southwest Monsoon 272
  Surface 271
  volume transports 272
 dissolved oxygen 58
 nutrients 62
 water masses
  Bottom, Deep 157, 273, 274
  Central 158, 273
  High Salinity Intermediate 273
  T-S characteristics 205
  Upper 273
 winds 237, 270–272
Instruments and methods 92–153
 expendable 102

profiling 100, 102, 111
 satellite 130–137
Interleaving 169
Intertropical Convergence Zone 259, 273
Irminger Sea 198
Isentropic analysis 143
Isobar, definition 118
Isobath, definition 44, 117
Isopleth, definition 7
Isopycnal 117, 118, 143, 148
ISOS 176
Isotherm, definition 7

Kelvin-Helmholtz billows 167
Knudsen
 Relations 66
 Tables 21, 22, 101
 titration 98
Kuroshio see North Pacific

Labrador
 Current 189, 196, 232
 Sea 5, 156, 189, 198, 215–219
Lagoon 278, 286, 287, 289, 290
LANDSAT 134
Langmuir circulations 42–44, 165
La Niña 262
Latitude, low or high, definition 37
Layer, deep scattering 25
Leeuwin Current 272, 273
Light
 in the sea 30–33
 measurement 125–127
 penetration 62, 63

Mediterranean Sea 5, 51, 67, 208–211
 outflow 167, 168, 186, 200, 202, 206, 209, 218, 219
 volume budget 67
 water masses 150
  Deep and Bottom 211
  Levantine Intermediate 158, 209, 210, 218
  Western Deep Water 209
MEDS 105
Meridional distributions 60, 61, 197, 200, 238, 239, 267
Meteor Expedition 3, 41, 93, 157, 195
Meter wheel 95
Microwave sensors 132, 133
Mixed layer 26, 44, 97, 165, 166, 172, 245
Mixing
 convective 165, 227
 double diffusive 168–170, 216
 lateral 143, 187

processes 55, 165–170
ratio 83
triangle 145
vertical 56, 212, 218, 268, 277, 283, 285, 291
MIZEX 232
MODE 3, 93, 94, 195
Monsoon 161, 237, 243, 247, 270–274, 277
Moorings 112–115

New Guinea Coastal Countercurrent 242, 243
NIMBUS 105
North Atlantic Ocean 184–208
  adjacent seas 208–218
  inflow and outflow 218, 219
  currents
    Antilles 185, 190
    Canary 185, 257
    Deep 201
    East Greenland 214, 216, 218, 220, 224, 232
    Florida 185–187, 190, 196
    Gulf Stream 185–190, 200, 206
    Gyre 196
    North Atlantic 185–188, 218
    North Atlantic Gyre 185, 186
    North Equatorial 186
    North Equatorial Counter 184
    Upper layer 122, 182
    West Greenland 216
    West Spitsbergen 223
    volume transports 190
  heat budget 70, 86–88
  water masses 195–207
    Arctic Deep 149, 150
    Arctic Intermediate 203
    Bottom 198, 200
    Central 201
    Deep 139, 150, 157, 196, 198–200
    Northeast Atlantic Deep 57–61, 198
    Sigma-4 200
    Slope 192
    Surface 51, 195
    T-S characteristics 201, 205–208
    Upper 155
    Vertical sections 61, 191, 197, 200, 201
North Pacific Ocean 235–255
  currents
    Alaskan Gyre 253
    California 253, 257
    Kuroshio 86, 157, 190, 252–254
    Kuroshio Extension 252, 268
    North Pacific Current 252, 253, 266
    North Pacific Gyre 235
    Oyashio 253, 268
    Upper 121, 236, 252–254

  volume transports 252, 254
  dissolved oxygen 56, 57
  heat budget 70, 84–87
  water masses
    Bottom 268
    Central 263, 266
    Deep 268, 269
    Equatorial 263
    Intermediate 242, 267, 268
    Subarctic 266
    Subtropical 242
    Surface 51
    Tropical Upper 203, 206
    T-S characteristics 204, 263, 264, 268
    Upper 263–265
    vertical section 60, 267
Norwegian
  Current 214, 218
  Gyre 214
  Sea 5, 214, 215
  Sea Deep Water 225
Numerical modelling 296
Nutrients 59–62, 213, 226, 239, 242, 251, 257, 265, 273, 279

Ocean, two-layer 56
Oceanic Common Water 269, 274
Okta, definition 73
Overturn 19, 55, 60, 215, 217, 227
Oxic 211
Oxygen, dissolved 57, 58, 100, 137, 138, 177, 179, 180, 195, 197, 198, 211, 212, 214, 215, 217, 218, 239, 242, 246, 256, 257, 265, 268, 269, 273, 274, 285, 295
  maximum 57
  minimum 57, 58
  type distribution 58
  units 300
Oyashio see North Pacific

Pacific Ocean 5, 6, 7, 121, 162, 163
  currents, water masses see under Equatorial, North, South Pacific
  heat budget 84–87
Pendulum day 161
Persian Gulf 158, 273, 275
Peru Current see South Pacific
Platforms 127–137
Polar regions 37, 57, 70, 146, 172
POLYGON 3, 195
POLYMODE 3, 195, 294
Polynya 231, 233–235
PSS-78 16
Precipitation 49, 65, 67, 78, 134, 258, 259, 263, 277
Precision, definition 12

Pressure 95, 300
Profile 35, 139, 140, 142
    definition 140
    stepped 166, 167
Pycnocline 40, 44, 52, 53, 56, 209, 211, 245
Pyranometer 74, 124

Radar
    altimeter 133, 135
    scatterometer 133
    synthetic aperture (SAR) 133
Radiation
    back or long-wave 68, 70, 71, 75–78
    measurement 124–127, 129, 134
    reflection 73, 74, 78
    scattered 71–73
    short-wave or solar 68, 70–75, 254
    sky 73
    theory 70–78
    thermometer 97
Radiometer 125
    microwave 132, 294
RAFOS floats 106
Red Sea 51, 158, 218, 273, 274, 275
    brine pools 274
Reference level 117, 118
Reflection of radiation 28, 74
Refraction 24, 27–29
Remote sensing 129–137
Residence time 67, 68, 186, 218, 226, 275, 290
Reverberation 25, 28
Richardson number 167
Ridge
    Lomonosov 220
    Mid-Atlantic 198, 200
    Mid-Ocean 9
    pressure 230, 231
Rings 192–194, 293, 294
River runoff 64, 66, 196, 211–213, 217, 228, 263, 273, 277, 281, 284–286
Ross Sea 156, 157, 177

Salinity 13–19, 66
    absolute 14, 299
    maxima, minima 146, 147, 206, 267
    measurement 98–101
    Practical 14, 16, 99, 299
    surface 49–51
    type distributions 36, 49–53
Salinometer
    in situ 99, 100
    electrical, inductive 99, 100
    refractometer 99
Salt fingering 168, 169
SAR (Synthetic Aperture Radar) 129, 133, 135, 295

Sargasso Sea 63, 187, 190, 192, 193, 195
Satellite
    altimeter 133, 135, 293
    geostationary 130
    images 188, 194, 248
    instruments 130–137
    navigation 136, 137
    types 129–137
    weather 36, 105, 129–132
Scale, Practical Salinity 99
Sea-ice 46, 77, 133, 135, 217, 226–235, 294
    albedo 231
    break-up, build-up 214, 232, 233
    distribution 231, 232
    freezing process 228, 229
    physics 227–231
    pressure ridges 230, 231
    properties 92, 155, 228–230
    rafting 230
    salinity 228, 229
    types 228, 231, 232
Sea mount 9, 253
SEASAT 132, 133, 187, 193, 294
Sea-water 13–33
    colour 32, 33
    equation of state 17, 21, 22
    freezing point 17–19
    maximum density 18, 19
    properties 95–102
    Standard 99, 100
Secchi disc 125, 126
    depths, distribution 62, 63
Section 35, 140, 142
    definition 140
Sediments 10, 11
Shadow zone 26
Shear, current, flow 117, 165, 167, 212
Shelf break 8, 289
Ship of opportunity 36, 94, 102
Shore 7
Sigma-4 49, 199, 200
Sigma-t 17, 18, 19, 21, 23, 143, 300
    definition 17
    type distributions 17–19, 23, 49, 53
Sigma-theta 20, 49, 56, 199, 217
Sill, sill depth 10, 284
Sinking
    near-boundary 156
    open-ocean 156, 211
SMMR 132
Snellius Expedition 47–49
Snow 77
SOFAR
    channel 27, 28, 106
    float 106
Solar
    constant 72
    energy, radiation 68, 72
Somali Current see Indian Ocean

Index

SONAR 24–28
Sound 23–29
  ambient 25
  channel 27, 28, 123
  diffraction 29
  refraction
    lateral 29
    vertical 27, 28
  scattering 25
  speed 17, 22, 25–29, 123
  speed minimum 26–29
Sounding 9, 10, 24, 25, 29
South Atlantic Ocean 181–184
  currents
    Benguela 181, 183, 257
    Brazil 181, 183, 196
    Falkland 176, 181, 196
    South Equatorial 181
    Upper Layer 182
    volume transports 181, 183
  heat budget 88
  water masses
    Bottom 183
    Central 205
    Deep 196, 198
    Sigma-4 200
    Surface 51, 196
    T-S characteristics 203–205
    Upper 183
  vertical sections 61, 197, 200, 201
Southern (Antarctic) Ocean 5, 173–181
  currents
    (Antarctic) Circumpolar 157, 175, 176, 178, 179, 256
    East Wind Drift 175
    Upper layer 174
    vertical section 178
    volume transports 175, 176
    West Wind Drift 175
  water masses
    (Antarctic) Bottom 59, 138, 150, 170, 177, 180, 183, 198, 200
    (Antarctic) Circumpolar 177, 179, 180
    (Atlantic) Deep 179, 180, 183, 198, 200
    (Antarctic) Intermediate 179, 180, 183, 185, 198–200, 206, 266, 273
    Shelf 177
    (Antarctic) Surface 155, 173–175, 177
  T-S diagram 178
Southern Oscillation (Index) 258–260, 261, 262
South Pacific Ocean 87, 255–263
  currents
    Bottom 269
    East Australian 256, 289
    Gunther 256
    Gyre (South Pacific) 5, 255
    Peru 176, 256, 257, 263
    Peru Countercurrent 246
    Peru-Chile Undercurrent 246
    Surface 236, 255, 256
    Upper layer 236
    volume transports 268
  heat budget 87
  water masses
    Bottom 200, 268, 269
    Central 263, 266
    Deep 268, 269
    Intermediate 269
    Subtropical 242, 243
    Surface 51, 263
    Tropical Upper 206
    T-S characteristics 204, 266
    Upper layer 263
    Vertical sections 60, 267
South-west Monsoon Current 270
Specific heat 23
Specific volume 20, 300
  anomaly 20, 21, 300
SST *see* Temperature
Stability, static (E) 49, 54–56, 84, 143, 146, 165, 168, 228, 263
Stagnation 58, 67, 68, 211–213, 219, 274, 285
Standard Sea-Water 16, 99, 100
Station
  anchor 35, 279
  oceanographic 35
STD 100
Steady state 65
Stefan's Law 70, 75, 96
Stick diagram 141, 142
Stokes drift 43
Stratification 287
Subantarctic
  Front 174, 175
  Surface or Upper Water 173, 174, 179
  Zone 173, 179, 180
Subarctic Convergence 266
Suboxic 212
Subtropical
  Convergence 173–175, 179, 181
  definition 37
Sverdrup, unit 138, 299
Synoptic approach 2, 4, 94, 122, 135

Tasman Sea 5
Temperature 13, 17, 34
  in situ 47–49
  maxima, minima 146, 147, 273
  measurement 96–98, 123
  potential 19, 20, 47–49, 196, 197
  sea-surface (SST) 37–40, 46, 47, 75–77, 128, 130–134, 254, 255, 258, 259, 261, 263
  type distributions 36–41
  variations 44–47

Thermocline 40, 41, 44, 45, 138, 156, 158, 193, 195, 241–243, 245, 246, 257, 258, 263, 265, 270, 274, 286, 291, 292
Thermostad 190, 241, 246
Thermometer
  electrical resistance 98
  radiation 97
  reversing 95, 96, 97
Thermosteric anomaly 21, 245, 246, 300
Tide 3, 276–278, 288
Time-series 35, 188, 295
  plots 140–142
TIROS 130, 134
Tongue-like distribution 147, 148, 198, 256
Topography, bottom 7–10, 176
Tracers 59–62, 107, 139, 269
Transparency 62, 63
  meter 126
Transport
  definition 64
  unit 138, 299
  volume 65, 118, 160, 175, 176, 183, 184
Trench 6, 9
Tritium 139
Tropical, definition 37
T-S characteristics
  Atlantic 149, 206, 207
  global 146, 147, 149–151
  Indian 205
  Pacific 204, 264
  Southern Ocean 178
T-S diagram 143–148, 294
  see also individual oceans
T-S-t diagram 152, 153, 287
T-S-V diagram 148–152
Turbidity 31, 100
  current 167
Turbulence 56, 79, 82, 161, 165–168, 212, 287
Two-layer model, ocean 56

Units, mixed, SI, definitions 298–300
Upper Layer 6, 166, 279, 280
  Water 42, 292
Upper zone, definition 40
Upwelling 42, 60, 61, 134, 160, 184, 196, 246, 251, 256
  coastal 257, 258, 262, 272, 276, 279, 280, 292

VACM, VMCM 109

Vapour pressure 80
Variations
  annual 287
  diel 287
  diurnal 35, 47
  seasonal 188, 196, 235, 246, 271, 274
  spatial 35, 75, 176, 242–245
  temporal 35, 44–47, 75, 176, 189, 213, 242–245, 253, 278, 292
Ventilation 68, 156
Vessel, research 127, 128
Vorticity 124, 157

Water
  budget 64–66
  mass, type, definitions 144
  masses, deep, upper 171, 172, 293
  pure, properties 12, 13
  sampling bottle 95, 96
  sea see Sea-water
  vapour 134
  vapour pressure 80
Wave 3
  equation 24
  equatorial trapped (Kelvin) 248, 262
  internal 56, 167, 283
  shelf 289
Weddell Sea 156, 157, 177
Western boundary deep flows 157, 158, 190
West Spitsbergen Current 223, 226
Westward intensification 164, 172
Wien's Law 70
Winches and wire 94, 95
Wind stress 84, 133, 158–165, 181, 230, 231, 245–257, 262, 272, 273, 279–281, 290
Winds
  Indian and Pacific Oceans 237
  Trade 49, 161, 164, 181, 185, 235, 237, 249–251, 258, 259, 261, 270
  westerly 161, 164, 175
WOCE 90, 294, 296
World Ocean 70, 88–91

XBT 36, 94, 102
XCTD 102

Zonal distribution 37
Zones, Deep, Thermocline, Upper, definitions 40